MODERN DEVELOPMENTS
IN
POWDER METALLURGY

Volume 5: Materials and Properties

MODERN DEVELOPMENTS IN POWDER METALLURGY
Edited by Henry H. Hausner

1966: Proceedings of the 1965 International
Powder Metallurgy Conference

Volume 1: Fundamentals and Methods
Volume 2: Applications
Volume 3: Development and Future Prospects

1971: Proceedings of the 1970 International
Powder Metallurgy Conference

Volume 4: Processes
Volume 5: Materials and Properties

MODERN DEVELOPMENTS
IN
POWDER METALLURGY

Proceedings of the 1970 International Powder Metallurgy Conference, sponsored by the Metal Powder Industries Federation and the American Powder Metallurgy Institute.

Editor
Henry H. Hausner

Adjunct Professor, Polytechnic Institute of Brooklyn
and Consulting Engineer
New York, N. Y.

Volume 5
Materials and Properties

ℙ PLENUM PRESS · NEW YORK–LONDON · 1971

These two-volume Proceedings of the 1970 International Powder Metallurgy Conference, published under the title *Modern Developments in Powder Metallurgy*, also comprise Volume 26 of the series *Progress in Powder Metallurgy*, published by the Metal Powder Industries Federation.

ISBN-13: 978-1-4615-8965-5   e-ISBN-13: 978-1-4615-8963-1
DOI: 10.1007/978-1-4615-8963-1

Library of Congress Catalog Card Number 61-65760

CONTENTS

NEW P/M MATERIALS

DISPERSION STRENGTHENING

P/M REFRACTORY MATERIALS AND
SPECIALTY PRODUCTS

OXIDES AND NUCLEAR MATERIALS

P/M PROPERTIES AND TESTING

NEW P/M MATERIALS

PROPERTIES OF HOT PRESSED TITANIUM ALLOY

POWDERS FOR CRYOGENIC APPLICATIONS *

Gerald I. Friedman
Whittaker Corporation/Nuclear Metals Division

John M. Kazaroff
NASA/Lewis Research Center

INTRODUCTION

This program was performed to evaluate the strength and toughness of hot pressed titanium alloy powders at room and at cryogenic temperatures. Our objective was to determine how the mechanical properties of solid bodies formed from powder would compare with wrought specimens of the same size and with the same chemical analysis. Such information would prove valuable in determining the usefulness of powder techniques for the fabrication of parts that are difficult or costly to make by conventional techniques. The basis for this expectation was the belief that the powder process would produce fine-grained structures possessing a high degree of chemical homogeneity and property uniformity.

In the first part of the program titanium alloy powders made by five different techniques were compared. Following an evaluation based on powder and solid characteristics, one powder type was selected for use in the balance of the program. Rectangular blocks 6 x 6 x 3 inches of both alloys were pressed, and samples cut from the blocks were tested at room temperature, -320°F and -423°F.

TECHNICAL PROGRAM

Ti-6Al-4V and Ti-5Al-2.5Sn ELI grade (Extra-Low Interstitial Content) forged bars were purchased by Nuclear Metals. Sections

---

*This work was performed for NASA/Lewis Research Center under Contract NAS 3-10301.

of each bar were sent to Nuclear Materials and Equipment Corporation, Apollo, Pennsylvania, for conversion to powder by both the
hydride-dehydride process (Hyd) and by mechanical attrition (MA).
In the former process, powder is prepared by enclosing the titanium
bar in a sealed chamber which is then evacuated and back-filled
with hydrogen as the chamber temperature is raised. Whereas
titanium is capable of dissolving nearly 8 atomic percent of hydrogen at 600°F, the reacted mass of metal can retain only from 0.05
to 0.14 atomic percent hydrogen at temperatures below 250°F. The
excess hydrogen is therefore forced out of solution at these lower
temperatures and is present as a titanium hydride phase. At levels
above 200 ppm, the precipitated hydrogen embrittles the matrix, thus
making possible a relatively simple crushing operation. The powder
which results from crushing the titanium hydride is converted back
to the metallic state by a subsequent vacuum heat treatment.

In the mechanical attrition process, the forged bar was
machined to chips, which were then converted to powder in a hammer
mill. The machining and milling operations were performed in a
helium atmosphere.

Dominion Magnesium, Ltd., Toronto, Canada, prepared ten pounds
of Ti-6Al-4V powder (Dom) by the coreduction of titanium, aluminum
and vanadium compounds.

Ten pounds of Ti-5Al-2.5Sn powder (Penn) were procured from
the Penn-Nuclear Corporation, Penn, Pennsylvania, which used a
combination gas impingement and mechanical attrition technique to
produce powder from their own starting stock.

Ten pounds of each alloy powder were prepared at Nuclear
Metals by their Rotating Electrode Process (REP). Portions of the
same forged bars sent to Numec were extruded, cut into 10-inch
lengths and finish machined to 1-1/4 inch diameter rods. The metal
rods were loaded into an 8-foot diameter tank which was subsequently evacuated and back-filled with helium. By means of a glove
port in the tank wall, each bar was positioned in turn in the chuck
of a high-speed spindle inside the tank. When the electrode had
attained the desired rotational speed, an arc was struck between
the face of the titanium bar and a non-rotating tungsten electrode.
The combined action of the arc and the rotating electrode results
in the formation of spherical droplets which fly off from the electrode face. The droplets freeze in flight into spherical powder
particles and are completely solid before they drop to the floor of
the tank.

The five powder types are illustrated in Figure 1. Chemical
analyses of the eight powders are given in Table I, from which it
may be seen that the powders can be placed in three categories. In

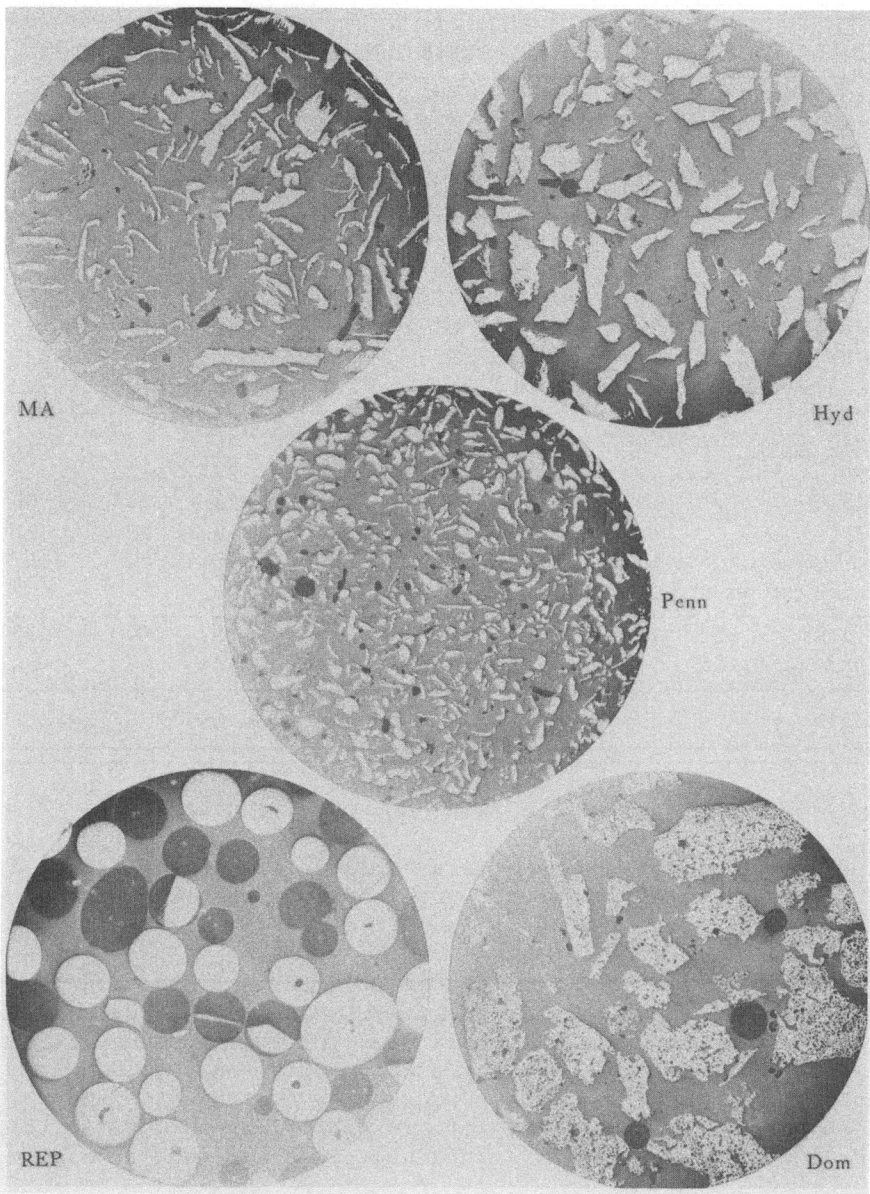

Figure 1. Titanium Alloy Powders, 25X.

the first group are the ELI powders, with very low oxygen contents
(REP). In the second group are powders with approximately twice
the oxygen in the first group (Hyd and MA). These powders are still
within specification for normal titanium alloys. Powders in the

TABLE I.   CHEMICAL ANALYSES FOR EIGHT TITANIUM ALLOY POWDERS

Sample	Al (%)	V (%)	Fe (%)	O (%)	C (%)	N (%)	H (%)
Ti-6Al-4V							
MA*	5.84	4.19	0.16	1560	130	189	98
REP*	5.67	4.30	0.15	750	67	146	58
Hyd*	5.94	4.32	0.21	1300	61	155	213
Dom	5.72	2.70	0.07	8100	940	187	2960
6" forged bar	6.31	4.27	0.11	700	270	70	16
MIL Spec,	5.50-	3.50-	0.25	1300	800	500	125
Ti-6Al-4V, ELI	6.75	4.50	(max)	(max)	(max)	(max)	(max)
Ti-5Al-2.5Sn							
REP**	5.25	2.49	0.20	750	27	106	39
Hyd**	4.93	2.41	0.04	1380	22	149	56
MA**	4.98	2.40	0.07	1220	42	191	90
Penn	5.16	2.45	0.20	2800	96	523	187
$6\frac{1}{2}$" forged bar	5.46	2.80	0.02	740	60	140	6
MIL Spec.,	4.25-	2.0-	0.25	1200	800	700	125
Ti-5Al-2.5SnELI	5.75	3.0	(max)	(max)	(max)	(max)	(max)

*Made from 6" forged bar   **Made from 6-1/2" forged bar

third group have very high oxygen contents (Penn and Dom).

Flow rate, apparent density and particle size distribution determinations were carried out in accordance with the appropriate ASTM specifications (Tables II and III).

These two tests illustrate a major difference between the irregular and the spherical powders. The angular-blocky co-reduced, hydride, and mechanically attrited powders are much more "fluffy" than the REP powder, and in fact occupy twice the volume for equal weights. The blocky-particle powders have a strong tendency to bridge over, which accounts for their poor flow characteristics in the Hall Flowmeter. The REP spherical powders flow very easily.

Each powder was then cold compacted into a steel can held in a packing die inside an inert gas glove box. The compacts were prepared by filling each can with powder at 1000 to 1400 lb/in^2. Each can contained approximately one pound of powder.

The compacts were then transferred to a welding box where the top lid, with steel evacuation tube attached, was joined to the can body. The cans were then evacuated and hot pressed at 1450,

TABLE II. DETERMINATION OF FLOW RATE AND APPARENT DENSITY			
Sample	Flow Rate (sec/50g)	Apparent Density (g/cc)	(% of theo.)
Ti-6A1-4V			
REP	24	2.72	61.5
MA	No flow	1.13	25.6
Hyd	44.5	1.65	37.2
Dom	No flow	1.29	29.2
Ti-5A1-2.5Sn			
REP	22	2.83	63.5
MA	No flow	1.52	34.2
Hyd	51	1.51	34.0
Penn	44	1.79	40.2

1650 and 1850°F under pressures of 50, 75 and 100 ton /in^2. Each compact was pressed individually using a floating-punch arrangement in a 2-inch diameter extrusion press liner. The extrusion press container, ram and hardened punches were maintained at 900°F. Transfer time from the billet-heating furnace to the container of the press averaged 10 to 15 seconds; application of ram pressure required approximately 5 additional seconds. Pressure was maintained on the billet for approximately 10 seconds.

The density of each compact was determined from measurements of weight and physical dimensions (Table IV). These data show:

1. The Dom Ti-6A1-4V compacts were all less than 98 percent dense. This is explainable in view of this powder's very high

TABLE III. TITANIUM POWDER PARTICLE SIZE DISTRIBUTION								
Screen Size (μ)	Fraction Retained on Screen (%)							
	Ti-6A1-4V				Ti-5A1-2.5Sn			
	MA	REP	Hyd	Dom	MA	REP	Hyd	Penn
250	13.8	36.3	28.0	83.8	23.6	21.0	24.6	0
177	27.2	40.9	21.6	7.1	30.8	49.5	22.1	0
125	24.2	13.1	14.4	3.9	21.9	17.5	17.6	24.2
88	17.2	7.4	15.6	2.6	13.2	9.3	15.2	41.8
63	7.0	1.5	8.4	1.8	4.3	1.7	8.2	15.7
44	5.6	0.4	8.9	0.1	3.2	0.5	8.9	8.9
Pan	5.1	0.4	3.1	0.7	3.0	0.4	3.4	9.4

TABLE IV. PERCENT THEORETICAL DENSITY OF HOT COMPACTED TITANIUM ALLOY POWDER BILLETS*

Temperature (°F)	1450			1650			1850		
Pressure (tsi)	50	75	100	50	75	100	50	75	100
Ti-6Al-4V									
MA	96.2	100.1	100.1	100.2	100.6	100.4	100.5	100.8	100.4
REP	96.9	99.3	99.4	99.4	99.9	99.9	99.7	99.9	99.8
Hyd	99.4	99.3	99.7	99.3	99.7	99.8	99.8	100.7	99.8
Dom	91.3	95.2	95.8	94.5	95.9	96.3	96.2	96.6	96.6
Ti-5Al-2.5Sn									
MA	95.8	99.6	100.0	97.5	99.3	99.6	100.0	100.0	99.2
REP	---	99.0	99.9	96.6	99.7	99.9	99.9	99.9	100.0
Hyd	95.2	99.4	99.8	96.3	99.6	99.8	100.2	99.8	99.9
Penn	96.1	99.8	99.8	98.6	99.7	100.3	100.1	100.0	100.0

*Based on a density of 4.424 g/cc for Ti-6Al-4V and 4.450 g/cc for Ti-5Al-2.5Sn

interstitial content, which would increase the difficulty of plastically deforming the powder particles.

2. No powder could be compacted to 98 percent density at 50 ton /in^2, 1450°F. Pressures of 75 and 100 ton /in^2 were sufficient to produce compacts of over 99 percent density at 1450°F, 1650°F and 1850°F.

The 1850°F/100 ton /in^2 compacts were then sectioned, providing samples for chemistry, heat treating, metallography and tensile testing.

Table V contains the results of chemical analyses for interstitial elements performed on the 1850°F/100 ton/in^2 compacts. Only the REP samples have low enough oxygen contents to be considered in the ELI category. In all other cases, the oxygen contents are higher than the approximately 1200 ppm established as the ELI limit. The progression of oxygen contamination during the conversion of these samples from bar to powder to pressed compact is significantly higher for the irregular powders than for the low surface area spherical powder particles. The oxygen content for the REP powders and most of the Numec powders is low enough to be included in the commercial range for ELI (REP) or standard grade (Numec) titanium.

High oxygen titanium powders are difficult to form because the oxygen in them is not uniformly distributed but is instead present in a heavy concentration on the surface of the powder particles. Surface oxygen levels can therefore be considerably higher than the nominal oxygen content for the powder. The high-oxygen surface layer is quite hard and impedes plastic deformation and surface welding of the powder particles. The direct consequence

TABLE V. CHEMICAL ANALYSIS OF HOT PRESSED TITANIUM ALLOY COMPACTS				
Sample	C (ppm)	O (ppm)	H (ppm)	N (ppm)
Ti-6Al-4V				
MA	650	1710	144	239
REP	200	900	90	234
Hyd	210	1570	210	148
Ti-5Al-2.5Sn				
MA	500	1640	258	155
REP	255	980	78	189
Hyd	150	3620	52	208
Penn	640	3530	183	591

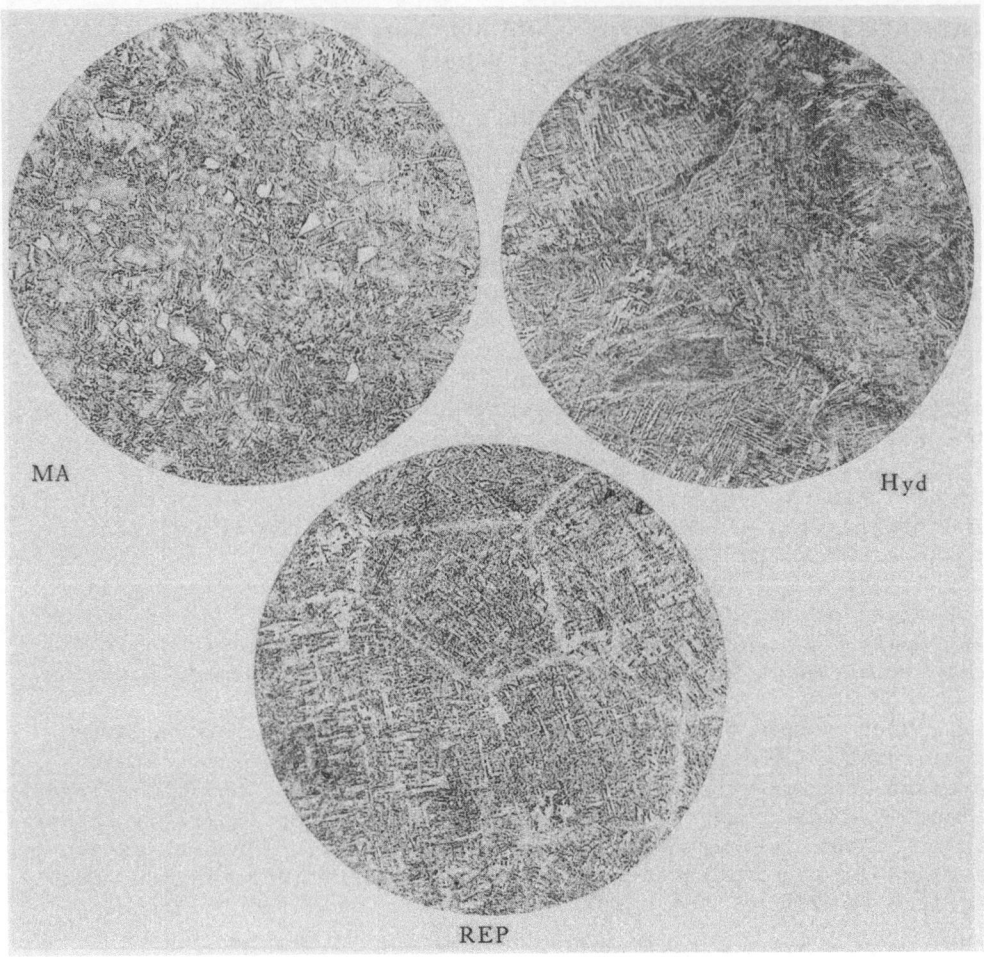

Figure 2. As-Pressed Ti-6Al-4V Compacts, 250X, etched.

of the hardened particle surface layer is a lowering of bond
strength between powder particles, and hence lower mechanical
properties for compacts in the as-pressed condition. High temperature annealing treatments can offset this condition somewhat by
diffusing oxygen away from the particle surface.

Photomicrographs of the as-pressed compacts are presented in
Figures 2 and 3. Prior particle boundaries are visible in some
of the samples. The structure of the Ti-6Al-4V samples is predominantly that of the acicular alpha phase in a framework of

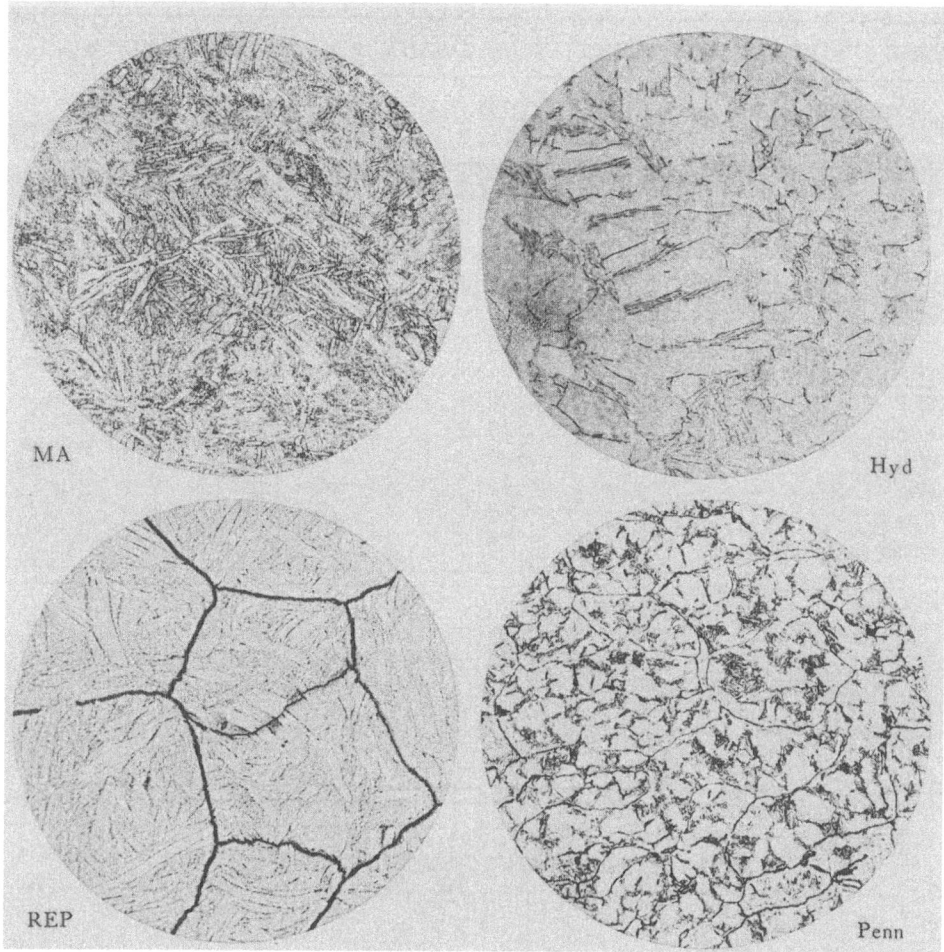

Figure 3. As-Pressed Ti-5Al-2.5Sn Compacts, 250X, etched.

prior beta grain boundaries. The Ti-5Al-2.5Sn alloy is single-
phase, and the structures observed are those of the alpha phase.

Tensile properties for the as-pressed compacts are listed in
Table VI. A comparison between the mechanical property data
presented in Table VI and military specifications shows that none
of the samples exhibited the minimum tensile elongation obtained
in conventionally processed material. A series of heat treatments
was therefore performed on the compacts made from the REP powders,
since they had the highest purity and therefore offered the
greatest potential in achieving increased values of ductility.

TABLE VI.   TENSILE PROPERTIES OF TITANIUM ALLOY POWDER COMPACTS

Sample	UTS (ksi)	0.2% YS (ksi)	Elongation (%)
Ti-6Al-4V			
MA	144.8	135.9	1.5
REP	145.1	134.1	7.5
Hyd	149.3	141.3	2.0
Spec., Ti-6Al-4V, ELI, Ann.	120	110	10
Ti-5Al-2.5Sn			
REP	130.7	130.7	4.0
Penn	129.9	-----	0
Hyd	126.3	120.3	5.5
MA	92.8	-----	0
Spec., Ti-5Al-2.5Sn, ELI, Ann.	105	100	10

The results of these heat treatments are tabluated in Table VII. Figures 4 and 5 illustrate microstructures associated with the heat treatment studies.

TABLE VII.   MECHANICAL PROPERTIES OF TITANIUM ALLOY COMPACTS AFTER VARIOUS HEAT TREATMENTS

Sample	Compacting Conditions Pres. (ton/in^2)	Temp. (°F)	Heat Treatment	UTS (lb/in^2 x 1000)	0.2%YS	Elong. (%)	RA (%)
Ti-6Al-4V							
1	75	1650	1650/2/WQ	157.5	138.6	7.3	26.4
2	75	1650	1650/4/WQ	159.3	148.7	7.7	29.2
3	75	1650	2200/4/AC	129.7	117.6	10.0	42.0
4	75	1850	1300/2/AC	140.0	131.5	10.4	37.4
Ti-5Al-2.5Sn							
1	100	1650	1550/4/AC	119.9	113.5	7.7	19.4
2	100	1850	1850/4/AC	125.1	115.0	11.4	25.2
3	75	1850	1775/4/AC	126.5	117.2	14.2	42.7

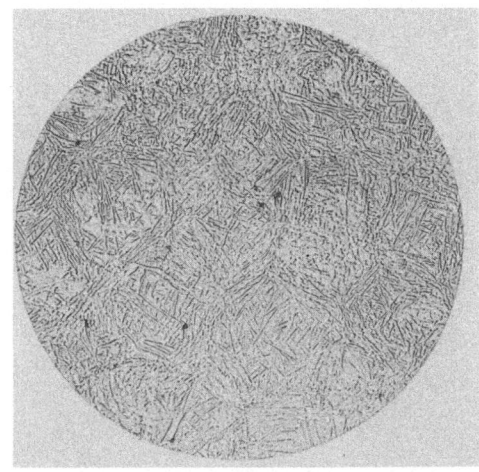

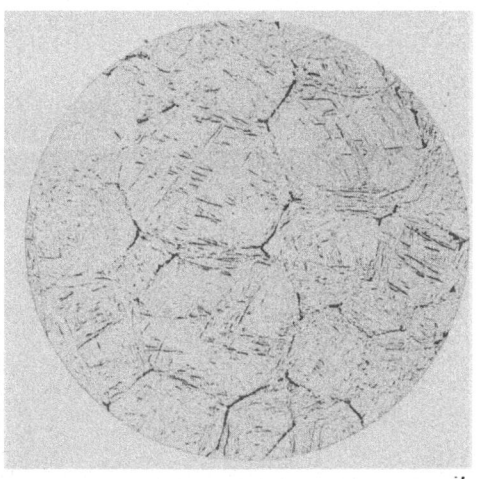

a. Pressed: 1650°F/75 ton/in^2 b. Pressed: 1650°F/75 ton/in^2
    Heat treated: 1650°F/2/WQ   Heat treated: 1650°F/4/WQ

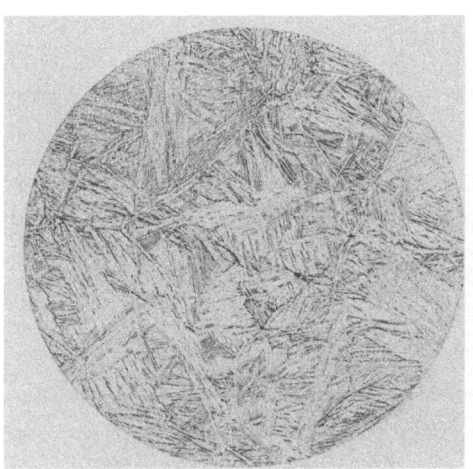

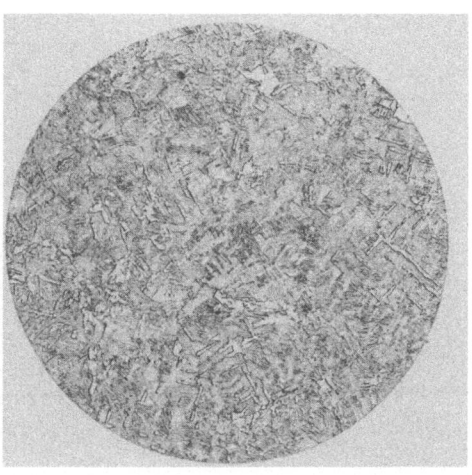

c. Pressed: 1650°F/75 ton/in^2 d. Pressed: 1850°F/75 ton/in^2
    Heat treated: 2200°F/4/AC   Heat treated: 1775°F/4/AC +
                                                            1300°F/2/AC

Figure 4.   REP Ti-6Al-4V, pressed and heat treated
            as indicated. Transverse sections,
            100X, etched.

Examination of the tensile results for the Ti-6Al-4V alloy
shows that tensile ductility, as measured by elongation and reduction in area, increases with an increase in compacting and
heat treatment temperature and time. The effect of the heat

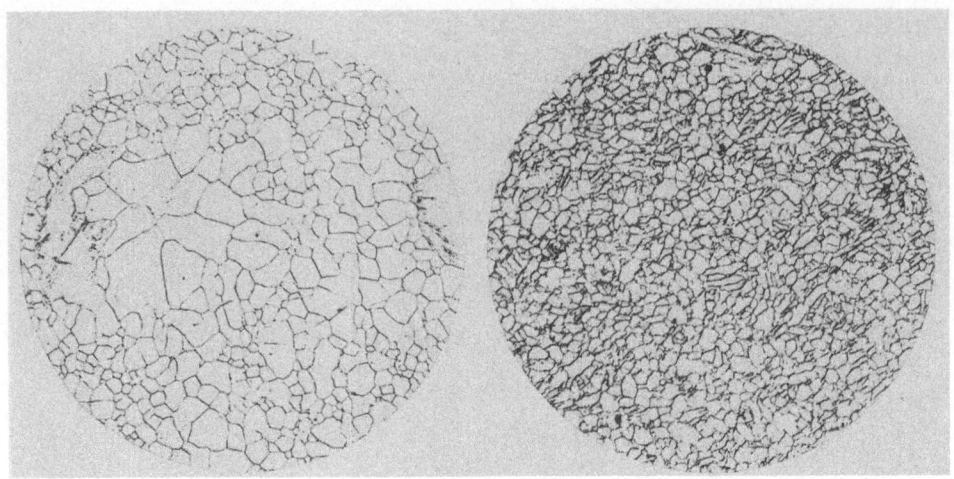

a. Pressed: 1650°F/100 ton/in^2 b. Pressed: 1850°F/100 ton/in^2
    Heat treated: 1550°F/4/AC   Heat treated: 1850°F/4/AC

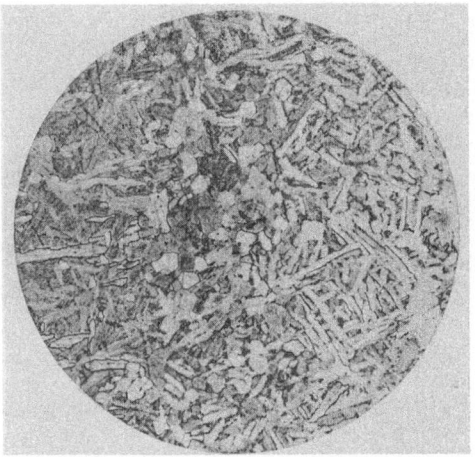

c. Pressed: 1850°F/75 ton/in^2
    Heat treated: 1775°F/4/AC

Figure 5.   REP Ti-5Al-2.5Sn, pressed and heat treated
            as indicated. Transverse sections, 100X,
            etched.

treatments would therefore appear to be that of enhancing the
disappearance, by diffusion, of the prior particle boundaries.
This effect is shown very clearly in Figures 4a and 4b in which
samples compacted at 1650°F, 75 ton/in^2 were annealed at 1650°F
for two and four hours respectively.

As the annealing temperatures are increased, further evidence of particle-boundary elimination is offered by the increase in grain size which occurs. Samples annealed at 2200°F possess adequate ductility, but their tensile strength has begun to fall off (as a result of greatly increased grain size) compared to the as-pressed condition. There is a large difference in grain size between the sample shown in Figure 4c, annealed at 2200°F, and the sample shown in Figure 4d, which was treated at 1775°F for a similar period of time. The microstructure shown in Figure 4c is one of acicular alpha and prior beta grain boundaries, obtained by heating entirely in the beta phase field. (The boundary between the alpha-plus-beta and the beta phase field lies at 1820°F for Ti-6Al-4V.) The sample annealed at 1775°F was therefore heated at the high end of the alpha-plus-beta field. The structure illustrated is that of primary alpha (white regions) and transformed beta (acicular alpha). The properties obtained by heating at this temperature offer a good balance between tensile strength and ductility.

Although Ti-6Al-4V is normally solution-treated for higher strengths, such a treatment is not applied to the ELI grade. This is because the lower oxygen content is specified to increase ductility, which would be lost in a strengthening heat treatment. Ti-5Al-2.5Sn, being a single-phase alloy, cannot be strengthened by heat treatment. The objective of any heat treatment for this alloy is therefore only to relieve working stresses, or (as in the present case) to diffuse away particle boundaries.

The result of the heat treatment - tensile test experiments was the development of a process producing adequate tensile strength along with good ductility in hot pressed REP titanium alloy powders. It was then possible to transfer these practices to the production of the larger test pieces, a pair of 6- x 6- x 3-inch blocks.

Ti-6Al-4V and Ti-5Al-2.5Sn electrode stock was extruded from the large forged bars. These extruded rods, approximately 1-1/8 inches in diameter, were then machined to electrode size and converted to powder via the Rotating Electrode Process, as described above. Chemical analysis and other pertinent data for the two alloy powders appear in Table VIII.

Using rectangular inserts in an extrusion press container, the powders were then formed into 6- x 6- x 3-inch blocks, in evacuated steel cans. The blocks were pressed at 1850°F, 73 ton/in^2, and annealed(1775°F/4 hr + 1300°F/2 hr). Tensile specimens were machined from the blocks parallel to and at right angles to the pressing direction. Duplicate sets of specimens were machined from the forged "parent" stock.

TABLE VIII. PROPERTIES OF FIFTY-POUND LOTS OF TITANIUM ALLOYS		
Property	Ti-6Al-4V	Ti-5Al-2.5Sn
Interstitial Analysis (ppm) C	196	40
O	730	510
H	58	34
N	175	268
Flow Rate (sec/50 g)	21.8	24.6
Apparent Density (% theo.)	60.5	63.8
Particle Size Distribution		
Screen Size (μ)	Fraction Retained on Screen(%)	
710	0	2.0
500	0	14.6
354	8.5	38.2
250	31.8	30.3
177	40.0	11.6
125	13.2	3.11
88	4.9	0.8
63	0.9	0.1
44	0.3	0
Pan	0.1	0

Each tensile specimen was instrumented with a pair of strain gages to compensate for any non-axial loading in the test machine. Tests were conducted at room temperature, -320°F (boiling point of nitrogen) and -423°F (boiling point of hydrogen). Tensile results are tabulated in Table IX. In most cases the powder specimens show properties equivalent to or better than those for the wrought specimens. The Ti-6Al-4V powder samples are stronger than their wrought counterparts at all temperatures, in both the longitudinal and transverse orientation, and they are more ductile than the wrought specimens except at -423°F where the longitudinal powder samples are slightly inferior.

In the case of the Ti-5Al-2.5Sn alloy, the transverse powder specimens are stronger and more ductile than their wrought counter-parts at the three test temperatures. The longitudinal powder specimens, however, show a lower yield strength at -320°F, lower tensile strength at the three test temperatures, and lower overall ductility at room temperature. At -320°F, the powder specimens show greater ductility as measured by reduction in area, but less ductility as measured by specimen elongation. The situation with respect to these two ductility criteria is reversed at -423°F. There are no obvious indications explaining why the longitudinal

TABLE IX. TENSILE PROPERTIES OF TWO TITANIUM POWDER ALLOYS, FROM ROOM TEMPERATURE TO -423°F

| Properties | Ti-6Al-4V, ELI | | | | Ti-5Al-2.5Sn, ELI | | | |
| | Longitudinal | | Transverse | | Longitudinal | | Transverse | |
	Powder	Wrought	Powder	Wrought	Powder	Wrought	Powder	Wrought
Room Temperature								
UTS (psi x 1000)	130	128	132	129	115	118	113	96.8
.2% YS (psi x 1000)	114	106	122	85.2	104	103	99.7	82.4
Elongation (%)	18.9	15.3	20.2	10.5	15.7	19.3	20.7	11.0
RA (%)	31.3	29.3	33.0	12.0	27.0	34.0	32.1	28.0
-320°F								
UTS (psi x 1000)	216	202	212	200	178	187	172	171
0.2% YS (psi x 1000)	188	174	189	170	157	170	149	149
Elongation (%)	11.7	12.0	10.8	9.6	10.0	17.0	16.0	11.3
RA (%)	17.3	11.3	13.0	11.7	18.6	16.8	18.5	14.3
-423°F								
UTS (psi x 1000)	237	225	237	223	197	214	203	198
0.2% YS	197	193	212	166	174	159	160	152
Elongation (%)	3.8	5.3	7.0	6.9	7.1	6.8	11.0	8.7
RA (%)	11.6	12.4	10.6	5.8	13.0	15.1	15.6	15.3

powder specimens do not show the superior mechanical properties, as compared to the wrought samples, that is evidenced by the other powder specimens.

Analyses of samples taken from the tensile specimens compared with the analysis of forged bars and powder indicate that very little change in chemistry had taken place.

CONCLUSIONS

This program demonstrated that of five titanium powder-making processes investigated, only the Rotating Electrode Process was capable of producing ELI grade titanium alloy powder. The oxygen content of compacted bodies made from the REP Ti-6Al-4V and Ti-5Al-2.5 Sn powders was 1000 ppm or less. In comparison, the compacts made from powder produced by the hydride-dehydride process, the mechanical attriting of chips, fluid-energy (gas) mill attriting of chips, and chemical reduction, had oxygen contents ranging from 1600 to over 8000 ppm.

All powders, except those made by chemical reduction, could be compacted to full density at a temperature/pressure combination of 1650°F and 75 ton/in^2. The chemically reduced powder could not be compacted to over 96.6 percent density at 1850°F and 1000 ton/in^2, the highest temperature/pressure combination investigated.

The as-compacted powder samples possessed only limited ductility, with elongation values ranging from 0 to 7.5 percent. The REP powder samples developed over 10 percent elongation and 40 percent reduction in area after a four-hour anneal at 1775°F.

Blocks hot-pressed from the spherical REP powders had tensile properties equivalent to or better than those obtained from wrought bar. In particular, the powder Ti-6Al-4V block specimens were stronger and more ductile than the wrought bar samples at all test temperatures, in both longitudinal and transverse directions. The powder Ti-5Al-2.5Sn samples showed this same superiority in the transverse direction only. Longitudinal powder specimens of this alloy were either weaker or less ductile than corresponding wrought samples.

IMPROVED SINTERING PROCEDURES FOR ALUMINUM P/M PARTS

J. H. Dudas* and C. B. Thompson**

*Group Leader, Atomizing and Powder Metallurgy Alcoa
   Research Laboratories, Alcoa Technical Center
**Development Engineer, Metal Working Division, Alcoa
   Process Development Laboratories, Alcoa Technical Center

ABSTRACT

This paper discusses sintering practices for the production
of aluminum P/M parts in nitrogen and dissociated ammonia and in
vacuum. The influence of sintering atmosphere, dew point and
temperature on the mechanical properties and dimensions of aluminum P/M alloys 601 AB and 201 AB are presented. Typical aluminum
microstructures are illustrated.

Nitrogen atmosphere produces highest strength in P/M parts
from aluminum, is moderate in cost and does not require the pur-
chase of a gas generator or dryer to achieve a high purity, dry
product. Since no special handling procedures are required, nitrogen can be readily used in both batch and continuous sintering
furnaces. Tensile strengths of aluminum P/M parts sintered in
nitrogen range from 20,000 to nearly 50,000 psi, depending on alloy
and heat treatment.

Dissociated ammonia is readily available in most P/M plants,
but care in handling should be exercised for aluminum sintering
because temperatures are not high enough to assure hydrogen self
ignition. Although the cost of dissociated ammonia compares favorably with nitrogen, the product needs to be adsorbent dried to
remove undissociated ammonia and obtain a low atmosphere dew point.
Mechanical properties of parts sintered in dissociated ammonia are
slightly lower than in nitrogen but satisfactory for most aluminum P/M applications. Tensile strengths up to 42,000 psi are
achieved.

Vacuum sintering is a suitable method for the production of high strength aluminum P/M parts. Neither inert atmosphere nor gas drying equipment is needed. Tensile properties approaching nitrogen sintered values are obtained in high density P/M parts with a mechanical pump vacuum of 200 microns pressure. Although separate presintering is required to remove lubricant, it may be conducted in air rather than nitrogen.

Both batch and continuous furnaces are suitable for production sintering of aluminum P/M parts. While batch furnaces have lower investment costs and moderate atmosphere consumption, production rates are low to medium. Continuous furnaces provide high production rates, but have higher equipment costs and require increased atmosphere flow. Both furnace types provide strong, ductile aluminum P/M parts for a variety of applications if atmosphere dew point and sintering temperature are controlled.

INTRODUCTION

Widespread interest has developed in the production of aluminum powder metal (P/M) parts because of their combination of high strength, light weight, corrosion resistance and high thermal and electrical conductivity. The production and performance of such parts have been discussed in several recent technical papers (Ref. 1-3). Previous problems of cold welding and galling of the aluminum powder to the tools and die walls during compaction have been solved by improved compositions having both controlled particle size and special admixed lubricants.

Successful sintering of aluminum P/M alloys requires proper selection of sintering conditions such as atmosphere, dew point and temperature. During the past year, a detailed investigation was conducted at Alcoa to delineate the effect of such conditions, especially atmosphere, on the properties, dimensions and microstructure of aluminum P/M alloys 601 AB and 201 AB sintered in both batch and conveyor furnaces. The results are presented in this paper.

MATERIALS AND TEST PROCEDURES

Sintering tests were conducted employing commercial aluminum P/M alloys, 601 AB and 201 AB, the nominal compositions of which are listed in Table I. Both alloys are premixed powders containing aluminum with alloying powder additions of copper, silicon and magnesium plus 1-1/2 weight percent lubricant.

TABLE I

NOMINAL COMPOSITIONS OF ALUMINUM P/M ALLOYS
601 AB AND 201 AB

Element	601 AB	201 AB
Copper	0.25%	4.4%
Silicon	0.6	0.8
Magnesium	1.0	0.5
Lubricant	1.5	1.5
Aluminum	Grade 1202	Grade 1202

Sintering atmospheres investigated included nitrogen, dissociated ammonia and vacuum. Atmosphere dew points were measured
by observation of condensation on the outside surface of an internally cooled stainless steel cup. Sintering temperatures were
monitored by a thermocouple inserted directly in the parts during
processing.

Sintered tensile properties were determined using flat powder
metal tension test bars (MPIF Standard 10-63). A description of
the thermal treatments used is given in Table II. Dimensional
changes were measured on 3-in. long x .4-in. wide x 5/8-in. thick
compacted bars.

TABLE II

THERMAL TREATMENTS FOR SINTERED
ALUMINUM P/M ALLOYS

Temper	Description
T1 As-Sintered	Cooled from sintering temperature to room temperature at an uncontrolled rate; refer to Figure 2 for approximate cooling rates.
T4	Heat treated 30 min either at 970 F (601 AB) or 940 F (201 AB) in air, cold water quenched and aged 4 days minimum at room temperature.
T6	Heat treated 30 min either at 970 F (601 AB) or 940 F (201 AB) in air, cold water quenched and aged 18 hr at 320 F.

SINTERING FURNACES

Both batch and conveyor atmosphere furnaces are suitable for aluminum sintering as long as provisions exist for lubricant burn off, sintering and cooling. Full muffle construction is required to maintain a high purity, low moisture atmosphere with minimum gas flow.

Batch Atmosphere Furnaces

Batch sintering furnaces provide an economical means for prototype and low to medium production sintering of aluminum parts. Investment costs for such furnaces are low and atmosphere flow requirements are usually less than for conveyor furnaces because door openings can be sealed easily.

Figure 1 shows a box-type batch furnace used for experimental aluminum sintering. The furnace is fitted with a 12 x 12 x 18-in. inconel muffle. An internal positive pressure of 1/2 to 1 psi achieved with a nitrogen flow of 60 to 100 cfh maintains a -40 to -60 F furnace atmosphere dew point with the entering gas at -70 to -80 F dew point.

FIGURE 1
BOX-TYPE BATCH FURNACE

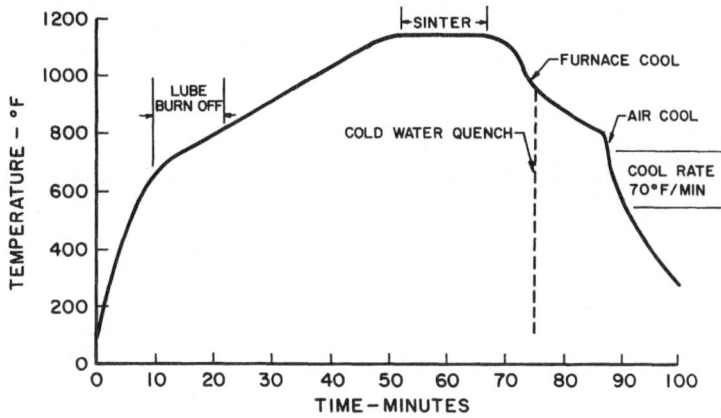

FIGURE 2
HEATING CYCLE FOR BATCH SINTERED ALUMINUM P/M PARTS

A typical heating cycle for this furnace is shown in Figure 2. Lubricant is expelled from the parts at 650 to 800 F after which the temperature is raised to 1150 F for sintering. The heating rate is not critical. After sintering, parts are normally cooled to 800 F by an air blast directed around the outside rear of the furnace muffle. The parts are then removed and air cooled to room temperature. Cycle time in the furnace is approximately 90 minutes. If heat treatment of the parts is desired, they are cooled in the furnace to the solution heat treating temperature (970 F for 601 AB) from which they are cold water quenched and either naturally or artificially aged to the T4 or T6 temper.

Many types and sizes of commercial batch furnaces are available for sintering aluminum P/M parts, some with capacities up to 100 lb of aluminum per hour. The designs include bell, elevator, pit and box-type furnaces.

Continuous Atmosphere Furnaces

Continuous conveyor sintering furnaces are widely used in the P/M industry for high volume parts production. The 35-ft long continuous hump-back conveyor furnace used in this investi-

FIGURE 3
CONTINUOUS HUMP-BACK CONVEYOR FURNACE

gation (Figure 3) was designed specially for sintering aluminum
using atmospheres of nitrogen or dissociated ammonia.

In operation, parts are conveyed in anodized aluminum trays
through the 6-in. wide x 4-in. high muffle opening at a pre-
selected belt speed between 2 and 8 ipm. Nitrogen curtains at
each end purge air from the parts and assist in maintaining a low
furnace dew point. A typical sintering cycle is illustrated in
Figure 4. After the lubricant is volatilized at 650-800 F, the
parts cool to approximately 570 F in the section separating the

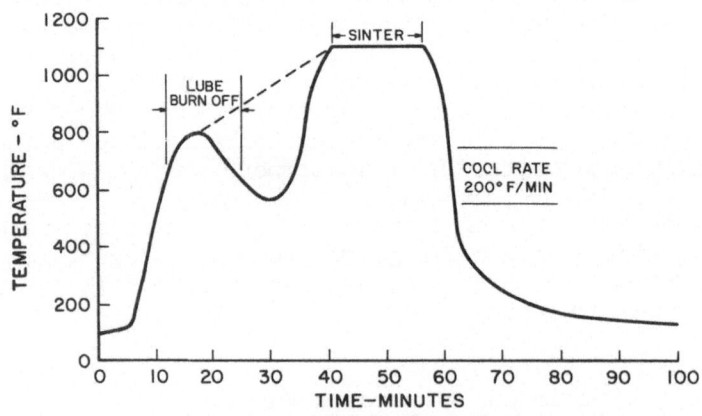

FIGURE 4
HEATING CYCLE FOR CONTINUOUS SINTERED ALUMINUM P/M PARTS

preheat and high heat chambers. To eliminate this inefficiency, the furnace is being modified to allow continued heating of the parts to the sintering temperature (dashed line - Figure 4). The rate of part heating has not been found to influence sintered properties.

The parts next proceed into the 90-in. long sintering chamber which is heated by nickel-chromium ribbon elements mounted above and below the inconel muffle. Temperature uniformity within ±5 F across the belt is achieved through adjustment of 12 saturable core reactors which proportionally control 12-in. individual heating sections along the sintering chamber. Sintered parts are cooled quickly to 300 F in the 24-in. long water-jacketed section, then slowly cooled to approximately 150 F in the non-water cooled inclined tunnel.

The hump-back design is a well established aid to atmosphere savings with "lighter than air" atmospheres such as dissociated ammonia, which contains a high percentage of hydrogen. With nitrogen, which has about the same density as air, the hump-back feature is not required nor is it desirable. An atmosphere flow of 500 to 700 cfh has been required with nitrogen but only 400 to 500 cfh with dissociated ammonia.

The production rate of sintered aluminum P/M parts in conveyor furnaces depends upon belt width and speed as well as part shape and size. Belt loading is rarely a limitation because of aluminum's light weight and lower sintering temperature. Sintering rates of 40 lb per hr have been achieved in the 6-in. wide furnace (2 lb per ft--4 ipm belt speed). On a per piece basis this is equivalent to 120 lb per hr of iron parts in the same furnace. Furnace manufacturers are prepared to manufacture continuous aluminum furnaces with conveyor widths up to 36 in. capable of production rates up to 500 lb of aluminum P/M parts per hour.

Vacuum Furnaces

Aluminum P/M parts may also be sintered in conventional cold-wall or hot-wall vacuum furnaces. Inert atmosphere or gas drying equipment is not needed. Roughing pump vacuums of about 200 microns appear suitable for higher density aluminum parts; hence, vacuum cold traps and diffusion pumps are not required.

FIGURE 5
8-IN. DIAMETER HOT-WALL VACUUM FURNACE

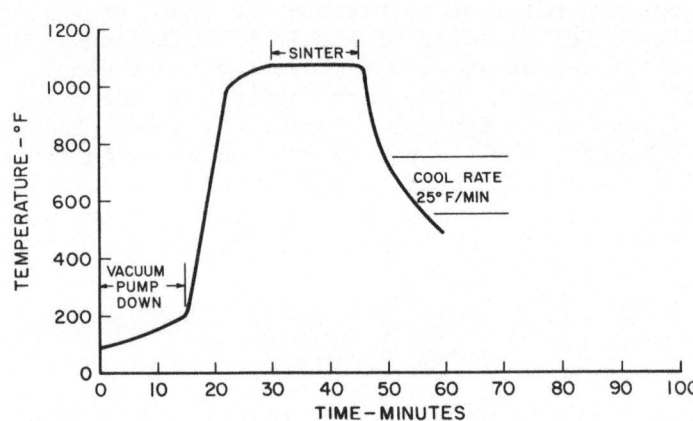

FIGURE 6
HEATING CYCLE FOR VACUUM SINTERED ALUMINUM P/M PARTS

An 8-in. diameter hot-wall vacuum furnace used for experimental aluminum sintering is shown in Figure 5; a typical sintering cycle in Figure 6. The parts, which are presintered sepa-
rately to remove lubricant, are heated directly to the sintering

temperature once pump down in complete. After sintering, they are furnace cooled to approximately 500 F before the vacuum is released.

Vacuum sintering, which is gaining increased acceptance in the production of stainless steel P/M parts, should not be over-looked as a potential method for sintering aluminum. Various sizes of batch and semi-continuous vacuum furnaces are available.

SINTERING ATMOSPHERE

Atmosphere sintering studies were conducted in nitrogen and dissociated ammonia in the continuous conveyor furnace shown in Figure 3 and vacuum sintering tests in the furnace illustrated in Figure 5. The effect of atmosphere and sintering parameters was related to properties, dimensions and microstructures of aluminum P/M alloys.

Nitrogen

Nitrogen at high purity and low dew point is especially suited for sintering aluminum P/M parts because of its ready avail-ability and moderate cost. Special handling is not required nor is a generator and adsorbent dryer needed to convert it to a dry, gaseous form. Bulk liquid nitrogen with a guaranteed purity of 99.995% and a dew point of -80 F will cost 15 to 75 cents per 100 cu ft, depending upon quantity consumed and delivery location. The highest sintered strength in both 601 AB and 201 AB parts is achieved in nitrogen.

In batch furnaces, nitrogen flow requirements are usually less than those for either hydrogen or dissociated ammonia be-cause larger nitrogen molecules are more easily contained at door and retort seals. A conveyor furnace using nitrogen should cost less than one using hydrogen-containing gases because a straight-through muffle can be used rather than a furnace of hump-back design.

Production rates for continuously sintered aluminum parts in nitrogen were varied from 10 to 40 lb per hour. Atmosphere flow of 500 to 700 cfh produced consistently well-sintered parts in simulated production runs that lasted 4 to 6 hours. Sintering time varied with part section thickness. Parts up to 1/4-in. thick sintered in 10 to 15 min whereas those 1 to 2-in. thick required 30 to 40 minutes. Extended sintering times caused some surface nitriding of parts but had no significant effect on ten-sile properties. Cooling rate from the sintering temperature varied with part size and belt speed, causing some variation in

as-sintered (T1) strengths. However, heat treating to the T4 or
T6 temper raised properties to a consistent acceptable level.

Table III lists the tensile properties of alloys 601 AB and
201 AB sintered in nitrogen. Tensile strengths of 601 AB parts
were 20,000 to 35,000 psi, depending on density and thermal treatment. As sintered (T1) parts had moderate strength and high ductility for easy repressing. The heat treated and naturally aged
T4 temper provided the best combination of strength and ductility
whereas highest strength was obtained in the fully heat treated
T6 temper. Alloy 201 AB extended the strengths to approximately
49,000 psi in the 95 percent density parts that were heat treated
to the T6 temper. This composition is specially suited for powder
metal applications requiring the highest aluminum strengths.

<u>TABLE III</u>

PROPERTIES OF 601 AB AND 201 AB
<u>ALLOYS SINTERED IN NITROGEN</u>

Green Density			Tensile Strength	Yield Strength	Elongation Percent
Percent	g/cm^3	Temper	psi	psi	in 1 in.
601 AB Alloy[a]					
90	2.42	T1	20,100	12,700	5.0
		T4	24,900	16,600	5.0
		T6	33,600	32,500	2.0
95	2.55	T1	21,000	13,700	6.0
		T4	25,600	17,000	6.0
		T6	34,500	33,400	2.0
201 AB Alloy[b]					
90	2.50	T1	24,200	21,300	3.0
		T4	29,200	24,300	3.0
		T6	38,500	38,000	1.5
95	2.64	T1	30,300	25,700	3.0
		T4	36,700	28,700	2.5
		T6	48,800	46,700	2.0

a - Sinter 10-30 min at 1150 F (-40 to -60 dew point).

b - Sinter 10-30 min at 1100 F (-40 to -60 dew point).

Dissociated Ammonia

Dissociated ammonia is used in many P/M plants for sintering brass and bronze parts and may be conveniently available for aluminum. The atmosphere used in this investigation was generated from premium grade anhydrous ammonia with a guaranteed minimum purity of 99.99 percent. It was dissociated in a 500 cfh dissociator at 1750 F. The dissociated product, 75 percent hydrogen--25 percent nitrogen, had a dew point of -60 F and contained 75 to 90 ppm of undissociated ammonia. A 500 cfh dual tower adsorbent dryer further dried the gas to a dew point of -80 to -100 F and reduced the undissociated ammonia to less than 1 ppm. The cost of dissociated ammonia may vary from 20 to 50 cents per 100 cu ft, depending on quantity consumed and size of generator employed. This compares favorably with nitrogen.

Since dissociated ammonia contains high concentrations of flammable hydrogen, care must be exercised in its handling, particularly for aluminum where sintering temperatures are not high enough to assure self ignition upon contact with air. One precaution is to purge the furnace with an inert gas such as nitrogen prior to introducing dissociated ammonia. Electric ignitors are installed in the burn-off stand pipe to provide positive ignition of the exiting hydrogen. Although some hydrogen escapes at the furnace doors, it is quickly diluted to a noncombustible concentration with nitrogen from the end curtains and with air. Burning of hydrogen at the furnace doors is not recommended because the resultant moisture may back diffuse into the furnace and raise the dew point.

Conditions and production rates for aluminum P/M parts sintered in dissociated ammonia were similar to those sintered in nitrogen; however, lower atmosphere flow rates were possible because of the hump-back furnace design. A dissociated ammonia flow of 400-500 cfh was satisfactory to maintain a -40 to -60 F furnace dew point during simulated production tests.

Tensile properties of 601 AB and 201 AB alloys sintered in dissociated ammonia are listed in Table IV. Both tensile strength and ductility were lower than for nitrogen sintered parts but satisfactory for most aluminum P/M applications. Strength levels up to 30,000 psi were obtained in 601 AB alloy and up to 42,000 psi for 201 AB alloy. For 601 AB parts, relative tensile strengths were 65 to 85 percent of those produced in a nitrogen atmosphere; 201 AB parts were 85 to 95 percent as strong as those sintered in nitrogen.

TABLE IV

PROPERTIES OF 601 AB AND 201 AB
ALLOYS SINTERED IN DISSOCIATED AMMONIA

Green Density			Tensile Strength	Yield Strength	Elongation Percent
Percent	g/cm^3	Temper	psi	psi	in 1 in.
601 AB Alloy[a]					
90	2.42	T1	13,500	11,000	2.5
		T4	15,700	12,700	3.5
		T6	23,100	---	1.0
95	2.55	T1	17,600	12,600	3.5
		T4	21,200	14,300	5.0
		T6	30,100	29,700	1.5
201 AB Alloy[b]					
90	2.50	T1	23,300	20,500	2.0
		T4	28,800	23,700	2.5
		T6	35,800	---	0.5
95	2.64	T1	25,200	22,000	2.0
		T4	32,000	26,100	3.0
		T6	41,800	41,600	1.0

a - Sinter 10-30 min at 1150 F (-40 to -60 dew point).

b - Sinter 10-30 min at 1100 F (-40 to -60 dew point).

The lower properties of parts sintered in dissociated ammonia
appear related to the presence of hydrogen and/or undissociated
ammonia in the sintering atmosphere. The adverse effect of hydrogen to cause gassing of aluminum has been well documented
(Ref. 4); also, it is known that ammonia will react with aluminum
to liberate hydrogen. When 100 percent ammonia vapor was used as
a heat treating atmosphere for aluminum, the tensile strength of
2024 alloy sheet was reduced by 29 percent and the elongation by
82 percent (Ref. 5).

Gas analysis performed during this investigation indicated
the level of undissociated ammonia in the sintering atmosphere
was less than 1 ppm at the dryer but it reformed to 8-15 ppm in
the furnace. It appears then that both hydrogen and undissociated
ammonia can be a source for hydrogen absorption by the liquid
phase at the sintering temperature. This can lead to increased

porosity in the compact during solidification and reduce the tensile properties of P/M parts.

Vacuum

Vacuum studies were aimed solely at establishing sintering parameters for alloys 601 AB and 201 AB and not to develop production data. Sintering times for 1/4 to 1/2-in. thick specimens were comparable to those for parts sintered in either nitrogen or dissociated ammonia, but sintering temperatures in vacuum had to be 25 F lower to prevent excessive distortion and melting.

The effect of presintering in air or nitrogen to remove lubricant prior to vacuum sintering was examined using a burn-off time of 15 min at 700-800 F. Figure 7 shows that 85 percent density 201 AB parts presintered in nitrogen developed significantly higher properties than presintered in air parts, but at 95 percent

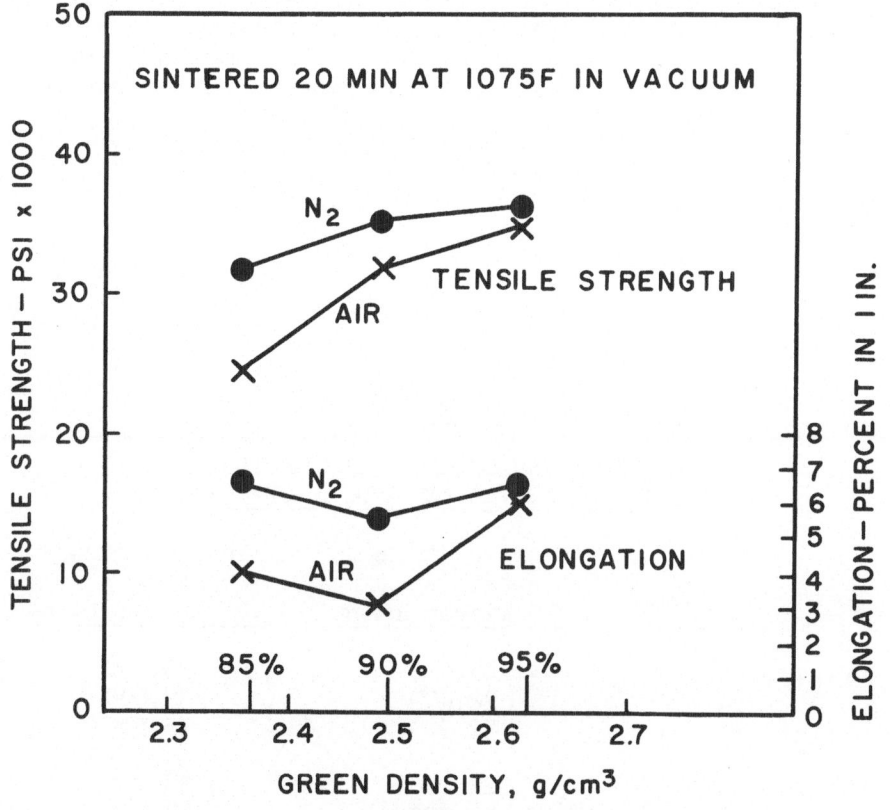

FIGURE 7
EFFECT OF PRESINTERING ATMOSPHERE ON
PROPERTIES OF 201 AB - T4 SINTERED IN VACUUM

density strength and ductility were comparable for both atmospheres. Although air presintering appears practical for higher density parts, the time should be kept as short as possible to prevent excessive oxidation.

The influence of vacuum level on sintered properties of 601 AB and 201 AB is shown in Table V. A low pressure of .01 microns produced substantially stronger parts at 85 percent density but only marginally stronger parts at 90 and 95 percent density. This is significant because most mechanical roughing pumps can easily achieve a vacuum level of 50 to 200 microns, thereby eliminating need of diffusion pumps and minimizing pump-down time.

Average mechanical properties of vacuum sintered 601 AB and 201 AB alloys developed over a range of vacuum levels, sintering times and presintering conditions are listed in Table VI. In all cases, strength and ductility were higher than for parts sintered in dissociated ammonia. Compared to nitrogen values, 601 AB vacuum sintered parts were 80 to 95 percent as strong. Alloy 201 AB parts, on the other hand, had higher sintered strengths in vacuum at 90 percent density and were only slightly lower than nitrogen values at 95 percent density. Elongations for vacuum sintered 201 AB parts were the highest obtained.

TABLE V

EFFECT OF VACUUM LEVEL ON TENSILE
STRENGTH OF 601 AB AND 201 AB SPECIMENS[a]

Alloy	Density, %	Tensile Strength psi, at 200 Micron Range	Tensile Strength psi, at .01 Micron Range
601 AB	85	25,500	30,200
	90	32,400	32,600
	95	32,900	35,500
201 AB	85	27,000	38,200
	90	43,900	44,100

a - All specimens heat treated to T6 temper after sintering.

TABLE VI

PROPERTIES OF 601 AB AND 201 AB
ALLOYS SINTERED IN VACUUM

Green Density			Tensile Strength	Yield Strength	Elongation Percent
Percent	g/cm^3	Temper	psi	psi	in 1 in.

601 AB Alloy[a]

90	2.42	T1	16,300	9,900	4.5
		T4	20,300	13,200	4.0
		T6	32,300	30,600	2.0
95	2.55	T1	19,000	11,600	5.0
		T4	23,400	14,400	7.0
		T6	33,300	31,800	2.0

201 AB Alloy[b]

90	2.50	T1	26,800	20,700	4.0
		T4	35,000	27,100	5.5
		T6	43,000	41,600	2.0
95	2.64	T1	26,700	21,200	4.0
		T4	36,300	26,900	6.5
		T6	45,300	42,000	2.0

a - Sinter 10-30 min at 1125 F (.01-200 micron vacuum)

b - Sinter 10-30 min at 1075 F (.01-200 micron vacuum)

The slightly lower properties for most vacuum sintered parts probably result from magnesium loss by vaporization at elevated temperatures and a higher porosity than nitrogen sintered parts. Chemical analyses of several specimens revealed that the magnesium content of the part surface was reduced by 25 to 50 percent during vacuum sintering. Future tests will determine if magnesium vaporization can be minimized by back filling the furnace with a partial pressure of nitrogen or argon.

Dimensions

Dimensions of sintered aluminum P/M parts are affected by compact density, sintering atmosphere, temperature and dew point. The effect of green density and atmosphere on sintered dimensions

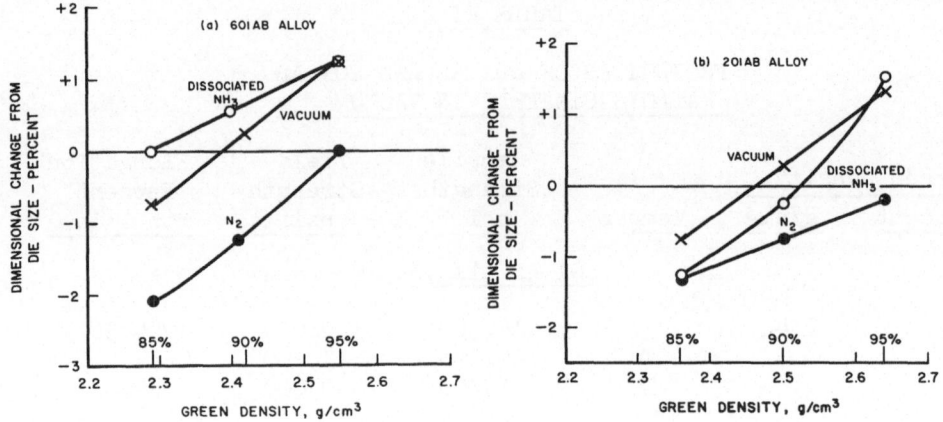

FIGURE 8
EFFECT OF GREEN DENSITY AND ATMOSPHERE ON SINTERED
DIMENSIONS OF 601 AB AND 201 AB ALLOYS

of 601 AB and 201 AB alloys are illustrated in Figures 8(a) and
8(b). Dimensions increased with increased green density in all
atmospheres. There was either shrinkage or no growth in 85 percent green density parts whereas high density compacts exhibited
growth when sintered in dissociated ammonia and vacuum. Nitrogen
sintered parts experienced shrinkage over the full range of densities except for 95 percent density 601 AB where no change was
noted.

These dimensional changes were quite consistent as long as
the sintering temperature was constant and the dew point in the
furnace was at least -40 F. Higher than normal temperatures
caused excessive shrinkage and distortion or even melting in extreme cases. Lower than normal temperatures produced parts having
increased dimensions and reduced properties. Thus, for example,
lowering the sintering temperature of 201 AB alloy from 1100 to
1060 F caused 95 percent density specimens sintered in dissociated
ammonia to change from 0.25 percent shrinkage to 1.0 percent growth.
At 1080 F, 0.25 percent growth was observed. Properties were less
affected, although a reduction in tensile strength of 2 to 10 percent was observed, depending on thermal treatment. High dew points
in the furnace resulted in excessive part expansion and a significant reduction in properties. The relationship of atmosphere dew
point to properties and dimensions is shown in Ref. 1 (Figures 4
and 7).

Microstructure

Representative samples of 201 AB and 601 AB were examined metallographically after sintering in nitrogen, dissociated ammonia and vacuum. No significant differences were observed in the general microstructure that could be related to atmosphere, although higher porosity was noted in specimens sintered in dissociated ammonia and vacuum. This was expected, however, since these specimens exhibited growth after sintering. The microstructure of 201 AB and 601 AB is typified by the photomicrographs for dissociated ammonia sintered specimens illustrated in Figures 9 and 10.

Figure 9 shows at 500X the well bonded particles of sintered 201 AB alloy. Recrystallized grains can be noted throughout the structure. A finely dispersed constituent can be observed in the center of many grains. This appears to be the original dendritic structure normally seen in atomized aluminum particles that was not completely dissolved during sintering. Intermetallic phases of Al-Cu eutectic and Al-Cu-Mg-Si constituent can be noted along particle boundaries. Similar phases are frequently observed in the as-cast structure of 2XXX series aluminum alloys, like 2014.

The microstructure of sintered 601 AB alloy at 500X is shown in Figure 10. This composition contains less alloying ingredients than 201 AB and, therefore, has less constituent dispersed throughout the microstructure. Disrupted oxide films can be noted along many of the original aluminum particle boundaries and evidence of recrystallization can be observed in some particles.

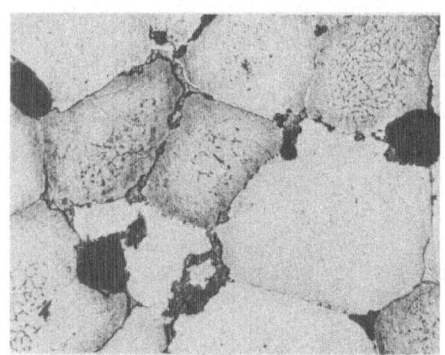
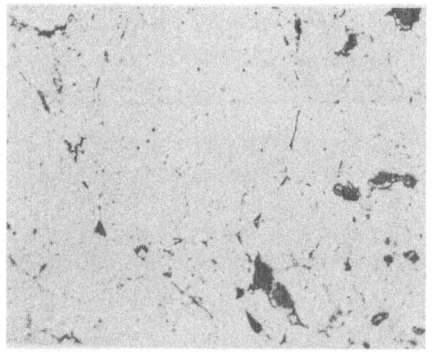

FIGURE 9 - 201 AB ALLOY   FIGURE 10 - 601 AB ALLOY
PHOTOMICROGRAPHS SHOWING MICROSTRUCTURE OF SINTERED 201 AB
AND 601 AB ALLOYS

Mag. 500X (Red. 50%)

SUMMARY

Aluminum P/M parts can be production sintered in various types of furnaces and atmospheres. Selection of sintering furnace depends upon economic considerations and production rates desired. Batch furnaces have lowest investment costs and are adequate for low to medium production whereas continuous furnaces are more costly but provide higher production rates.

Strong, well-sintered P/M parts can be obtained in atmospheres of nitrogen, dissociated ammonia and in vacuum. Atmosphere selection depends upon facilities available within individual plants plus property requirements. Highest strengths are produced in nitrogen followed by vacuum and dissociated ammonia. Reproducible dimensions can be achieved with proper attention to compact density, sintering temperature, dew point and atmosphere.

REFERENCES

1. J. H. Dudas and W. A. Dean, "The Production of Precision Aluminum P/M Parts," International Journal of Powder Metallurgy, Vol. 5, April, 1969.

2. P. F. Mathews, "Effects of Processing Variables on the Properties of Sintered Aluminum Compacts," International Journal of Powder Metallurgy, Vol. 4, October, 1968.

3. J. H. Dudas and K. J. Brondyke, "Aluminum P/M Parts - Their Properties and Performance," Technical Paper No. 700141, Society of Automotive Engineers, Inc., Two Pennsylvania Plaza, New York, New York, 10001.

4. K. R. Van Horn (Editor), Aluminum Vol. I, pp. 26-28, American Society for Metals, Metals Park, Ohio, 1967.

5. K. R. Van Horn (Editor), Aluminum Vol. III, pp. 313-314, American Society for Metals, Metals Park, Ohio, 1967.

EXPERIMENTAL MANUFACTURE OF INCONEL ALLOY 718

COMPRESSOR ROTOR BLADES FROM METAL POWDER PREFORMS

Bernard Triffleman, F. C. Wagner, Keki K. Irani

Curtiss-Wright Corporation

Buffalo Facility

The objective of the program was to develop methods for making aircraft quality parts from forged Inconel Alloy 718 metal powder preforms. Aircraft quality parts were defined as those having room-temperature tensile and fatigue properties, and 1200 $^\circ$F. tensile and 1300 $^\circ$F. stress-rupture properties meeting aircraft specifications.

Basic decisions made on this program were:
    a) to use, wherever possible, standard production type equipment for the consolidation of the metal powders.
    b) to design metal powder preforms which would fit into present forging shop practices.
It was deemed that this approach would enable us to scale up the processes in the shortest possible time.

To design a forging preform, it was necessary to determine the amount of reduction in area required in forging so as to obtain a specimen meeting minimum properties. In order to make forging blanks it was necessary to learn how to press and sinter billets to a density where internal oxidation or nitrification would be minimal during the hot forging step. Previous experience at Curtiss-Wright had shown that a minimum density of 92% of theoretical was necessary to achieve these latter goals.

Since it had been shown that inert gas atomized powders gave the lowest oxygen and interstitial element content, it was decided to use these types of powders for the program. Several lots of powders (identified as Lots A, B, C) were obtained from two vendors, (Vendor A and Vendor B).

The first task was to set specifications for the metal powders. Since powders are a different type of raw material than commercial bar stock of the same alloy, a new specification had to be written.

Chemical specifications for commercial bar stock were reviewed. Commercial bar stock for aircraft forgings is purchased to inclusive specifications which includes, among other things, a fairly wide chemical specification plus a series of physical tests. In receiving a great many heats of acceptable bar stock meeting all the physical tests, it was found that the various elements of the alloy fell within a rather narrow range. This range was then selected for our chemical specifications. Oxygen and nitrogen are not specified in commercial bar stock but vacuum melting is specified and analyses of various bar stock heats gave contents of 0.01% O_2 and 0.007% N_2 and therefore these elements were specified to the lowest limit which the powder manufacturers would accept at the time.

Table 1 shows the range of C, O_2, and N_2 specified and the analyses of these elements on various lots of powders purchased from several manufacturers.

The size specifications for the powders were made on the basis of previous experience for a good pressing powder and also what could be obtained from the powder manufacturers at the time. In commercial operations, apparent density, tap density, and flow rate specifications also have to be set but for this initial work these requirements were omitted. The requirements and actual results are shown in Table 2 along with apparent density, tap density, and flow rates.

A study showed that the most economical method of pressing the powders into shapes was to use mechanical (or hydraulic) presses or isostatic presses at ambient temperatures. While most of the literature (1), (2) has indicated that spherical, low oxygen content atomized powders were not pressable at ambient temperatures, it was

TABLE 2

PHYSICAL PROPERTIES OF VARIOUS INCO 718 POWDERS

	C-W Spec.	Vendor A Lot A	Vendor A Lot B	Vendor A Lot C	Vendor B Lot A	Vendor B Lot B
1) Screen Analysis - %						
+ 50	---------	0.00	0.00	0.00	0.00	------
-50 + 100	2 Max	0.83	9.22	1.71	0.52	------
-100 + 200	40 ± 10	42.81	42.49	42.80	29.72	------
-200 + 400	20 ± 10	31.36	26.53	30.01	30.66	------
-400	40 ± 10	25.00	21.76	25.48	39.10	------
2) Apparent Density - g/cc	--------	4.13	4.14	3.82	4.57	4.61
3) Tap Density - g/cc	--------	5.26	5.13	4.85	5.26	5.32
4) Hall Flow Rate - Sec.	--------	16.0	14.2	16.5	None	11.0

decided to undertake the pressing program using special binders
which would not leave residues after sintering.

It was found that the first lot of powder (Vendor A, Lot A)
could be pressed both mechancially and isostatically with and without binders in test dies. Densities obtained vs pressures for both
types of pressings are shown in Figure 1.

Green strength vs pressure for mechanically pressed compacts
are shown in Figure 2 and green strength vs pressure for isostatically pressed powders are shown in Figure 3. Thus it was established
that these powders could be pressed at ambient temperature and would
have sufficient strength to be pressed into more complicated shapes.
While some of the other lots of powders could not be pressed without
binders, all lots could be pressed with binders.

Two inch and smaller billets were pressed at 40 T.S.I. mechanically and at 30 T.S.I. isostatically and sintered in a vacuum
furnace at 2100 - 2200 $^{\circ}$F. for 2 to 8 hours. Densities were achieved which varied between 92 and 98% of theoretical density.
Figure 4 shows a typical microstructure of a sintered billet.

A forging die was designed and the configuration of this die
is shown in Figure 5. Billets were machined so that reductions in
height of 10, 20 and 70% would be achieved with about 0.5% in excess
for flash. These reduction ratios were chosen to give some idea
of the reductions that may be required for complete densification
and for obtaining wrought properties. A 70% reduction in height
represented a limit that would make forging preforms uneconomical
since no forging operations could be eliminated using this reduction in area.

Billets were forged to 70% reduction in thickness at 1900 $^{\circ}$F.,
2000 $^{\circ}$F., 2100 $^{\circ}$F. and 2200 $^{\circ}$F. on a 1500 pound hammer. The billets
forged at 2000 $^{\circ}$F. and 2100 $^{\circ}$F. were all crack free while those
forged at 2200 $^{\circ}$F. cracked severely. The billets forged at 1900 $^{\circ}$F.
exhibited small cracks. A photograph of the 2100 $^{\circ}$F. forgings are
shown in Figure 6. The remaining 10 and 20% billets were therefore
forged at 2000 $^{\circ}$F. and 2100 $^{\circ}$F. Along with the sintered billets,
billets machined from commercial bar stock were forged at 10%, 20%
and 70% R.A. in the same dies.

The forgings were checked for hardness, density, C, O_2, N_2
and microstructure. The densities varied between 99.5% of theoretical for the 10% R.A. billets to 100% for the 70% R.A. billets.
The C, O_2, and N_2 contents are shown in Table 3 along with comparable figures of the powder and in the as-sintered state. It can
be seen that there was some increase in both oxygen and nitrogen
content.

TABLE 1
INTERSTITIAL ANALYSES ON
VARIOUS LOTS OF INCO 718 POWDERS

	C	O_2	N_2
C-W Spec.	0.03-0.05	0.030-Max.	0.010-Max.
Vendor A, Lot A	0.042	0.014	0.0098
Vendor A, Lot B	0.075	0.017	0.0095
Vendor A, Lot C	0.060	0.020	0.0090
Vendor B, Lot A	0.078	0.019	0.018
Vendor B, Lot B*	0.096	0.029	

*Remelt of Inco 718 Scrap

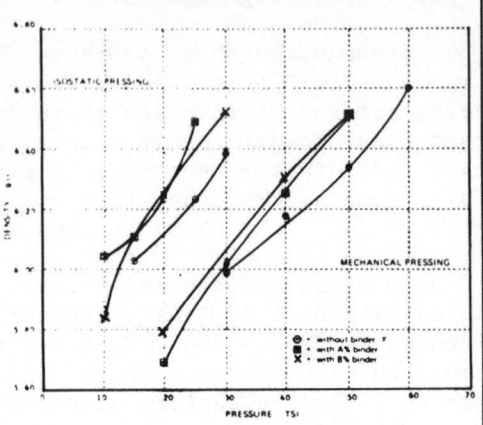

FIGURE 1
VENDOR A INCO 718 POWDER LOT A
PRESSURE vs DENSITY

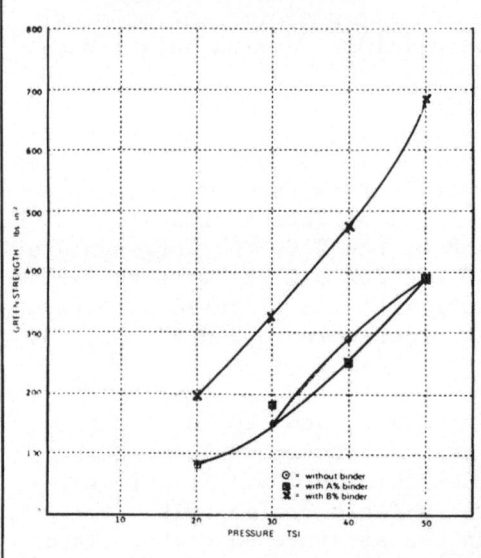

FIGURE 2
VENDOR A INCO 718 POWDER LOT A
MECHANICALLY PRESSED BARS
PRESSURE vs GREEN STRENGTH

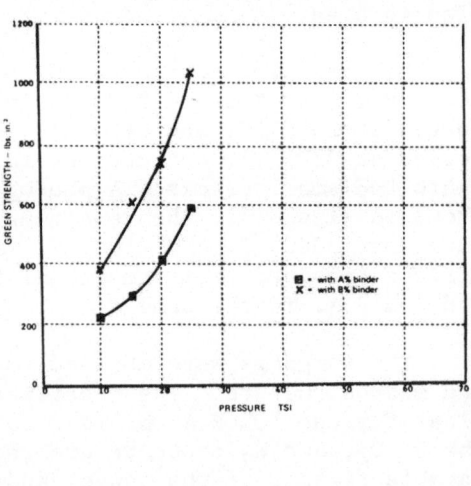

FIGURE 3
VENDOR A INCO 718 POWDER LOT A
ISOSTATICALLY PRESSED BARS
PRESSURE vs GREEN STRENGTH

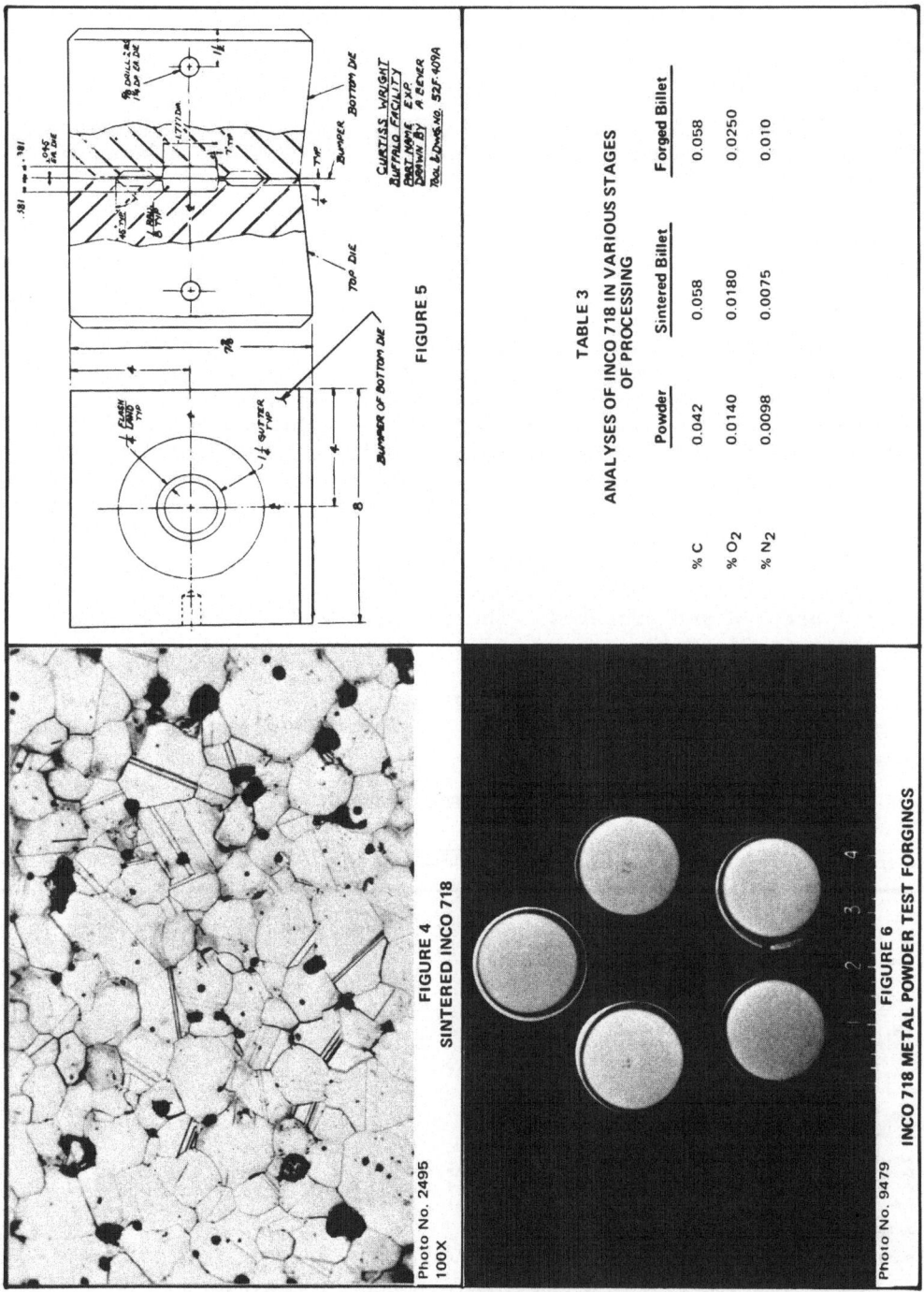

FIGURE 5

TABLE 3

ANALYSES OF INCO 718 IN VARIOUS STAGES
OF PROCESSING

	Powder	Sintered Billet	Forged Billet
% C	0.042	0.058	0.058
% O_2	0.0140	0.0180	0.0250
% N_2	0.0098	0.0075	0.010

FIGURE 4
SINTERED INCO 718

Photo No. 2495
100X

FIGURE 6
INCO 718 METAL POWDER TEST FORGINGS

Photo No. 9479

Ambient temperature and 1200 °F. properties were run on the 70% R.A. forged billets and no significant difference in properties were observed between the forged powder billets and the forged commercial bar stock billets.

The forgings were given the standard heat treatment (1 hour @ 1750 °F., AC, followed by another hour @ 1750 °F., AC, 8 hours @ 1325 °F., Cool 100 °F/hr. to 1150 °F., 8 hours @ 1150 °F., AC). Test specimens were cut from the heat treated pieces and these were tested for typical customer requirements. The results are shown in Table 4. It should be noted that except for some of the ductilities, all other physical specifications were met.

At this point, there was confidence that a blue print part could be made from a metal powder preform. It was reasoned that by proper design of the forging preform and dies porosity could be avoided in the portions of the preform where the reduction in area would be small and/or where the metal movement would not be great. Porosity in the test forgings probably accounted for the low ductility noted in some cases. Such porosity would also be detrimental to fatigue properties.

Isostatic molds were designed and powder preforms were isostatically pressed and sintered. The sintered and peened preforms are shown in Figure 7.

The sintered preforms were subjected to the following basic forging operations: 1. Block 2. Semi Finish Forge 3. Finish Forge 4. Coin.

TABLE 4

PROPERTIES OF 2000 °F. FORGED AND HEAT-TREATED* METAL POWDER BILLETS OF INCO 718

	Customer Specification	Vendor "A" Lot A		Vendor "B" Lot A		Commercial Bar Stock HT # 6720	
HT Properties		10% R.A.	70% R.A.	10% R.A.	70% R.A.	10% R.A.	70% R.A.
Hardness-Rc	32 min.	42-43	44-45	41-41	43-44	41-42	43-44
Grain Size	3 or finer	9-10	8-10	9-10	8-10	2-5	7-8
Y.S. - K.S.I.	150.0 min.	170.2	176.3	148.5	164.9	171.8	169.0
U.T.S. - K.S.I.	180.0 min.	193.5	207.3	182.5	208.4	192.4	209.2
El. - %	12 min.	10.4	13.0	7.4	15.6	15.5	15.8
R.A. - %	15 min.	19.4	24.9	7.9	33.5	26.6	37.4
1200 °F. Properties							
Y.S. - K.S.I.	125.0 min.	141.6	153.9	126.5	168.8	143.0	147.8
U.T.S. - K.S.I.	145.0 min.	156.1	173.4	145.3	171.6	153.9	169.1
El. - %	10 min.	10.1	9.7	5.8	10.0	14.8	17.0
R.A. - %	15 min.	18.5	23.7	8.9	20.7	31.3	45.3
1200 °F. Stress-Rupture at 100 K.S.I.							
Smooth							
Hours	25 min.**	99.0	84.8	218.1	68.7	91.9	119
El. - %	----	2.6	7.5	3.9	4.4	----	3.0
R.A. - %	----	7.9	20.1	9.9	17.8	----	1.7
Notched							
Hours	25 min.**	361	116.0	364.2	113.1	162.3	138.4

* 1 hour @ 1750 °F.,AC; 1 hour @ 1750 °F., AC; 8 hours @ 1325 °F., FC to 1150 °F.; 8 hours @ 1150 °F., AC.
** Test may be discontinued after 70 hours. Below 70 hours the notched specimen must last as long as the smooth specimen.

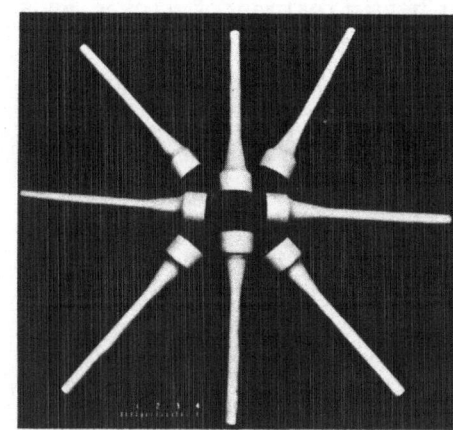

Photo No. 9478

FIGURE 7
VENDOR B INCO 718
FORGING PREFORMS
SINTERED AND PEENED

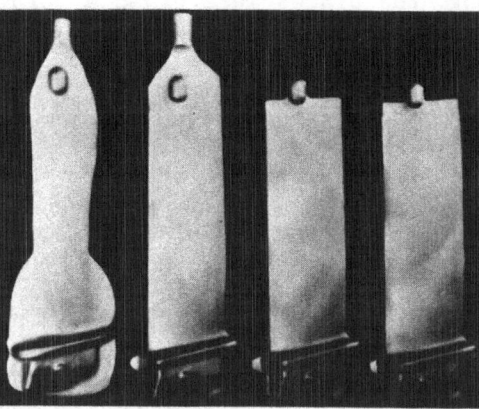

Blocked   Semi Finished   Finished   Coined
Forged   Forged

FIGURE 8
COMPRESSOR ROTOR BLADES
IN VARIOUS STAGES OF FORGING

Photo No. 9483
Part No. 231818
Lot A

FIGURE 9
VENDOR B INCO 718
FORGED AND COINED
SOLUTIONED & AGED

100X
Photo No. 2959
Part No. 231818

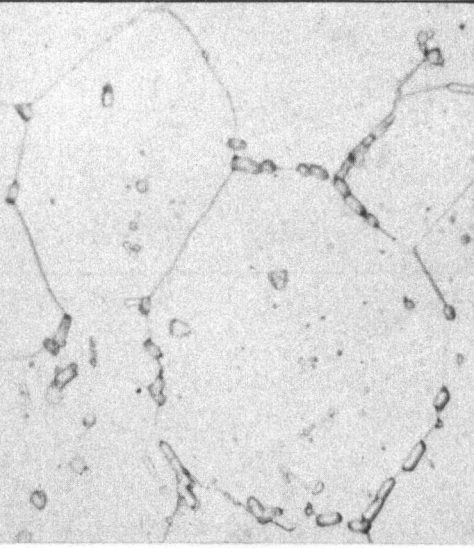

FIGURE 10
VENDOR B INCO 718
FORGED AND COINED
SOLUTIONED & AGED

1000X
Photo No. 2960
Part No. 231818

Figure 8 shows a part in all the forging stages. After coining, the blades were given a solution heat treatment followed by a precipitation heat treatment. Photomicrographs of the solution and aged pieces are shown in Figures 9 and 10.

A few blades were selected for mechanical physical testing while the remaining blades were machined, ground, and inspected. The finished rotor blades are shown in Figure 11.

In forging the blades, three different lots of powder were used as raw material to study the reproductility of the process. The test specimens were cut from the butt end of the blade where the preforms receives the least amount of work. It was presumed that if these test results were satisfactory, then the airfoil which receives a great deal of more work would also be satisfactory. The results of the test on the heat treated blades are shown in Table 5.

Minimum requirements were met for blades made from all three lots of powder. For the most part, the test results fell within the range obtained from commercial bar stock.

The majority of the blades forged in this series were made from Vendor B, Lot A powder. These blades were fatigue tested. Blades made from commercial bar stock were fatigue tested to serve as a control. The fatigue tests were conducted at room temperature. Testing was performed by a "step technique" with the initial stress step at 24 K.S.I. Upon completion of 2×10^6 cycles, the stress level was increased in 4 K.S.I. increments until cracking. Cracking was detected by a change in blade tip amplitude with time. Stresses were measured using a strain gage. A summary of the calibration data is shown in Table 6. The comparison point chosen was the tip amplitude of vibration at 19,500 P.S.I. The fatigue data are summarized in Table 7. The forged powder blades exhibited approximately 20% higher fatigue strength. An examination of the blade fractures after testing showed them to be normal fatigue fractures. Metallographic examination of specimens taken through the fatigue nuclei showed the absence of any material or processing defects.

The final step was to check the blades in an official military accepted 150 hour engine parts qualification test. This engine test is used for releasing new materials and parts for production. The blades have passed this most important test.

CONCLUSIONS

A method of processing Inco 718 powders was developed which resulted in fully dense aircraft engine compressor blades meeting specification properties, including improved fatigue properties

TABLE 5

PROPERTIES OF FORGED AND HEAT TREATED INCO 718 ROTOR BLADES

H. T. Properties	Customer Specification	Vendor "B" Lot A	Vendor "B" Lot B	Vendor "A" Lot A	Commercial Bar Stock (20 Lots)
Butt Hardness — Rc	39 min.	41	44	43	41.5—45.0
Air Foil Hardness — Rc	39 min.	41	44	43	
Butt Grain Size	5 or finer	7—8	7—8	5—6	5—8
Air Foil Grain Size	5 or finer	7—8	7—8	5—6	5—8
Y.S. — K.S.I.	150.0 min.	169.7	186.0	181.7	160—194
U.T.S. — K.S.I.	180.0 min.	201.7	208.4	199.4	186—213
El. — %	10 min.	16.2	16.4	15.2	13.2—23.4
R.A. — %	12 min.	35.4	33.5	42.4	30.1—49.7
1200°F Properties					
Y.S. — K.S.I.	125.0	132.2	154.9	145.6	131—164
U.T.S. — K.S.I.	140.0	165.5	170.6	157.1	146—178
El. — %	10 min.	11.3	11.1	14.5	12.3—20.5
R.A. — %	12 min.	33.9	32.8	43.5	31.6—47.1
1300°F Stress-Rupture 75 K.S.I. Load					
Smooth					
Hours	23.0 min.*	37.9	93.6	158.6	52—513
El. — %	5.0 min.	5.1	10.0	11.6	5.1—32.3
R.A. — %	—	6.8	17.0	22.6	
Notched					
Hours	23.0 min.*	37.9	93.6	158.6	

*In a combination Smooth and Notched specimen, fracture shall occur in the smooth section.

FIGURE 11

COMPRESSOR ROTOR BLADES MADE FROM METAL POWDER PREFORMS

Photo No. 9270
Part No. 231817
Part No. 231818
Lot A

TABLE 6

FATIGUE TEST CALIBRATIONS SUMMARY

Characteristic	Forged Powder Blades	Forged Bar Stock Blades
Max. Amplitude @ 19,500 P.S.I.	0.281"	0.281"
Min. Amplitude @ 19,500 P.S.I.	0.260"	0.255"
Max. Frequency, CPS	240	237
Min. Frequency, CPS	230	230

TABLE 7

FATIGUE DATA ON FORGED POWDER AND WROUGHT BLADES

Characteristic	Forged Powder Blades	Forged Bar Stock Blades
Max. (K.S.I.)	56.8	49.6
Min. (K.S.I.)	41.2	30.4
Ave. (K.S.I.)	50.4	41.7

as compared with conventional wrought blades.[*]

REFERENCES

1. Kortovich, C. S. - "Close Tolerance Forgings From Powder Met-
    allurgy Preforms"; Air Force Materials Laboratory Technical
    Report TR-69-181 June, 1969.

2. Moyer, K. H. - "Development of An Improved Manufacturing Pro-
    cess For The Production of Higher Purity Superalloy Powders";
    Air Force Materials Laboratory Interim Engineering Progress
    Report IR-9-182-CIV July, 1967.

3. Wagner, H. J. and Hall, A. M. - "Physical Metallurgy of Alloy
    718"; DMIC Report 217 June, 1965.

4. Hall, R. C. - "The Metallurgy of Alloy 718"; Transactions of
    the ASME September, 1967.

5. Barker, J. F.; Ross, E. W. and Radavich, J. F. - "Long Time
    Stability of Inconel 718"; Journal of Metals, Vol. 22 No. 1
    January, 1970.

6. Allen, M. M.; Athey, R. L. and Moore, J. B. - "Application of
    Powder Metallurgy to Superalloy Forgings"; Metals Engineering
    Quarterly, Vol. 10-No. 1 February, 1970.

*Latest information of the authors: These blades have passed an
 official military accepted 150 hr engine parts qualification test.

ELEVATED TEMPERATURE MECHANICAL PROPERTIES OF A DISPERSION

STRENGTHENED SUPERALLOY

J. S. Benjamin and R. L. Cairns

The International Nickel Company, Inc.

Sterling Forrest, New York

ABSTRACT

A new process called mechanical alloying has been used to produce composite superalloy powders containing an intimate dispersion of refractory oxide particles. Consolidated materials made from these powders exhibit dispersion strengthening and gamma prime precipitation hardening.

High temperature creep and stress rupture properties of a nickel base alloy containing 19.7%Cr, 2.3%Ti, .88%Al, .067%Zr, .007%B, .06%C, 1.12%Al_2O_3 by weight with 2.25 volume % yttria dispersoid are presented. Two hardening mechanisms each influence the high temperature mechanical properties of this material in a complex interaction. At lower temperatures the rupture strength behavior is similar to that of ordinary nickel base superalloys, while at high temperatures it corresponds to the behavior found in dispersion strengthened nickel.

The apparent activation energy for creep was found to be 153 kcal/mole over the temperature range 0.55 to 0.6 Tm and 202 kcals/mole between 0.65-0.8 Tm. These energies were compared with values obtained for a dispersion-free, wrought superalloy of similar composition (103 kcals/mole between 0.55-0.6 Tm) and stress relief annealed TD Nickel bar (133 kcals/mole between 0.7-0.75 Tm). Although creep of the dispersion strengthened superalloy clearly showed two regimes of behavior, the activation energy found below 0.6 Tm was still considerably higher than that for conventional

hardened superalloys. This difference was also evident from the
slopes of stress rupture plots which were flatter for the dispersion strengthened superalloy than for conventional alloys. High
stress exponents were obtained over both temperature ranges.

INTRODUCTION

Improvements in the properties of nickel base superalloys are
being sought to meet the needs of advancing technology particularly
for applications in gas turbines and other aerospace fields.
Materials are required which combine high creep strength at intermediate and elevated temperatures with good high temperature corrosion resistance.

Dispersion strengthening has offered one of the best means
for increasing the strength of metals at temperatures from 0.7 Tm
to the melting point. While this concept has been exploited since
1910 with the invention of thoriated tungsten(1) the first dispersion strengthened structural load bearing system was sintered
aluminum powder SAP(2). Further advances were made with the application of dispersion strengthening to higher melting point metals
such as copper and nickel(3,4). Although mechanical working techniques were also developed to give improved high temperature
strength these dispersion strengthened metals were limited by low
strength at intermediate temperatures and lack of corrosion resistance. Some success has been achieved in producing a relatively
corrosion resistant dispersion strengthened 80Ni-20Cr alloy but the
problem of low intermediate temperature creep strength still re-
stricts the usefulness of this material.

The most obvious way to overcome these deficiencies would be
to combine the corrosion resistance and intermediate temperature
strength of a conventional γ' hardened nickel base superalloy with
the high temperature strength and stability characteristics of a
dispersion strengthened metal. There are many suitable powder
metallurgy techniques for producing dispersion strengthened metals
and dispersion strengthened simple solid solutions. The problem
of including reactive γ' forming elements such as aluminum and
titanium in a dispersion strengthened alloy powder without causing
severe contamination or destroying the required oxide dispersion
has, until recently, defied solution.

Now a new process called mechanical alloying has been developed which will produce composite superalloy powders containing an
intimate dispersion of refractory oxide particles(5). Consolidated
materials made from these powders exhibit both dispersion strengthening and γ' precipitation hardening.

EXPERIMENTAL PROCEDURE

Materials

This paper presents properties of a nickel-base superalloy (nominal matrix composition essentially Ni, 19Cr, 1.2Al, 2.4Ti) dispersion strengthened with 2.5 volume percent of yttrium oxide. The creep rupture behavior of this material is compared with that of dispersion free superalloy bar of similar composition produced by conventional melting and working and of commercial stress-relief annealed TD Nickel bar*.

Dispersion Strengthened Superalloy

The dispersion strengthened superalloy used in this study was made by the dry high energy mechanical alloying technique(5). In this process a blend of raw material powders(Table I) was processed to composite powder particles in a Model 10-S Szegvari Attritor Grinding Mill**. This machine is a high energy driven ball mill in which the charge of powder and balls is held in a stationary, vertical, water cooled tank and agitated by impellers radiating from a rotating central shaft.

TABLE I

RAW MATERIAL POWDERS USED IN PRODUCTION OF DISPERSION STRENGTHENED SUPERALLOY

Component	Particle Size
Type 123 Carbonyl Nickel	$4-7\mu$
Chromium	-200 mesh
Ni/Al/Ti master alloy	-200 mesh
Ni/Zr master alloy	-200 mesh
Ni/B master alloy	-200 mesh
Y_2O_3	200-300 Å

Key features of the mechanical alloying process are the high energy milling and the omission of any surface active agent other than the air sealed into the tank with the charge. This practice actually promotes particle welding in contrast to conventional ball milling in which welding is inhibited by use of liquids and other surfactants.

---

* Registered trademark of Fansteel Metals Division
**Registered trademark of Union Process, Inc.

Mechanically alloyed powder was packed in mild steel cans
which were evacuated and sealed by fusion welding. Consolidation
was accomplished by hot extrusion.

Prior to machining, specimens were heat treated as follows:
2 hours at 2325°F or 2400°F in argon, air cool; 7 hours at 1975°F,
air cool; then 16 hours at 1300°F, air cool. The first treatment
produced a coarse grain structure while the last two treatments were
conventional nickel-base superalloy solution and aging treatments.

TD Nickel. The TD Nickel used was 1/2" diameter stress relief
annealed barstock containing 2.4 weight percent ThO_2 dispersoid.
This was purchased from E.I. DuPont de Nemours and Company, Inc.,
and was tested in the as-received condition.

20Cr-1.2Al-2.4Ti Nickel-Base Superalloy. The conventional
nickel-base superalloy was 1" diameter bar extruded at 2150°F from
a cast 3-1/2" round billet. Specimens were heat treated 7 hours at
1975°F in air, air cool, followed by 16 hours at 1300°F in air, air
cool.

Compositions of the three materials are shown in Table II.

Testing Procedure

Creep rupture samples of each material were tested in air
according to ASTM Specification E-139-66T at various stresses and
temperatures. Temperatures were controlled within 2°F and creep
extensions were measured with a sensitivity of 50 microinches.

EXPERIMENTAL RESULTS AND DISCUSSION

Structure

Structures of the three materials are compared in Figure 1.
The structure of the heat treated dispersion strengthened superalloy consisted of coarse cylindrical grains of irregular cross
section elongated in the extrusion direction (Figure 1a). The
macrograph in Figure 1b shows the coarse grain structure clearly.
The TD Nickel bar had a fine grained fibrous structure (Figure 1c).
The longest grain dimensions were parallel to the bar length and,
by inference, parallel to the major working direction. The wrought
superalloy bar had generally equiaxed grains of mixed sizes
(Figure 1d).

The different longitudinal grain structures of the heat treated
dispersion strengthened superalloy and the TD Nickel are further
illustrated by the replica electron micrographs shown in Figure 2.
The distribution of carbides in the dispersion strengthened superalloy within the grains and at grain boundaries can be seen in

TABLE II

COMPOSITION OF MATERIALS

Material	Ni	Cr	Al	Ti	C	B	Zr	O	Y_2O_3	ThO_2	Al_2O_3*
Dispersion Strengthened Superalloy	Bal	19.7	0.88	2.30	.06	.007	.067	.78	1.18	--	1.12
Conventional Superalloy	Bal	19.1	1.17	2.48	.077	.007	.07	--	--	--	--
TD Nickel	Bal	.003		<.001	.0027	--	--	--	--	2.4	--

* Estimated from oxygen content

Figure 2a. Identification of dispersoid and γ' is difficult due to the complex nature of the etched surface. The transverse grain dimension of the TD Nickel is 1 to 2 microns compared to a value of 50 to 100 microns in the dispersion strengthened superalloy determined from Figure 1a. Thoria can be seen in the TD Nickel as can differential etching patterns delineating the grains.

Figure 3 shows transmission electron micrographs of these materials. In the coarse grained dispersion strengthened super-alloy (Figure 3a and b) there are four populations of particle species discernible. The γ', present as a fine semi-transparent precipitate of from 50Å to 150Å diameter, is seen best in Figure 3a. The remaining coarser dispersion consists of Y_2O_3, Al_2O_3 and car-bide particles. The larger particles are probably MC carbides of up to 3000Å in size. Some of these display a polygonal outline suggesting a cubic shape. The finer particles are of Y_2O_3 and Al_2O_3, with sizes of from 1000Å down to 100Å well within the range required for dispersion strengthening. At a given size the yttria particles are more opaque due to the higher atomic number of yttrium compared to aluminum. Some of these particles also display polygonal shapes. The dislocation substructure in the dispersion strengthened superalloy is seen best in Figure 3b in which dislo-cations pinned by dispersoid particles are labeled as well as other features.

The ThO_2 dispersoid particles can be seen in the fine grained TD Nickel in Figure 3c. The population of very small semi-trans-parent particles is believed to be due to contamination of the foil. The balance of the structure consists of high angle grain boundaries and dislocations, some of which are pinned by thoria particles.

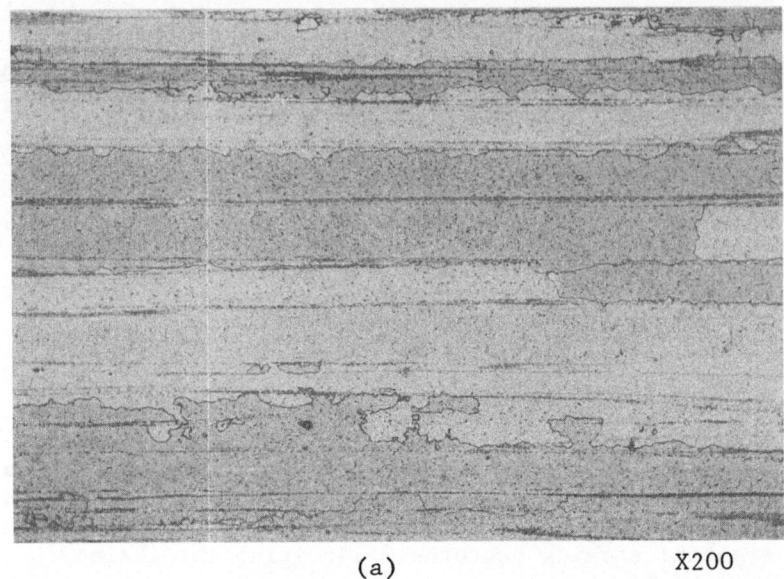

(a)   X200

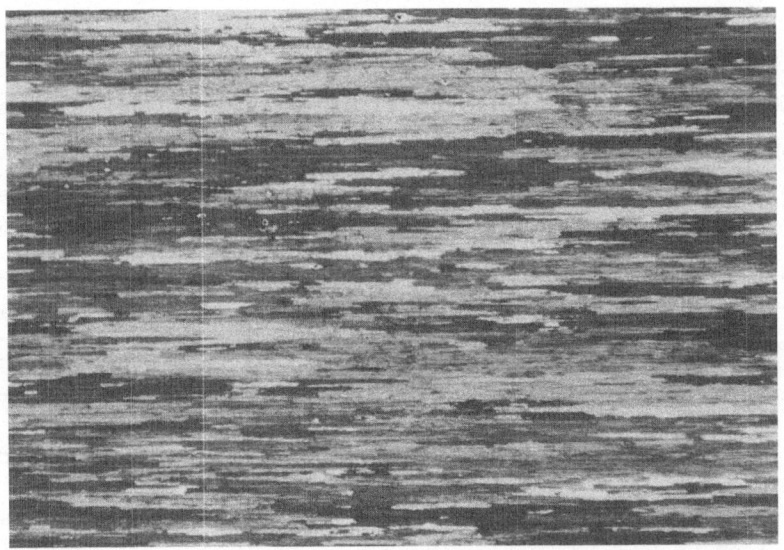

(b)   X10

FIGURE 1.   (a)   Longitudinal Microstructure of Dispersion
                  Strengthened Superalloy
            (b)   Longitudinal Macrostructure of Dispersion
                  Strengthened Superalloy

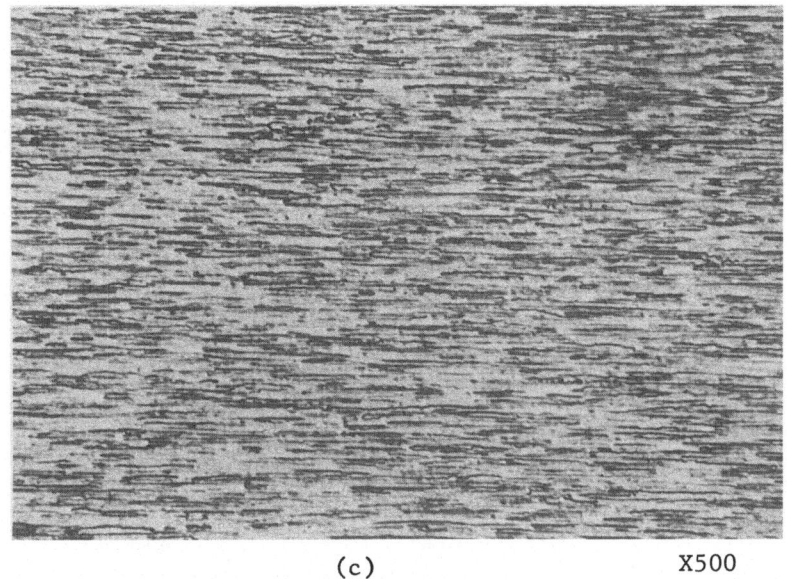

(c)   X500

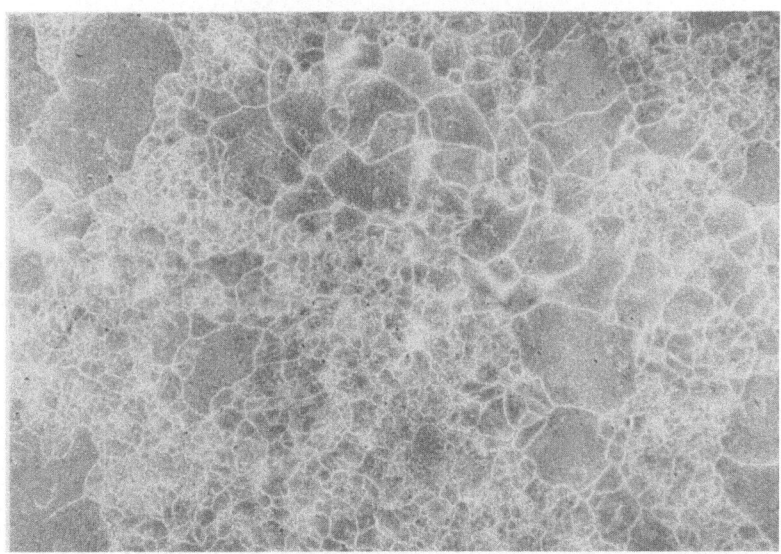

(d)   X100

FIGURE 1. (continued)
        (c) Longitudinal Microstructure of Stress Relief
            Annealed TD Nickel
        (d) Longitudinal Microstructure of Cast and Wrought
            Nickel Base Superalloy

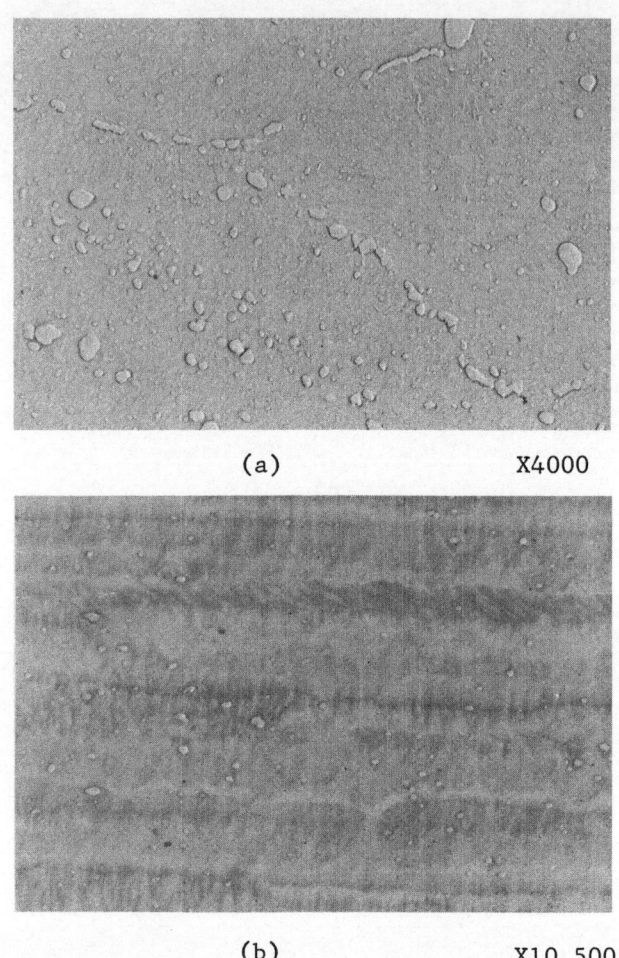

(a)   X4000

(b)   X10,500

FIGURE 2. Electron Replica Photomicrographs Comparing
           Longitudinal Structures of Dispersion
           Strengthened Superalloy (a), with Stress
           Relief Annealed TD Nickel (b)

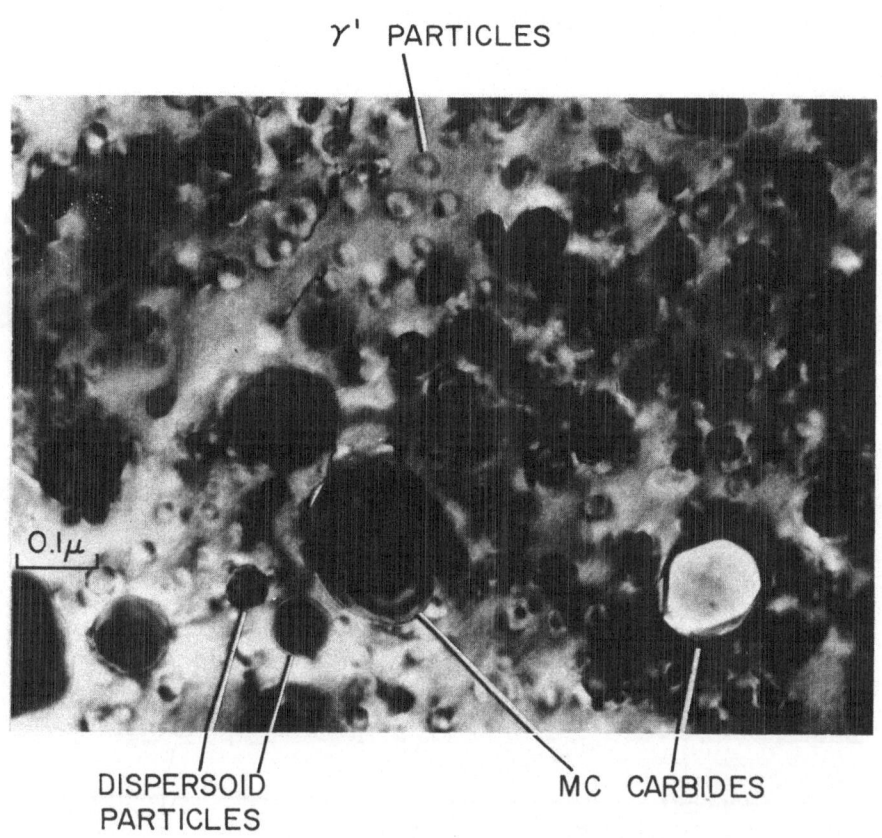

FIGURE 3a. Transmission Electron Micrograph of
Dispersion Strengthened Superalloy

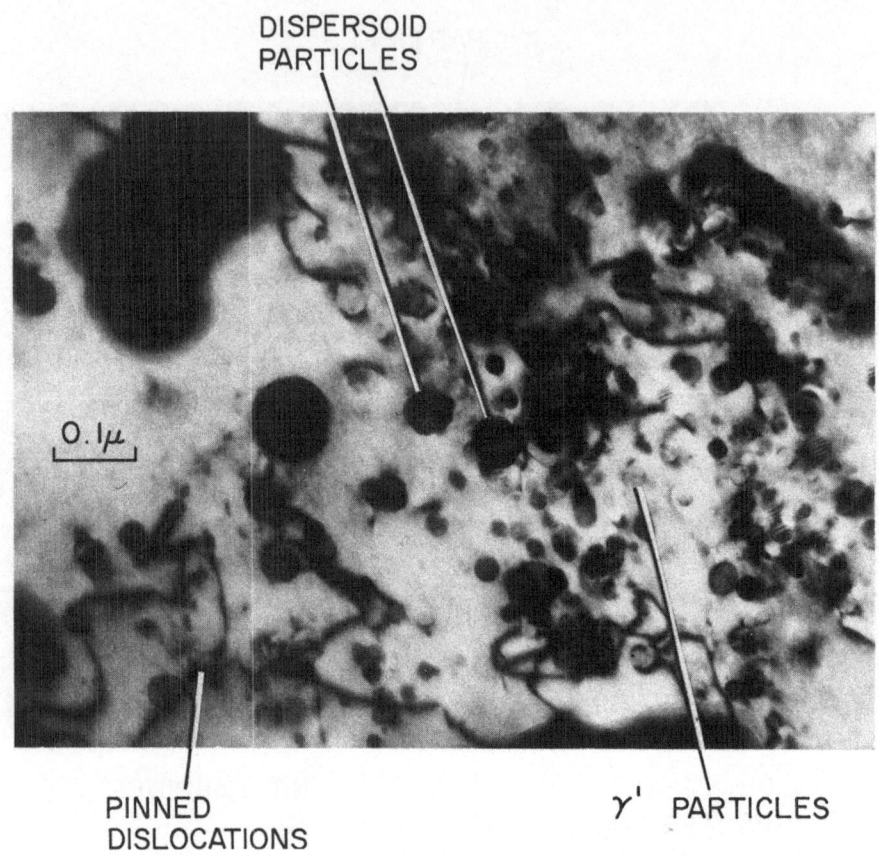

FIGURE 3b. Transmission Electron Micrograph of
            Dispersion Strengthened Superalloy

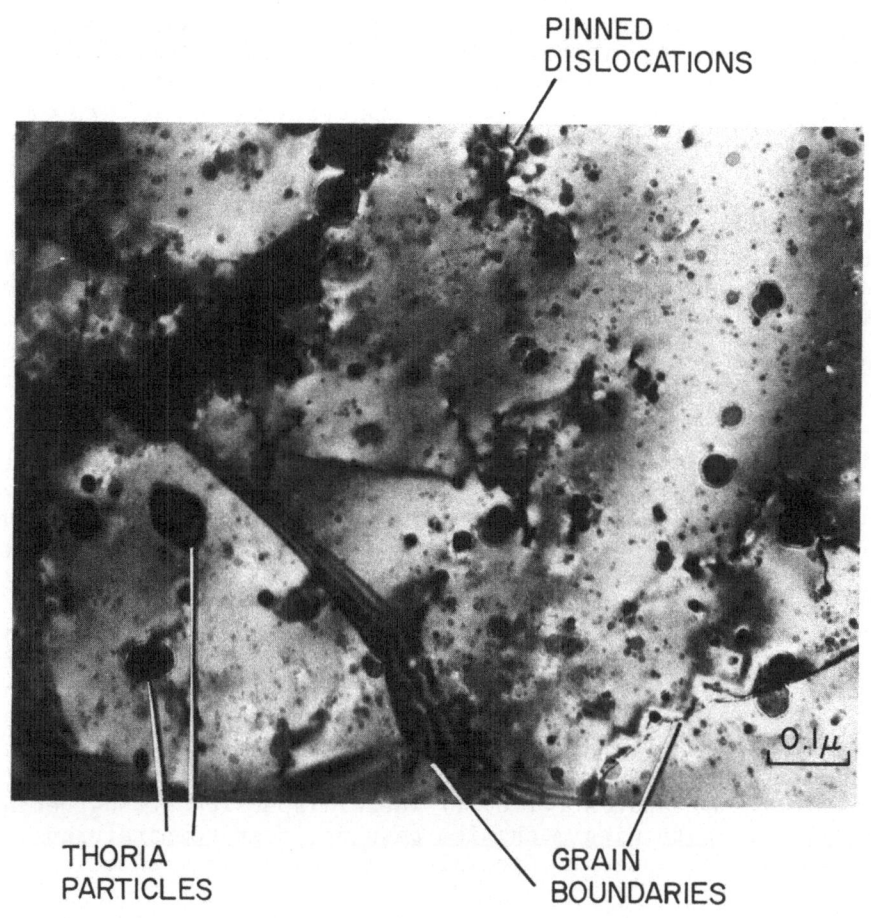

FIGURE 3c. Transmission Electron Micrograph of Stress
Relief Annealed TD Nickel

Tensile Properties

The elevated temperature tensile strength of the dispersion strengthened superalloy clearly shows two strengthening regimes. Data for material similar to that described in this paper have been reported elsewhere(5). These are replotted in Figure 4 together with data for TD Nickel(6). Data for NIMONIC*80A(6), a commercial nickel base superalloy with a composition similar to the matrix of the dispersion strengthened superalloy and to the wrought superalloy used in this study, is also plotted. Dispersion strengthening appears to have little influence on the tensile strength below about 1600 °F where the similarity between the tensile strengths of the dispersion strengthened superalloy and NIMONIC 80A is most striking. At temperatures above 1600 °F the predominant influence of the dispersion is equally apparent in comparison with the properties of TD Nickel.

Stress Rupture Properties

Stress rupture lives of the dispersion strengthened superalloy are presented in Figure 5. Tests were run at various temperatures ranging from 1200°F to 1900°F. It is interesting to note that the slopes of rupture stress versus rupture time plots are less at higher temperatures (1500°F to 1900°F) than at lower temperatures (1200°F to 1400°F). This is different from the behavior of conventional nickel-base superalloys where the slopes of rupture plots increase with increasing temperature. This behavior also suggests that more than one strengthening mechanism is operating and again reflects the increasing influence of dispersion strengthening at temperatures above about 1500°F. Rupture elongations shown in Table III are typical for the material. The ductility was slightly less at high temperatures (>1500 °F) where dispersion strengthening is the main strengthening mechanism than at lower temperatures (1200°F to 1400°F) where γ' precipitation hardening predominates.

The 100 hour and 1000 hour rupture strengths are plotted as a function of temperature in Figure 6. For clarity of presentation the curves for the experimental alloy are linked by hatched lines. Also included are typical values for TD Nickel and NIMONIC 80A(6). The influence of two different strengthening mechanisms in the dispersion strengthened superalloy is again apparent.

At low temperatures (below 1400°F) the rupture strength follows the curve for the γ' strengthened NIMONIC 80A closely. At high temperatures (above 1500°F) the rupture strength follows the curve for the dispersion strengthened TD Nickel. At intermediate temperatures the two hardening mechanisms appear to augment one another producing a smooth transition of slope in the rupture strength curve. Because the rupture life curves of the dispersion

---

* Registered trademark of The International Nickel Co., Inc.

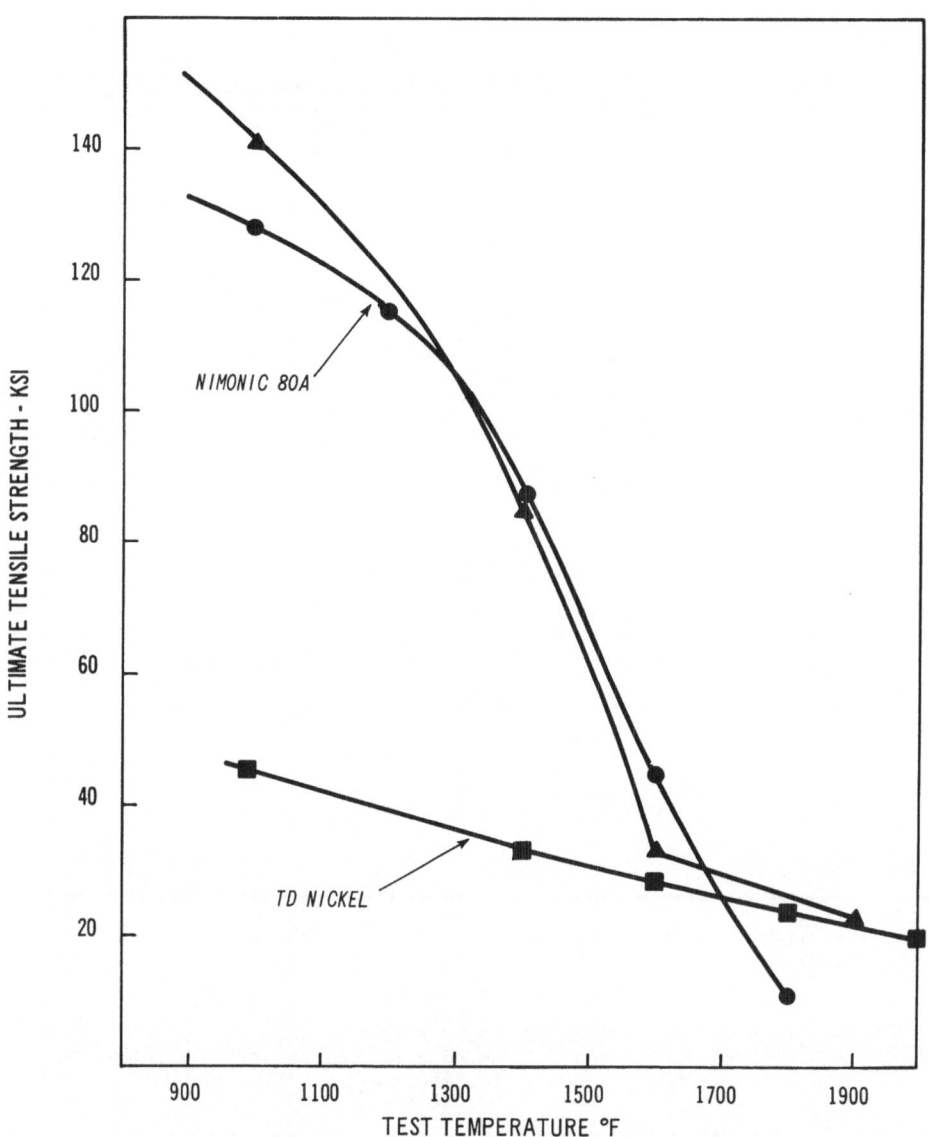

FIGURE 4 – ULTIMATE TENSILE STRENGTH OF DISPERSION STRENGTHENED
SUPERALLOY AS A FUNCTION OF TEMPERATURE.

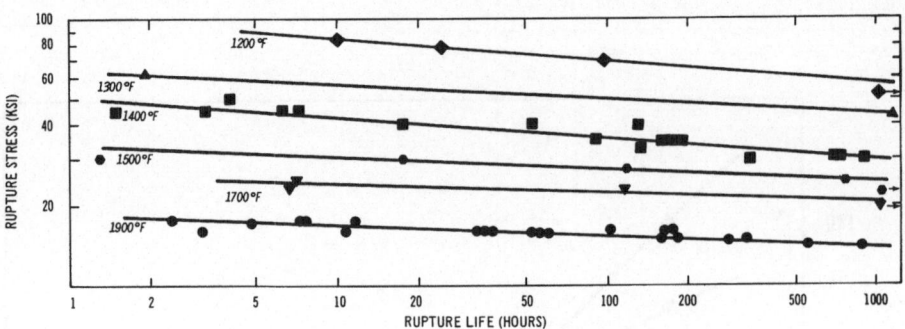

FIGURE 5 - STRESS RUPTURE TEST LIVES OF DISPERSION STRENGTHENED SUPERALLOY.

TABLE III

CREEP RUPTURE TEST DATA

Test Temperature °F	Test Stress ksi	Minimum Creep Rate Sec-1	Rupture Life Hours	Elong.%	Reduction of area %
Dispersion Strengthened Superalloy					
1900	14	1.724×10^{-7}	882.7	3.0	0.7
1900	17	1.197×10^{-4}	4.8	6.0	13.3
1700	19	6.67×10^{-7}	311.0*	---	---
1700	23	1.17×10^{-3}	1.0	12.0	25.0
1500	22.5	2.89×10^{-7}	1390.0*	---	---
1500	30	1.614×10^{-3}	1.3	31.0	38.9
1300	42	7.51×10^{-8}	1201.8	7.0	16.0
1300	63	5.03×10^{-4}	1.9	14.0	31.2
1200	52.5	6.23×10^{-8}	1652.5*	---	---
1200	85	1.168×10^{-3}	1.0	23.0	28.6
1800	19	2.71×10^{-5}	22.7	5.0	11.6
1600	19	4.68×10^{-8}	817.7*	---	---
1230	60	1.751×10^{-5}	67.9	8.5	18.1
1260	60	8.34×10^{-5}	13.0	12.5	21.8
T.D. Nickel					
1500	26	7.78×10^{-8}	777.3	4.0	12.4
1500	30	1.04×10^{-5}	12.4	9.6	24.5
1700	20	4.34×10^{-8}	1200.6*	---	---
1700	26	4.73×10^{-5}	1.1	7.0	28.1
1900	15.5	1.502×10^{-7}	840.0*	---	---
1900	20	7.12×10^{-6}	6.1	5.0	3.6
1770	20	1.557×10^{-7}	464.4	3.5	12.3
1840	20	9.67×10^{-7}	109.3	1.6	13.0
1800	20	5.06×10^{-7}	63.3	4.8	17.1
1870	20	1.668×10^{-6}	35.5	3.6	15.1
1740	20	1.390×10^{-7}	200.6+	---	---
1840	20	7.12×10^{-7}	69.9+	---	---
Cast and Wrought Superalloy(1)					
1200	67	1.35×10^{-7}	647.0*	---	---
1200	80	1.223×10^{-6}	147.0*	---	---
1300	45	2.50×10^{-7}	669.6*	---	---
1300	60	1.069×10^{-6}	170.3*	---	---
1500	26	2.28×10^{-6}	99.1	34.0	51.6

* Test discontinued.
\+ Specimen broke outside gauge length.

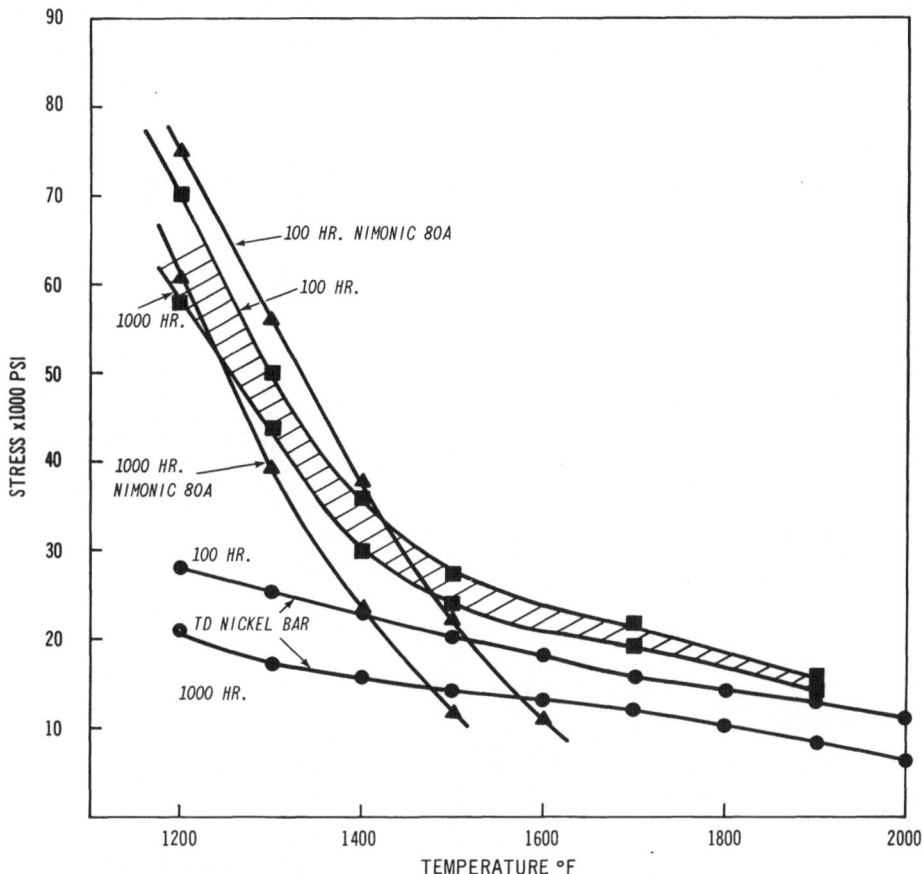

FIGURE 6 - 100 HOUR AND 1000 HOUR RUPTURE STRESS FOR DISPERSION STRENGTHENED
SUPERALLOY COMPARED WITH DATA FOR NIMONIC 80A AND TD NICKEL.

strengthened superalloy are flatter than those of conventional
superalloys the former has significantly greater rupture strength
at longer times. This is seen by comparing the 100 and 1000 hour
curves with those for NIMONIC 80A in Figure 6.

Dependence of Steady State Creep Rate on Temperature and Stress

Creep data for the three materials are given in Table III.
It was found that these creep results could be represented by an
empirical expression of the form:

$$\dot{\epsilon}^m \; \alpha \; \frac{\sigma_m}{T} \cdot \exp{}^{-\left(\frac{Q_c}{RT}\right)}$$

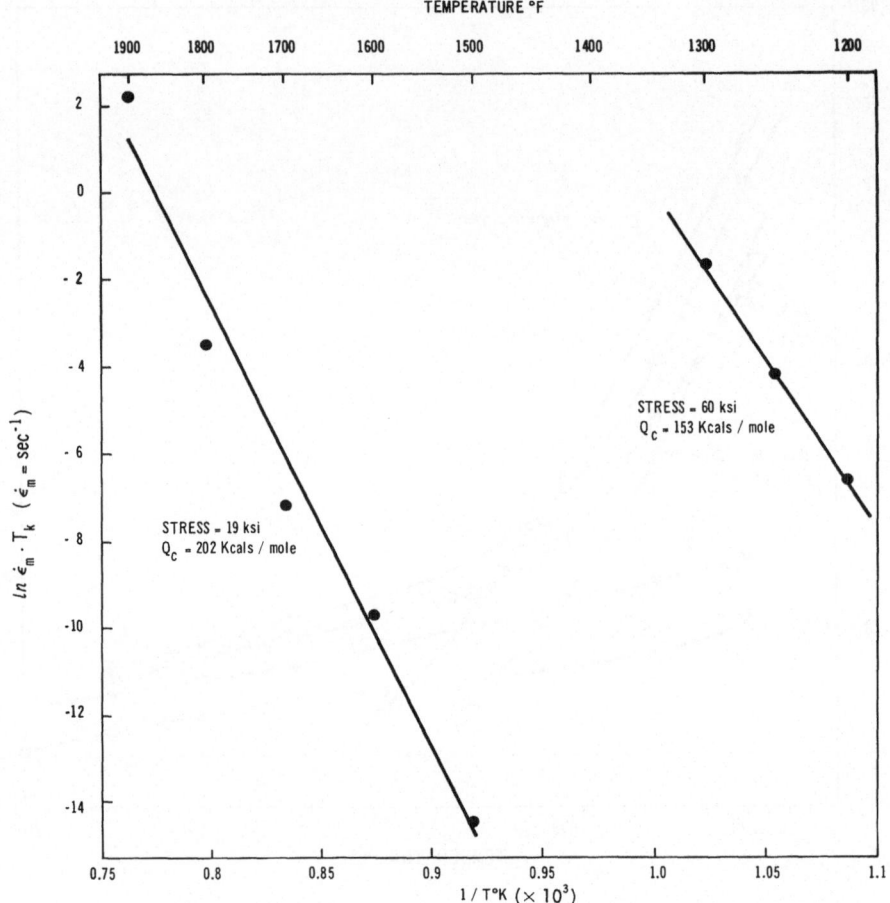

FIGURE 7 - TEMPERATURE DEPENDENCE OF THE STEADY STATE CREEP RATE OF THE DISPERSION STRENGTHENED SUPERALLOY.

The apparent activation energy, Q_c, for creep in the dispersion strengthened superalloy was found by least squares analysis of the data for two stress levels and two temperature ranges. A value of 153 kcal/mole was found between 1200°F and 1300°F at a stress of 60,000 psi while 202 kcal/mole was obtained between 1500°F and 1900°F at a stress of 19,000 psi. The data are plotted as variations of the function $\ln(\dot{\epsilon}_m \cdot T_k)$ with $\frac{1}{T_k}$ in Figure 7. The best fits are also shown.

Using the activation energies, the stress exponents were similarly calculated for each temperature range. This exponent, m, was found to be 20.6 between 1200°F and 1300°F where Q_c=153 kcal/mole and 32.8 between 1500°F and 1900°F where Q_c=202 kcal/mole.

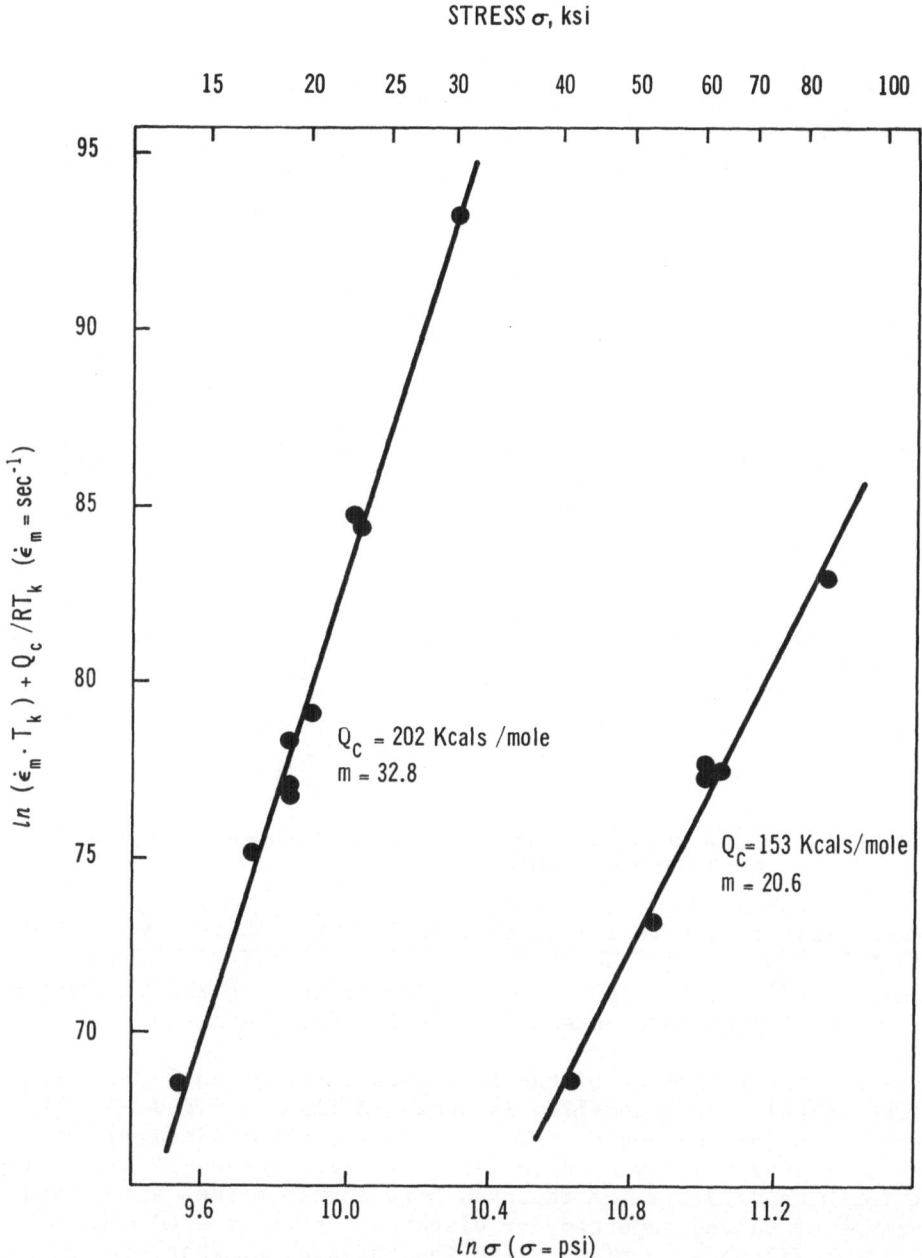

FIGURE 8 - STRESS DEPENDENCE OF TEMPERATURE COMPENSATED CREEP RATE OF DISPERSION STRENGTHENED SUPERALLOY.

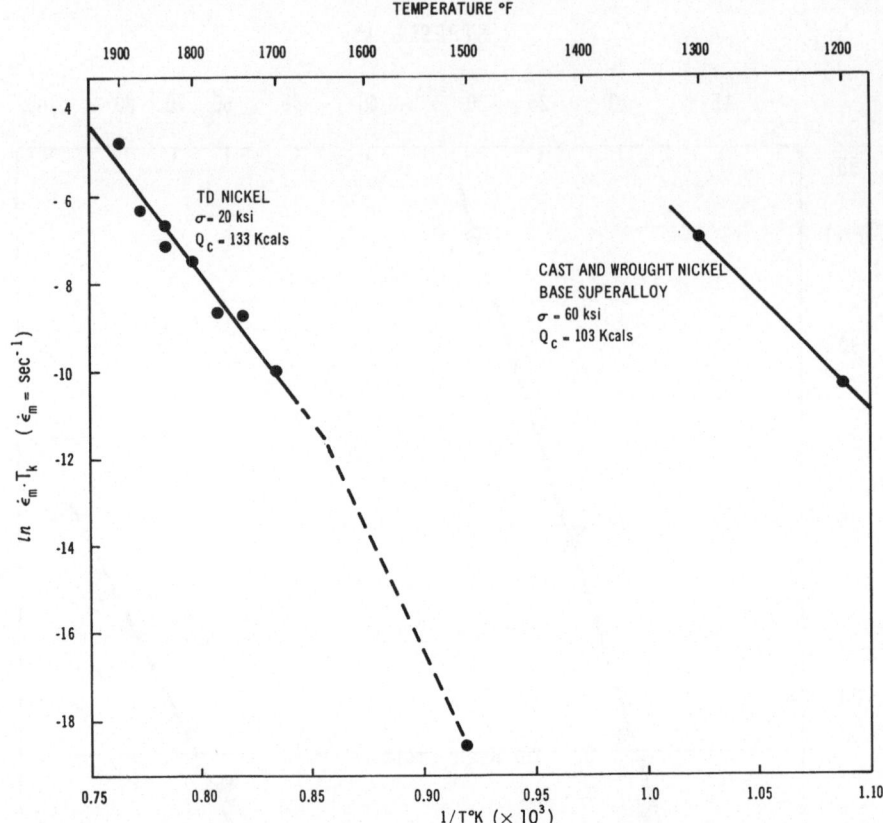

FIGURE 9 - TEMPERATURE DEPENDENCE OF THE STEADY STATE CREEP RATE OF TD NICKEL AND THE CAST AND
WROUGHT NICKEL BASE SUPERALLOY.

The closeness of fit is illustrated in Figure 8 where the function $\ln (\dot{\epsilon}_m \cdot T_k) + \frac{Q_c}{RT}$ is plotted as a function of $\ln (\sigma)$ for both sets of data. The data in Figure 8 were consistent with the assumption that Q_c was stress independent for the present tests.

    A similar treatment of the test data obtained on TD Nickel and the nickel-base superalloy is shown in Figures 9 and 10. The values of activation energy and stress exponent obtained on the three materials are compared in Table IV. The values of 133 kcal/ mole for Q_c and 23.2 for m reported here for TD Nickel are within the range of values reported for dispersion strengthened nickel and nickel alloys of various histories but are somewhat low for stress relief annealed material(7,8). The value of Q_c=103 kcals/ mole obtained on the wrought nickel-base superalloy compares favorably with the value of 97.6 kcals/mole obtained on NIMONIC 80A by

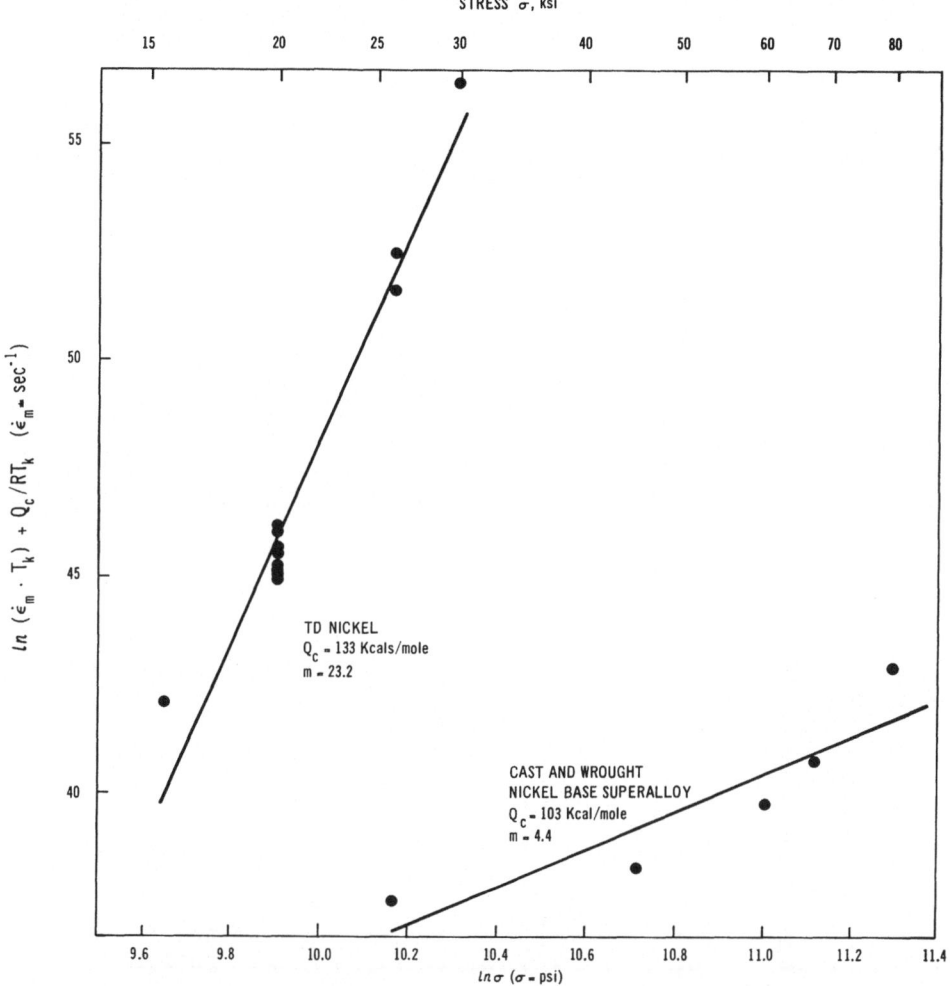

FIGURE 10- STRESS DEPENDENCE OF TEMPERATURE COMPENSATED CREEP RATE OF TD NICKEL AND CAST AND WROUGHT NICKEL BASE SUPERALLOY.

Sidey and Wilshire(9), and is well within the range 83-125 kcals/ mole cited by Heslop(10) for this alloy. The value of m=4.4 is typical of values normally obtained for polycrystalline metals.

The high values of activation energy Q_c and stress exponent, m, obtained for the two dispersion strengthened materials cannot be rationalized in terms of current creep theories and should be considered empirical quantities. High values are, however, charac- teristic of dispersion strengthened materials processed to have high stored energies.

TABLE V

ACTIVATION ENERGIES, STRESS EXPONENTS AND ELEVATED TEMPERATURE STRENGTHS OBTAINED IN STUDIES ON DISPERSION STRENGTHENED NICKEL BASE MATERIALS

Material	Material Condition And Final Processing	Activation Energy Qc kc cals/mole	Stress Exponent	Stress Range and Temperature Range of Data Measurements ksi °F	Empirical Creep Equation	Elevated Temperature Tensile Strength ksi	Reference
TD Nickel	1/2" bar cold swaged 95%. Stress relief annealed 1 hr/1846°F. (Fine fibrous grain structure)	190 above 1112°F	40	15-24/1112-2012	$\dot{\epsilon}_m = A\sigma^m \cdot \exp\frac{-Q}{RT}$	15 at 2192°F	7
TD Nickel	Extruded bar, no subsequent processing	175-193	-	-	-	4 at 2000°F	8
	As received bar annealed 1 hr/2300 F	160-220	-	-	-	18.5 at 2000°F	8
	Extruded bar rolled 47% Recrystallized at 2300 F	190-330	-	-	-	20 at 2000°F	8
TD Nickel	0.020" Sheet Cross Rolled 95% Recrystallized at 2372°F	235	119	13-17/1292-1832	$\dot{\epsilon}_m = A\sigma^m \exp\frac{-Q}{RT}$	11 at 2192°F	11
		235	15	9-13/1292-1832	"	-	11
Various Ni-ThO$_2$ Alloys 1.12 wt. % ThO$_2$	0.020" Sheet Cross Rolled From Powder Compacts With Intermittant Anneals at 2192°F. Recrystallized at 2192°F. (Procedure designed to make isotropic material.)	64	7	2-14/1292-1742	$\dot{\epsilon}_m = A\sigma^m \exp\frac{-Q}{T}$	-	15
3.0 wt. % ThO$_2$		64	7		"	-	15
4.97 wt. % ThO$_2$		64	7	-		-	15
Ni-Cr-2ThO$_2$	0.040" Sheet as received. Coarse elongated twinned Grains.	92	24.3 Longitudinal / 9.7 Transverse	14-19/1508-1900 / 4-12.5/1508-2100	$\dot{\epsilon}_m = A\sigma^m \exp\frac{-Q}{RT}$	-	17 / 17
Various Ni-Cr-ThO$_2$ Alloys 0.0 wt. % Cr	0.020" Sheet Cross Rolled With Intermittant Anneals at 2192°F. Recrystallized at 2192°F. (Procedure designed to make isotropic material.)	64	8	4/1472-1652	$\dot{\epsilon}_m = A\left(\sigma\right)^m \left(\frac{1}{E}\right) \exp\frac{-Q}{RT}$	-	16
13.5 wt. % Cr		65	6.3	7/1472-1652	"	-	16
22.6 wt. %		74	7.2	8/1472-1652	"	-	16
33.7 wt. %		78	6.7	8/1472-1652	"	-	16

TABLE IV

CALCULATED VALUES OF ACTIVATION ENERGY Q_c, AND STRESS EXPONENT m OBTAINED FROM CREEP TESTS

Material	Activation Energy Qc kcals/mole	Stress Exponent "m"	Temperature Range °F	Range of Test Stress ksi
Dispersion Strengthened Nickel Base Superalloy	202 (19 ksi stress) 153 (60 ksi stress)	32.8 20.6	1500-1900 1200-1300	15-30 40-85
TD Nickel	133 (20 ksi stress)	23.2	1700-1900	15-30
Nickel Base Superalloy	103 (60 ksi stress)	4.4	1200-1300	25-80

The complexities of the interaction between precipitation hardening and dispersion strengthening, indicated by the stress rupture results of the dispersion strengthened superalloy, are certainly reinforced by these creep findings. It is clear that, while the low temperature strength levels of the dispersion strengthened superalloy are typical of precipitation hardenable materials, the creep characteristics, as indicated by the apparent activation energy and stress exponent, are more typical of values obtained for dispersion strengthened materials.

Thermomechanical Processing History Grain Structure and Properties of Dispersion Strengthened Materials

Grain Size. A wide variation in apparent activation energies and stress exponents has been reported in the literature for both experimental and commercial dispersion strengthened materials with differing thermomechanical processing histories (see Table V). These results have been invoked to support various theories proposed to explain all or part of the creep behavior of dispersion strengthened materials.

1. Pinning of dislocations by dispersoid particles accounting for the strength of coarse grained and recrystallized materials(11).

2.   Stabilization by dispersoid particles of a fine grain size or subgrain structure produced during fabrication, and accounting for the strength of fine grained commercial materials processed to contain a high stored energy of cold work. In such cases the total structure is supposed to control dislocation mobility and the strength cannot be related to dislocation theories of dispersion strengthening(11).

3.   The creep behavior of TD Nickel has also been attributed to grain boundary sliding at high temperatures(7).

4.   Other workers(8,12) have argued that elevated temperature strength is limited by grain boundary failure mechanisms and concluded that high strength could best be realized by developing coarse grained elongated structures, thereby minimizing transverse grain boundary area. This effect was also held to account for anisotropic mechanical properties in TD Nickel.

The various factors contributing to dispersion strengthening cannot be separated simply by comparing materials in the fine grained ("as-received") condition, with materials given additional mechanical processing and annealing treatments to produce a coarse grain size (recrystallization) as has been attempted. The differences between these groups of materials are more complex than the terms "as-received" and "recrystallized" would indicate. Grain growth in dispersion strengthened materials is inherently affected by factors other than temperature alone. The as-received structure of commercial TD Nickel resists change at temperatures up to 2500 F. Various studies(8,13,14) have shown that different types of cold working operations may produce widely varying response to such heat treatment after comparable amounts of strain.

Working Direction and Grain Shape. Extruded thoriated nickel bar is very weak but exhibits a high activation energy (see Table V) (8). Unidirectional post extrusion working designed to produce a fine fibrous grain structure and leading to improved creep rupture properties is used for commercial thoriated nickel. In this condition the material is strong and has both a high activation energy and stress exponent for creep. Similarly, TD Nickel rolled only in the extrusion direction prior to recrystallization had both high activation energy and high elevated temperature tensile strength(8).

In another study(11) Wilcox et al subjected thoriated nickel to a cold strain perpendicular to and in the original extrusion direction before recrystallization. This material had plate shaped grains and moderate elevated temperature strength but high activation energy and stress exponent. In contrast, the dispersion strengthened alloys used in (15,16) were intentionally worked by cross rolling techniques and had disc or plate shaped coarse grains. These alloys had low activation energies and low stress exponents in addition to being relatively weak.

SUMMARY

The working of the dispersion strengthened superalloy in the present study was limited to a single extrusion operation. The material was then heat treated to obtain a coarse grain structure elongated in the extrusion direction. It is notable that high values of strength, activation energy Q_c and stress exponent "m" were obtained. The creep and rupture strengths of the coarse grained alloy in the present study were much higher than those of cross rolled, recrystallized thoriated nickel and thoriated alloys cited and are comparable to those of as-received TD Nickel and of unidirectionally worked, recrystallized TD Nickel(8).

Different combinations of microstructure, strength level, activation energy and stress exponent can be obtained depending on the details of the complete processing history. It is not clear whether microstructural features are primary factors required for high strength or are merely secondary results of thermomechanical processing history. It is possible that microstructural features do not play the same role in all dispersion strengthened metals, solid solutions and complex alloys. This reinforces the conclusion that a satisfactory theory explaining elevated temperature creep behavior of dispersion strengthened materials will not be obtained until some means is found to account for the different effects of thermomechanical processing history, matrix composition and grain structure.

CONCLUSIONS

1.   Dispersion strengthening has been combined with precipitation hardening in a nickel base superalloy.

Elevated temperature properties of this material represent a superposition of dispersion strengthening upon precipitation hardening in a desirable combination.

2.   Constant-life, rupture-stress curves show two regimes of behavior dominated by precipitation hardening below 1500°F and dispersion strengthening above 1500°F.

3.   Creep properties are controlled primarily by dispersion strengthening. At temperatures above 1500°F the material exhibits apparent activation energies corresponding to values found in stress relief annealed TD Nickel. At intermediate temperatures where rupture life appears dominated by precipitation hardening the apparent activation energy and stress exponent for creep remain significantly higher than those of conventional superalloys.

REFERENCES

1.   W. D. Coolidge "Ductile Tungsten", Proc. AIEE 1910,
p. 961.

2.   R. Irmann, "Sintered Aluminum with High Strength at
Elevated Temperatures", Metallurgia 49, 1952, p. 125.

3.   C. J. Leadbeater and discussion by E. Gregory, "Some
Developments in Sintered Structural Parts", Symposium on Powder
Metallurgy, 1954, ISI Special Report No. 58, pp. 149-159 and
p. 357.

4.   G. B. Alexander et al, U.S. Patent No. 2,972,529,
February 21, 1961.

5.   J. S. Benjamin, "Dispersion Strengthened Superalloys
by Mechanical Alloying", submitted for publication in Metallurgical
Transactions.

6.   "High Temperature, High Strength, Nickel Base Alloys",
The International Nickel Co., Inc. 2nd Edition, June 1968.
Note: The values for the properties of TD Nickel are given only
for reference and represent an average of the numbers reported in
the open literature and in manufacturer's pamphlets for varying bar
sizes of this alloy.

7.   B. A. Wilcox and A. H. Clauer,"Creep of Thoriated Nickel
Above and Below 0.5Tm," Trans. AIME 236, 570 (1966).

8.   G. S. Doble, L. Leonard, L. J. Ebert, "The Effect of
Deformation on Dispersion Hardened Alloys", Final Report NASA
grant NGR 36-033-094, November, 1967.

9.   D. Sidey and B. Wilshire, "Mechanisms of Creep and
Recovery in NIMONIC 80A, Metal Science Journal 3, 1969, p. 56.

10.   J. Heslop, "Creep Fracture in Nickel-Chromium Base Creep
Resistant Alloys", Jrnl. of Inst. of Metals, 91, 1962-63, p. 28.

11.   B. A. Wilcox and A. H. Clauer, "High Temperature Deformation of Dispersion Strengthened Nickel Alloys", NASA CR-72367,
February 29, 1968.

12.   R. W. Fraser, D.J.I. Evans, "Oxide Dispersion Strengthening", Bolton Landing Conference, 1966, p. 375.

13.   M. C. Inman, P. J. Smith, "Oxide Dispersion Strengthening", Bolton Landing Conference, 1966, p. 291.

14.   M. Van Heimendahl, R. A. Huggins, Trans. AIME, 233, 1965, p. 1076.

15.   A. H. Clauer and B. A. Wilcox, "Steady State Creep of Dispersion Strengthened Nickel", Metal Science Journal, $\underline{1}$, 86, 1967.

16.   B. A. Wilcox and A. H. Clauer, "Creep of Dispersion Strengthened Nickel-Chromium Alloys", Metal Science Journal, $\underline{3}$, 1969, p. 26.

17.   B. A. Wilcox, A. H. Clauer, and W. S. McCain, "Creep and Creep Fracture of a Ni-20Cr-2ThO$_2$ Alloy", Trans. AIME, $\underline{239}$, 1967, p. 1791.

NEW DEVELOPMENTS IN SUPERALLOY POWDERS

S. H. Reichman and J. W. Smythe

Federal-Mogul Corporation

Metal Powder Division Research

Abstract

The advantages of a P/M process for Ni-based superalloys are shown in terms of better homogeneity than cast alloys. By using low interstitial-atomized powders densified by direct extrusion, superplastic structures are developed. The results of a thermo-mechanical process are presented through which large grained or even single crystal structures can be fabricated. Finally, an alloy series made specifically for P/M processes is presented. These alloys have no carbon as atomized, densified or formed. The resultant solution-annealed grain sizes are ASTM 1 to 0. These alloys are subsequently carburized to stabilize the grain boundaries.

Introduction

Within the next year or so there will be wide acceptance of P/M superalloy components for high temperature turbine applications. Pratt and Whitney and General Electric are currently testing P/M superalloy turbine discs for large aircraft engines, and Curtiss-Wright is testing P/M superalloy turbine blades. The primary reason for the impending success of P/M superalloys has been the production of low interstitial-atomized superalloy powders combined with sophisticated densification techniques.

The P/M approach for Ni-based superalloys offers several important advantages over cast and/or wrought superalloys. These alloys for use at high temperatures are complex, consisting of a solid solution-hardened nickel-based matrix with a gamma-prime

$\left\{Ni_3 \; (Al, \; Ti)\right\}$ dispersed precipitate phase and complex metal carbides, both of which are stable to high temperatures. The alloys can contain alloying additions of Al, Ti, Mo, W, Ta, Th, etc.; and the combination of these high and low melting temperature alloying elements, especially in the newer alloys, results in severe segregation in castings. Typical compositions of some Ni-based superalloys are shown in Table 1. Furthermore the cast alloys, because of the high temperature hardening nature of the constituents, cannot be hot worked without severe cracking developing. Typically the gamma-prime phase does not re-solution to quite close to or even above the incipient melting temperature of the alloy, and usually the carbides do not re-solution until incipient melting has taken place. Therefore, the hot working capacity of the alloy is severely limited for both forming and homogenization processing. These aspects become quite important when one considers large discs (up to three feet diameter) where property uniformity is very important.

Atomized superalloy powder is inherently homogeneous and avoids the problems of property uniformity. The low interstitial content (less than 100 ppm O) is necessary to avoid insoluble interstitial compound formation in prior powder boundaries of the densified material. The presence of these compounds (particularly in a continuous network) will result in an extremely brittle material with poor high temperature properties.

The densification method utilized in the following discussion is direct extrusion of vacuum canned powder to fully dense rod. This technique insures good cross-section uniformity and prior powder (particle) boundary elimination at high extrusion ratios (greater than 6:1).

The topics to be discussed in this paper will include: superplastic forming of P/M superalloys in which low stress - high ductility forming is performed, thermo-mechanical processing of P/M superalloy material in order to result in large grain sizes, and an alloy development series is shown in which alloy compositions are tailored specifically for powder processing.

TABLE 1

Nominal Compositions of Ni-Based Superalloys

Alloy	Cr	Co	Al	Ti	Mo	W	C	V	B	Ni
IN-100	9.5	15	5.5	5.0	3.0	--	0.2	.95	.015	Balance
U-700	15.0	18	4.5	3.5	5.0	--	0.15	--	.03	Balance
IN-713C	12.5	--	6.1	0.8	4.2	2.0 Cb	0.12	--	.012	Balance

Superplasticity [1][2]

Superplasticity is commonly referred to as the ability of an alloy to suffer large neck-free deformations at temperatures above one half the absolute melting temperature. The deformation stress or flow stress is usually considerably lower than the flow stress for the non-superplastic form of the alloy at the same temperature. Furthermore, the deformation is highly rate sensitive and usually follows the relationship:

$$\sigma = \varkappa \dot{\varepsilon}^m$$

where σ = Flow Stress
$\varkappa$ = Constant
$\dot{\varepsilon}$ = Strain Rate
m = Strain Rate Sensitivity

Most alloys which are superplastic have a value of m above 0.3 (m is usually below 0.1 for non-superplastic alloys).

It has been observed empirically that in order for an alloy to behave in a superplastic manner, it must, in addition to having a high strain rate sensitivity, be of extremely fine structure and that it must be microduplex in nature. The grain size for superplastic alloys is usually less than 10^{-4} inches in diameter. The microduplex structure (incorporating a stable second phase at the deformation temperature) is necessary in order to preserve the fine structure during deformation and to prevent grain growth. While the exact nature of the deformation is not precisely known, it is postulated that boundary mobility (either grain boundary or second-phase boundary) is responsible for the phenomenon. The above rules as previously mentioned are empirical and do not ensure that an alloy meeting the criteria will be superplastic.

Nickel-based superalloys do meet all of the above criteria. The structure consists of a gamma-phase matrix with the gamma-prime phase stable to high temperature; the strain rate sensitivity is quite high for this type of alloy, and via a P/M route an extremely fine grain structure can be achieved.

By starting with low interstitial (less than 100 ppm Oxygen) atomized powders and densifying by direct extrusion, a homogeneous (though somewhat anisotropic) product is produced. Low interstitial powder is necessary to insure that there are no continuous prior powder boundaries which would essentially act as large grain boundaries and become rate controlling in the deformation process. The extrusion parameters are picked so that the billet temperature is slightly below the recrystallization temperature of the alloy and the adiabatic heating during the extrusion process will cause recrystallization, but minimal grain growth. This fine structure with the attendant empirical superplastic structural features results in a material with the optimum superplastic characteristics for the particular alloy composition.

Figure 1 shows the as-extruded structure of P/M IN-100. The extrusion was carried out at 1900°F at a 10:1 ratio (a temperature slightly below the recrystallization temperature of 1950-2000°F for IN-100). The adiabatic heating caused recrystallization and a resultant grain size of approximately 2×10^{-5} inches. This material was then superplastically deformed at temperatures from 1700°F to 2000°F in tension and compression. As-machined and deformed bars are shown in Figure 2 showing the large neck-free deformations obtained for this alloy. Incidentally, IN-100 is used as-cast since it is too sensitive to cracking with even light hot forging. Elongations to 1000 percent have been observed in. P/M IN-100 and U-700. Figure 3 shows the isothermal superplastic deformation data for IN-100 and U-700. The slope of these curves represents the strain rate sensitivity, m . For IN-100, m = 0.5. For U-700, m is 0.42. It is evident that the deformations take place at stresses quite below the flow stress for as-cast IN-100 and U-700 at corresponding temperatures.

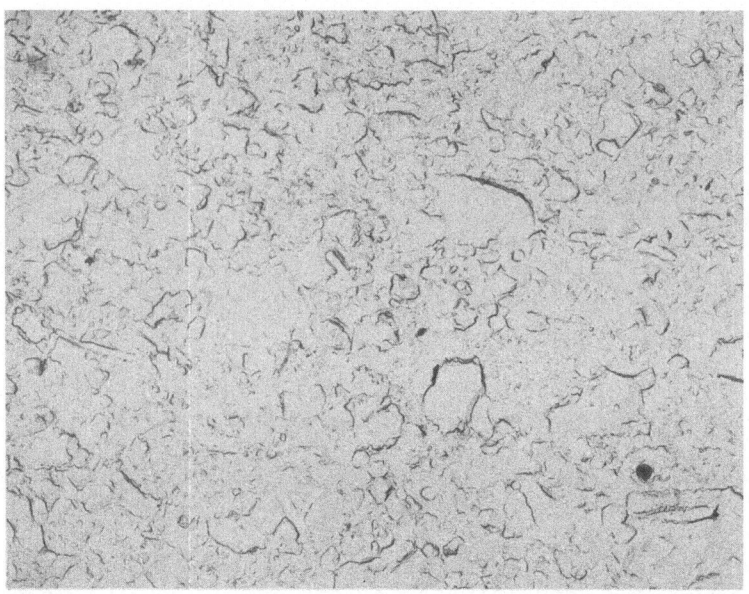

6000X   Kalling's Etch

Figure 1. Replica Electron Mocrograph of As-Extruded P/M IN-100 Alloy (1900°F, 10:1 Ratio)

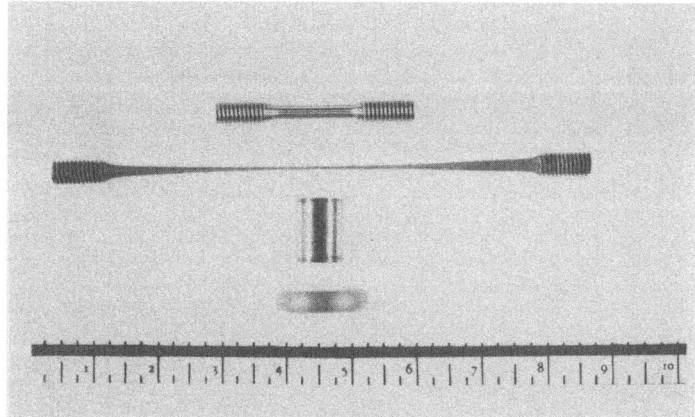

Figure 2. Deformed and Undeformed P/M IN-100
Tensile and Compression Specimen, Superplastically
Tested at 1850°F.

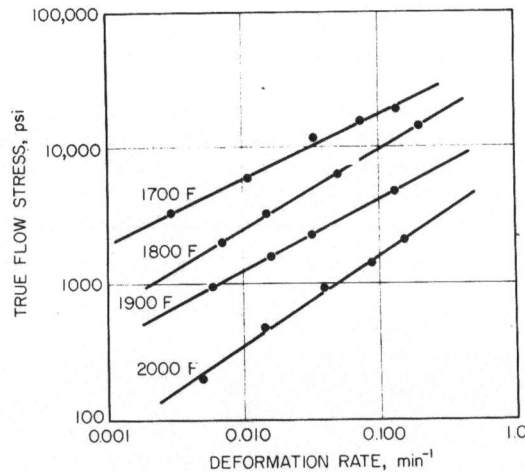

Figure 3a. Isothermal Superplastic Deformation
Relationships (Flow Stress Versus Deformation
Rate in Tension) IN-100 Extruded at 1900°F and 10:1.

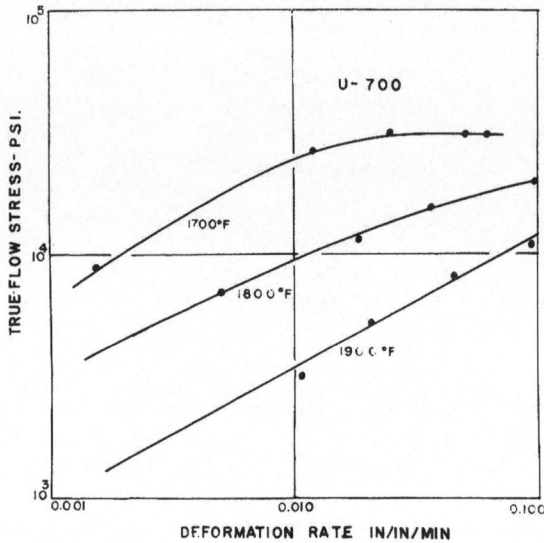

Figure 3b.   Isothermal Superplastic Deformation Relationships (Flow Stress Versus
Deformation Rate in Tension) U-700 Extruded at
1900°F and 10:1.

        Subsequent to superplastic deformation, the superalloy is given
a "solution-anneal" treatment to grow the grains and stop the alloy's
superplastic propensity. Figure 4 is the structure of P/M IN-100
superplastically deformed, at 1850°F, and solution annealed for
56 hours at 2270°F. This material is no longer superplastic and
displays properties comparable to cast IN-100 as shown in Figure 5.
Similar results are observed for U-700.

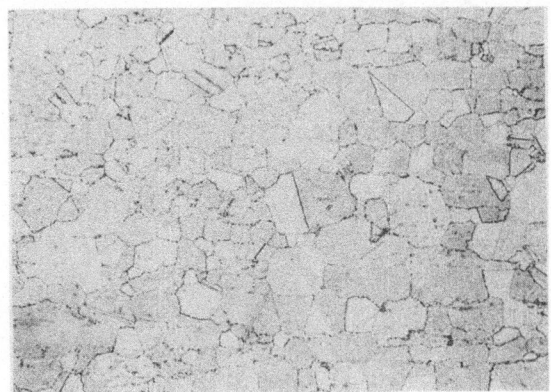

100X   Kalling's Etch

Figure 4. IN-100 P/M Alloy
Superplastically Deformed
and Subsequently Solution
Treated for 56 Hours at
2270°F.

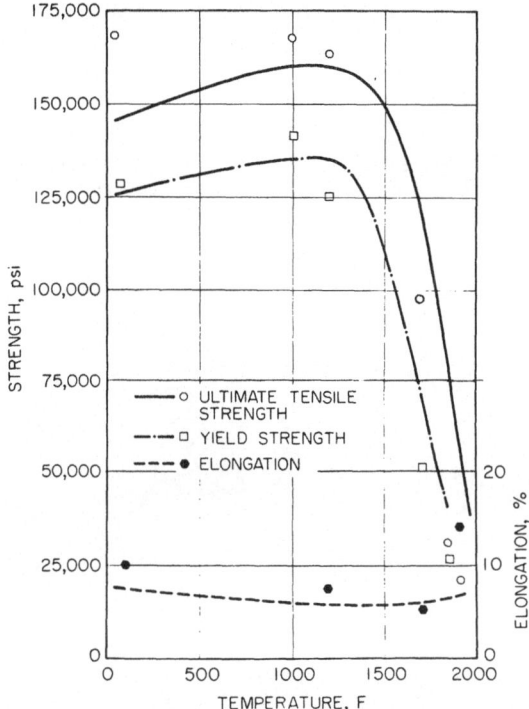

Figure 5. Mechanical Properties of P/M IN-100, Superplastically Deformed, Solution Treated for 56 hours at 2270°F and Tested in Tension at the Indicated Temperatures. The Data Points are for the P/M Alloy and the Line Data is Handbook Data for Cast IN-100.

Thermo-Mechanical Processing [2]

Due to the nature of a superalloy powder product with the unavoidable presence of prior powder particle boundaries, grain growth is usually limited to fairly small sizes. Even in a high purity P/M material with relatively clean prior powder boundaries, there are inevitantly interstitial impurities which are insoluble in the alloy matrix plus large amounts of undissolved carbides in the grain boundaries, making grain growth to sizes larger than ASTM 4-5 quite difficult with only a solution anneal treatment. The stress rupture properties of this material at high temperatures, eg. 1800°F, are somewhat lower than the cast alloy due to the relatively small grain size.

In order to attain larger grain sizes, a thermo-mechanical process (TMP) has been developed for P/M alloys which permits growth

to such large grain sizes. While the exact details of the process
are proprietary at this time, several examples of its application
are shown in Figure 6. A range of grain sizes has been obtained, up
to single crystals in 1/4-inch tensile bars. The mechanical properties
of TMP U-700 alloy are shown in Figure 7. The UTS and yield strength
to 1800°F are seen to be as-good or superior to cast and wrought
U-700 values shown in the same figure. Tensile ductility is observed
to be much higher than the values for wrought U-700.

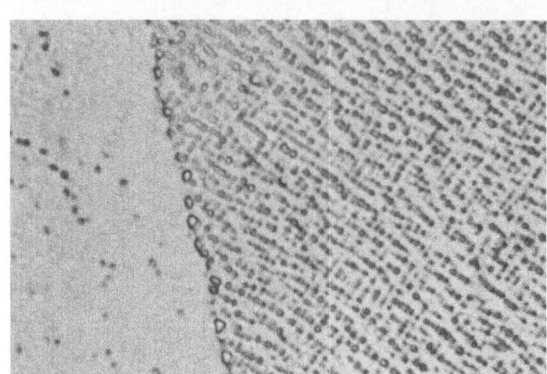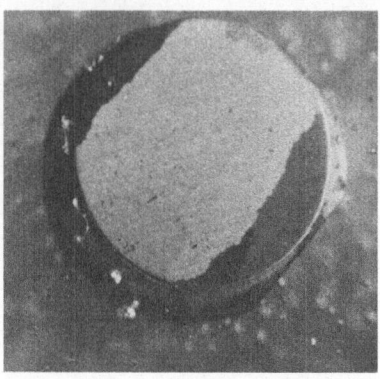

2500X

Figure 6. (a) Thermomechanically treated U-700 structure showing
large resultant grain size. The cross sectional diameter of the bar
is 0.150 inches. (b) Resultant microstructure of (a) showing the
gamma-prime structure and the carbide distribution in the grain
boundary.

Figure 6, showing the typical grain structure of this material,
reveals a directional macrostructure due to the TMP, but there is no
evidence of prior powder boundary persistence. More significant than
tensile properties of this material, the measured stress rupture life
of a single crystal tensile bar at 1850°F and 20,000 psi is 196 hours.
The specified life of wrought U-700 is less than one hour. The
increased life of the P/M product is due to the lack of grain boundaries in a single crystal, the preferred orientation from the TMP,
and the improved homogeneity of the powder product over the wrought
alloys.

The thermo-mechanical process can be applied to nickel-based
superalloys in which the gamma-prime phase solution temperature is
below the incipient melting point of the alloy. This includes most
of the currently used superalloys, but for more highly alloyed compositions another approach is necessary as will be discussed in the
next section.

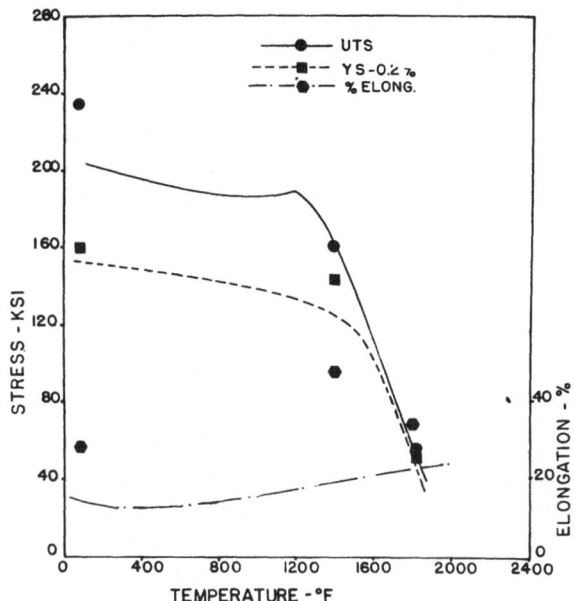

Figure 7. Mechanical
Properties of Thermo-
Mechanically Processed U-700
P/M Alloy. The Data Points
are for the P/M Alloy and
the Line Data is Handbook
Data for U-700.

Alloy Development

The foregoing discussions on thermo-mechanical processing and
superplastic forming have been based on taking cast alloy compositions,
applying P/M technology, and working towards bringing the P/M
product up to the cast alloy property levels. Perhaps the largest
problem area once a low interstitial, uniform structure is attained,
is increasing the grain size to cast levels in order to match the
high temperature stress rupture life (hence the TMP for P/M alloys).
IN-100 and U-700 alloy compositions that have been discussed above
are designed for castings and take into consideration the segregation,
catastrophic grain growth, coring, banding and dendritic structure
often found in highly alloyed cast superalloys. These features are
not found in a P/M product densified from atomized powder since the
powder is made from an homogeneous melt and resultant segregation
is virtually non-existent. One can therefore design a series of alloys
to take advantage of the features of atomized P/M products rather
than work around casting deficiencies.

We present a series of nickel-based superalloys atomized from
the molten state, containing essentially no carbon, and in the dense
product (for example by direct extrusion) no carbides at the grain
boundaries. The lack of carbides makes the grain boundaries highly
mobile and amenable to significantly more grain growth than the alloy
with carbides.

The alloy as densified can also have a superplastic structure. Solution annealing increases the grain size typically to ASTM 0-1 or even single crystal structure. In order to stabilize these large grains, the grain boundaries are carburized and the alloy further solution annealed to result in a P/M product which is easily formed (superplastically), the grain size easily increased to a size for good high temperature stress rupture life, and finally carburized to stabilize the boundaries and the alloy and further increase the stress rupture life.

Table 2 shows typical compositions for the as-densified P/M alloys. Alloy 1 as-densified is shown in Figure 8. The isothermal superplastic deformation behavior of this alloy at 1900°F is shown in Figure 9. The strain rate sensitivity is quite high compared, for example, to P/M IN-100. Solution annealing of the alloy for 72 hours at 2250°F results in ASTM 0-1 as shown in Figure 10. The grain size is quite uniform and equiaxed. There is no evidence of the prior powder boundaries and the grain boundaries of the material are free of carbides and are composed of discreet gamma-prime particles. The carbon content for this material is 40 ppm. Upon increasing the carbon content to 170 ppm, primarily in the grain boundaries, discreet carbides in the boundaries are evident as seen in Figure 11. Stress rupture life at 1800°F and 29,000 psi is 12 hours. This value of carbide content is apparently not sufficient to immobilize the boundaries and some growth took place during the tests. Testing is currently underway on material carburized to 1,000 ppm levels which should be sufficient to immobilize the boundaries.

TABLE 2

Low Carbon P/M Nickel-Based Superalloy Compositions

Alloy						Element %			
	Cr	Co	Al	Ti	Mo	W	V	B	Ni
1	9.5	15	5.5	5.0	3.0	--	--	--	Balance
2	15.0	18	4.5	3.5	5.0	--	--	--	Balance
3	12.5	--	6.1	0.8	4.2	2.0 Cb	--	--	Balance
4	--	--	6.0	--	--	2.0	1.4 Zr	--	Balance
5	6.0	--	6.0	--	4.0	4.0	8.0 Ta	2.5 Cb	Balance

The above results are extremely encouraging for powder products and indicate the direction that P/M research must take in order to become more accepted and successful for high temperature use in turbine applications. By this alloy modification process, very highly alloyed compositions can be utilized which could never be cast because of severe segregation and a total lack of workability.

Several of the newer alloy compositions such as WAZ-20 (with 20 w/o tungsten) or TAZ-8A (with 8 w/o tantalum) cannot be cast because of segregation, and are special P/M alloys.

1000X   Marble's Etch

Figure 8. As-Extruded Structure of Low Carbon Alloy No. 1 from Table 2. Extruded at 1900°F, 10:1.

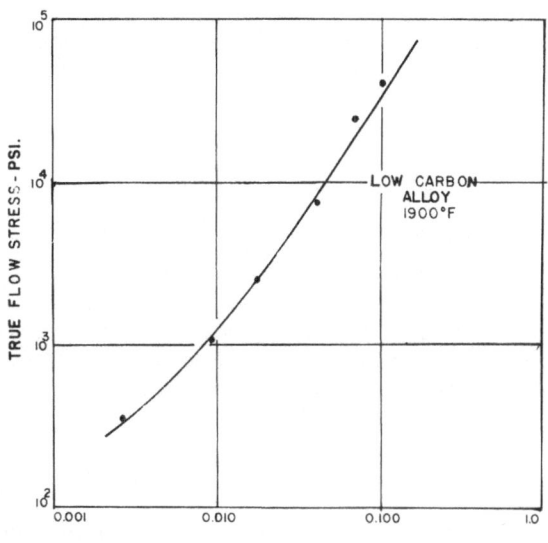

Figure 9. Isothermal Superplastic Deformation of Low Carbon Alloy No. 1 from Table 2, Deformed at 1900°F.

50X   Marble's Etch

Figure 10. P/M Low Carbon Alloy No. 1 from Table 2 Solution Treated for 72 Hours at 2250°F.

1000X   Marble's Etch

Figure 11. P/M Low Carbon Alloy No. 1 from Table 2 Carburized to 150 ppm Carbon.

Conclusions

1. A powder metallurgy process incorporating atomized low interstitial content powder results in a dense product with superior homogeneity than cast and wrought alloy forms.

2. By densifying the low interstitial content powders by direct extrusion, superplastic microstructures are achieved; and easy-to-form close tolerance parts can be made. The alloy is then made non-superplastic by a solution anneal treatment.

3. As-extruded or superplastically formed alloy parts can be thermo-mechanically processed (TMP) to a large resultant grain size.

4. An alloy series is developed for P/M processing which contains no carbon. The alloys can be superplastically formed and, subsequently, can be grown to large grain sizes by only a solution anneal and result in improved high-temperature stress rupture life. The alloys can also be carburized to put boundary stabilizing carbides in the structure.

References

(1)   S. H. Reichman and J. W. Smythe, Superplasticity in P/M IN-100 Alloy, International Journal of Powder Metallurgy, 6, 1, 1970, p.65.

(2)   S. H. Reichman, B. W. Castledine and J. W. Smythe, Superalloy P/M Components for Elevated Temperature Applications, Society of Automotive Engineers, 700140, January, 1970.

A CORRELATION OF MECHANICAL PROPERTIES OF SINTERED U-700 POWDER

WITH PARTICLE BOUNDARY MORPHOLOGY

Kenneth H. Moyer

Hoeganaes Corporation

INTRODUCTION

Difficulties in forging cast superalloys have caused the Air Force and engine manufacturers to concentrate their efforts to develop a powder metallurgy preform which would be suitable for forging.[1-7] As superalloys have developed, large concentrations of alloying additions have been made to improve high temperature properties. These increased additions have promoted segregation within the cast alloy. This segregation has narrowed the forging temperature range and resulted in alloys which are difficult to forge. Since a powder particle is an extremely small casting, segregation is restricted to each particle of powder. The fine particles with their microsegregation then facilitate complete homogenization of solubles. This in turn provides a more efficient utilization of alloying elements and implies that the optimum alloy composition for superalloy powders may be different than present alloys which are now fabricated by the cast/forge process. Similarly, it might be suspected that heat treatment response could be different for a powder compact because of the homogeneity of the powder and the increased degree of supersaturation.

Although superalloy powder compacts offer attractive advantages for forging, powder consolidation is still not free of problems. Unlike a solid ingot where the surface area is relatively low, the surface area of powder compacts is high. The abundant surface area renders the powdered alloys prone to oxidation; hence the oxide content of powder compacts is higher by several orders of magnitude than the oxide content of vacuum cast ingot. The influence of oxygen upon mechanical properties is still a matter of controversy, although most people believe that oxides are harmful to properties.[8,9] In addition, the stability of the complex oxides, which can form as a film about the powder particles, render the powders difficult to sinter.

The work which is reported includes part of the work for Air Force Contract AF 33 (615) 3994.[6] The purpose was to produce a superalloy powder with less than 300 ppm oxygen content and to consolidate the powder into a compact which was 93% of theoretical density. Sintering was one of the processes which was chosen to consolidate the required preforms. The mechanical properties of the powder compacts used in this study were lower than the mechanical properties of wrought material. However, it had been shown previously that the properties of powder compacts were comparable. Therefore, the balance of the work involved a study of why the mechanical properties were lower.

VACUUM SINTERING PROCESS

The temperatures which are required for sintering superalloy powders to densities greater than 90% of theoretical density are close to the melting point of the alloy and it is necessary to sinter in vacua.[10] Chromium, present in all nickel base superalloys, vaporizes at a pressure of 10^{-5} Torr, imposing a limit on the degree of vacuum which is practical. Hydropressed specimens (75/85% theoretical density) 1 inch diameter by 3 inches long were sintered at 2280° F for 6 hours at 10^{-5} Torr. The sintered density was 89% of theoretical density. A section of a specimen was analyzed for chromium, carbon, and oxygen. The chromium content was lowered to 14.38 w/o compared with an original chromium content of 15.16 w/o as a result of sintering. To alleviate the problem of vaporization all other compacts were sintered in vacua at a pressure of only 10^{-2} Torr. There was also a slight pick-up in carbon content from the polyvinyl alcohol binder which was required to hydropress the compacts but the oxygen content did not increase as a result of sintering.

The sintering parameters and the resulting densities of the sintered compacts are shown in Table 1. The data show that it is only possible to sinter U-700 powder within a narrow temperature range; at 2280° F the density is less than 90% of theoretical density whereas at 2330° F, eutectic melting occurs. The densities of the sintered fine powder were slightly higher than those of the coarse powder, as might be expected.

The microstructure of a sintered compact is shown in Fig. 1. Fine gamma prime is seen within the powder particles; carbides, which we will discuss further, are in abundance about the particle boundaries. A small amount of eutectic was found at some of the triple points.

Table 1.   Densities of 1 Inch Diameter x 3 Inch Long Billets
           Hydropressed With 0.1 w/o PVA Binder and Sintered
           At 10^{-2} Torr Pressure.

Material	Sintering Temperature, °F	Time, Hours	Density g/cc	% Theoretical Density
U700 Coarse	2330	1-1/2	Eutectic Melting	
U700 Fine	2330	1-1/2	Eutectic Melting	
U700 Coarse	2280	6	6.87	88.8
U700 Coarse	2300	6	7.45	94.3
U700 Fine	2300	6	7.60	96.2

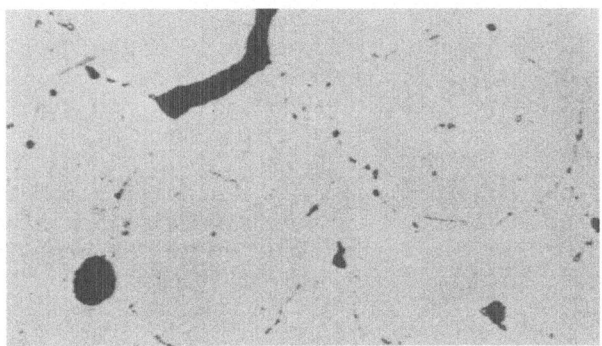

Figure 1.   Microstructure of U-700 Billet 1 Inch Diameter
            Using Fine Powder, Hydropressed at 100,000 psi
            And Sintered 6 Hours at 2300° F.   Mag. 250X
            Marbles Etch

The grain size of the sintered specimens averaged an ASTM grain
size of 6 after sintering for six hours at temperature. The grain
size was not influenced by the particle size distribution although
the coarse distribution recrystallized more readily than the finer
distribution. The porosity of the coarser sintered powder tended
to have less frequent large rounded pores whereas the finer sintered powder had more abundant smaller pores, triangular or rectangular
shaped.

The process was scaled up to manufacture billets which were
3 inches diameter by 6 inches long. The conditions for sintering
these larger billets are shown in Table 2. Some modifications had
to be made to the sintering process to obtain crack free billets.
Initially the green compacts were rested on flat plates of alumina

Table 2. Vacuum Heat Treating Procedures For Sintering at 2300° F

Distri-bution	Heat Treat. Time, Hours	Cooling Rate[1]	Support Material	Surface Appearance	% Theor. Density
Fine	6	Fast	Flat alumina plate	One large transverse crack	93.5
Fine	6	Fast	Flat alumina plate	Two large transverse cracks	93.5
Fine	12	Slow	Granular alumina	No Cracks	97.0
Coarse	6	Fast	Flat alumina plate	No Cracks	91.0
Coarse	12	Slow	Granular alumina	No Cracks	94.8
Coarse	12	Slow	Granular alumina	No Cracks	93.7
Coarse	12	Slow	Granular alumina	No Cracks	93.1

(1) Fast – Furnace power turned off at sintering temperature, cooled to 1400° F in approximately 45 mintues.

Slow – Controlled cooling to 1400° F in 2 hours with 30 minute holding intervals at 2050° F, 1800° F, and 1500° F.

for sintering. The sintered compacts contained large peripheral cracks which resulted from lack of support of the green compacts (billets numbers 1 and 2). Subsequent compacts were supported in a bed of fine granular alumina. The alumina bed helped to minimize tensile stresses but still a maze of fine surface cracks was observed after sintering. The first billets which were sintered were cooled rapidly after sintering at a rate of 675° F/hour within the temperature range of 2340° F to 1400° F. In order to eliminate the cracking problem, the cooling cycle was modified as follows:

1. Cool from the sintering temperature to 2050° F and hold for 30 minutes.
2. Cool to 1800° F and hold at temperature for 30 mintues.
3. Cool to 1500° F and hold at temperature for 30 minutes.
4. Furnace cool to ambient temperature.

One billet of the fine powder distributions and three billets of the coarse distribution were sintered with the above interrupted cool. No cracking was observed.

In addition to the modified cooling cycle, the sintering conditions required modification because these conditions developed for the 1 inch diameter specimens did not produce sintered billets which were at least 93% of theoretical density. To obtain a density which was a minimum 93% of theroetical density, the sintering time had to be increased from 6 hours to 12 hours.

A wafer was cut from two of the 3 inch diameter billets for chemical analysis (See Table 3). The carbon content did not increase appreciably as a result of adding the polyvinyl alcohol binder nor did the gas content increase as a result of processing.

Table 3. Chemical Analyses of Udimet 700, 2-1/2" Diameter Billets

Element	Specification w/o	Billets Made From Blend 104-105 Coarse w/o	Billets Made From Blend 106 Coarse w/o
C	0.03 to 0.09	0.083	0.081
Cr	14 to 16	15.19	15.22
Co	17 to 20	19.21	19.36
Mo	4.5 to 5.5	5.42	5.40
Ti	2.75 to 3.75	3.03	3.04
Al	3.75 to 4.75	4.65	4.78
B	0.025 to 0.035	0.0323	0.0320
Fe	1.0 max.	0.1	<0.1
Cu	0.1 max.	<0.05	<0.05
Zr	0.06 max.	<0.01	<0.01
Mn	0.15 max.	<0.05	<0.05
S	0.015 max.	0.009	0.007
P	0.015 max.	--	--
Si	0.2 max.	0.1	<0.1
Ni	Balance	Balance	Balance
O	0.03 max.	0.0228	0.0193
N	Low	0.0116	0.0121
H	Low	0.00023	0.00016

The microstructures of the billets were free of excessive precipitates, free of foreign matter, and the structure is similar to the structure observed in the one inch specimens. The pores of the sintered fine powder were smaller but more numerous, and usually found at triple points. In contrast, the pores in the sintered coarse powder were more massive and fewer. They were more randomly distributed throughout the structure. A discontinuous precipitate was seen within the powder particle boundaries and some incipient melting was observed at triple points. The original powder particle boundaries can be seen in the sintered structure.

MECHANICAL PROPERTIES

Tensile and stress rupture specimens were machined from the one inch diameter sintered material. The specimens were solution treated for 25 hours at 2300° F and given an argon quench. An oil quench rather than an air cool was used after aging treatments.

Figure 2 shows the microstructure of the powder compact after solution treatment and after aging. Large particles with the precipitate dispersed around the particle boundaries are seen in the solution treated compacts. Gamma prime is seen within the particles after aging. The grain size of the compacts was ASTM 4, the hardness after aging was 346 VHN and the density was 7.45 g/cm^3 or 94% of theoretical.

After Solution Heat Treatment   After Aging

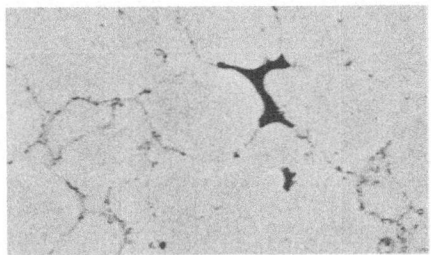 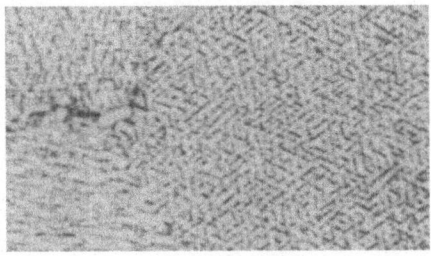

Figure 2.   Microstructure of Sintered Compacts.   Mag. 250X
            Glyceregia Etch

The ductility of the compacts was nil and the strength properties were inferior to the strength propertes of wrought material. The fracture mode of the specimens pulled in tension at room temperature was interparticle.

The stress rupture properties were also poor which further indicates a grain boundary problem. Initially the stress level for testing at 1400° F was 79 ksi. The stress had to be reduced to 40 ksi in order to obtain any rupture life. Similarly the stress level had to be halved to obtain any appreciable life at 1800° F.

GRAIN BOUNDARY AND FRACTURE SURFACE STUDIES

The mechanical properties were disappointing. This was especially so because powder manufactured by the atomizing process and fabricated by a proprietory process had yielded properties which were better than those properties normally found for wrought material.

Additional samples were sintered using the proprietory process which had yielded compacts with good properties. Powder compacts fabricated by this process were found to be cracked after processing. The billets were sectioned and the fracture was found to propagate along particle boundaries. The only difference between the two powders which were processed by the proprietory process was that the powder compacts which had the favorable properties had carbon content of only 0.03 w/o whereas the powder compacts which had the undesirable properties had a carbon content of 0.08 w/o.

Since the carbon had the greatest influence on the properties through strengthening the grain boundaries, the fracture surfaces were studied by using the electron microprobe, x-ray diffraction, electron microscopy, and electron diffraction techniques.

Carbon extraction replicas were taken from fracture surfaces using a 10% HCl alcohol solution. This solution dissolves the gamma and the gamma prime phases leaving only the MC, $M_{23}C_6$ and M_3B_2 precipitates normally found in the grain boundaries. The electron beam scan of these replicas for titanium, aluminum, molybdenum, and chromium indicated that within the particle boundaries at least three phases were present, a titanium rich carbide (MC), a molybdenum and chromium rich carbide ($M_{23}C_6$), and/or boride (M_3B_2) intermetallic, and an aluminum rich phase which fluoresced in the electron beam which indicated that it was an oxide. The titanium and the aluminum were widely distributed throughout the particle boundaries; molybdenum and chromium were found together and were found in isolated areas of the particle boundaries.

X-ray diffraction analysis of the same samples also detected a TiC phase, a complex $M_{23}C_6$ phase, a M_3B_2 boride, and very weak lines of alpha alumina.

Similar work was done with the electron microprobe on powder compacts that were fabricated by the proprietory process. Titanium, some molybdenum, and carbon were found to be prevalent in the particle boundary; aluminum, sodium, florine, oxygen, nitrogen and boron could not be detected. Cobalt, nickel and chromium were found to be depleted from large sections of the boundaries but were homogeneously distributed throughout the matrix. There was an adjacent area which looked like a pore at low magnification. When scanned at 1500X magnifications however, the area was found to contain a solid particle. This particle was composed of aluminum, silicon, carbon and oxygen.

Two particle morphologies were seen, fine globular particles and thin plates. These particles are shown in Figure 3. Electron diffraction patterns indicated that the globular particles were $M_{23}C_6$ type carbides whereas the platelike particles were MC carbides and/or M_3B_2 borides.

Fine Globular Particles   Thin Platelike Particles

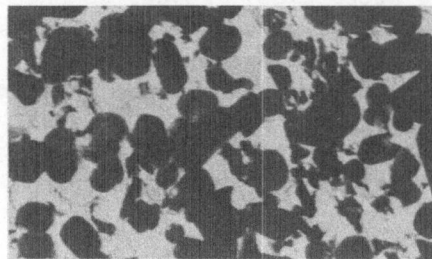 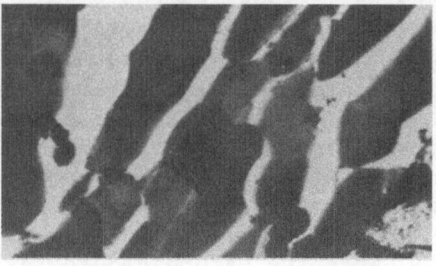

Figure 3.   Electron Microscopy Photomicrograph.   Mag.   4500X

Two stage replicas were prepared from the fracture surfaces. Electron micrographs of these replicas are shown in Figure 4.   In these studies the M_3B_2 boride could not be resolved.   A continuous boundary phase is present.   This phase may be the MC carbide phase and thus would provide an easy path for fracture.

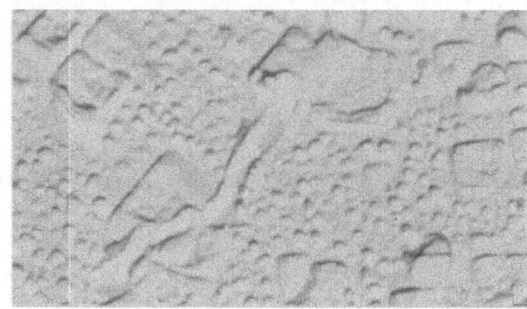

Figure 4.   Electron Microscope Photomicrograph of Fracture Areas
            Mag.   25,500X

In summary, a continuous phase probably MC carbide was seen to be concentrated within the particle boundaries.   Phases normally found within the alloy were present, but there was little evidence to show that a deleterious oxide phase was present, although a fine dispersion of Al_2O_3 was present.   It is believed that the Al_2O_3 present is not of sufficient size or concentration to form a continuous film which could affect the properties.   It is most probable however that the continuous platelike MC carbide did affect the mechanical properties.

The electron photomicrographs showed that the size and the morphology of the gamma prime of the heat treated alloy compared well with the size and morphology of the gamma prime measured in

wrought alloys. The size of the primary gamma prime was 2000/
3000 Å square, the secondary gamma prime was 100 Å square (Fig. 5).
Boesch & Canada report that the size of the primary gamma prime in
wrought U-700 is approximately 5000 Å square, secondary gamma prime
less than 1000 Å square.[11] With the exception of the continuous
film at the boundaries, the structure of the powder metallurgy alloy
compared well with the structure which is characteristic of wrought
alloys.

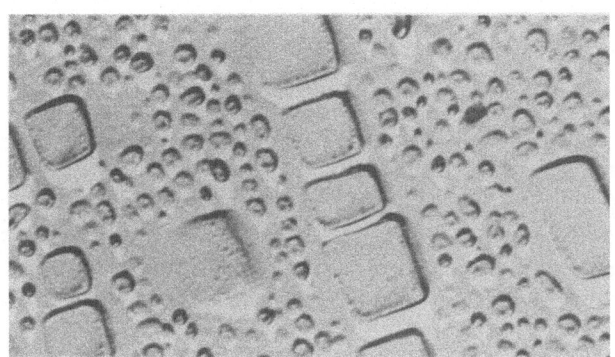

Figure 5.   Gamma Prime Morphology and Size of Sintered U-700
             Powder Compacts Mag. 25.500x

CONCLUSIONS

1. Hydropressed U-700 compacts can be sintered at 2300° F to a
    density of 93% of theoretical. The temperature range for
    sintering is narrow. At 2280° F, 93% theoretical density
    cannot be obtained; at 2330° F, incipient melting begins.

2. Mechanical properties and fracture studies suggest a weakened
    particle boundary condition. It is thought that a continuous
    platelike MC carbide is responsible for the weakened condition.
    Further fabrication studies should be made to distribute the
    carbides more favorably in the particle boundaries.

ACKNOWLEDGEMENTS

We wish to thank Mr. R.L. Kennard and Wright-Patterson Air Force
Base, United States Air Force for permission to publish this paper.
Most of the work which is reported was done under Air Force Contract
AF 33 (615) 3994. We wish also to thank Mr. Hoy McIntire and
Mr. Fred Joyce of Battelle Memorial Institute for their partici-
pation in the fabrication of the powder compacts.

REFERENCES

1. Borok, B.A., et.al, "Manufacture by Powder Metallurgy Techniques of Semi-Products for Subsequent Processing, "Soviet Powder Metallurgy and Metal Ceramics, November 1967, No. 11 (59).

2. Kortovich, C.S., "Close Tolerance Forging From Powder Metallurgy Preforms," Air Force Contract AF 33 (615) 5411.

3. Friedman, G.J. and Lowenstein, P., "Development of Processing Techniques for The Extrusion of Metal Powders," Air Force Contract AF 33 (615) 67C-1160.

4. Lyon, D., et.al, "Slipcast Superalloy Sheet," Air Force Contract AF 33 (615) 3998.

5. Ingram, J.F. and Durdaller, C., "Development of An Improved Manufacturing Process for the Production of Higher Purity Superalloy Powders," Phase I, Air Force Contract AF 33 (615) 3994.

6. Moyer, K.H., "Development of An Improved Manufacturing Process for the Production of Higher Purity Superalloy Powders," Phase II, Air Force Contract AF 33 (615) 3994.

7. Tracey, V.A., Poyner, G.T., and Watkinson, J.F., "Sintered Higher Temperature Alloys," Journal of Metals, May 1961.

8. Barker, J.T. and Calhoun, C.D., "AF 95 Powder Manufacturing Techniques" Interim Engineering Progress Report, October 1969, Air Force Contract F33615-69-1825.

9. Allen, M.M., Athey, R.L., and Moore, J.B. "Application of Powder Metallurgy to Superalloy Forgings," Pratt & Whitney Aircraft Report GP68-234, 27 November 1968.

10. Farrell, K. "Sintering of Atomized Superalloys and a Hardenable Stainless Steel," International Journal of Powder Metallurgy, 1, (3), 1965.

11. Boesch, W.J. and Canada, H.B., "Phases Present In the Wrought Superalloy Udimet 700," Journal of Metals, April 1968.

DISPERSION STRENGTHENING

THE STABILITY OF OXIDE DISPERSIONS IN IRON- AND NICKEL-BASED ALLOYS

H. Fischmeister and E. Navara

Dept. of Engineering Materials,

Chalmers University of Technology, Gothenburg, Sweden

INTRODUCTION

The service life of dispersion strengthened materials is in many cases limited by the coarsening of the dispersed phase which reduces the strength of the alloy by increasing the mean free path for dislocation movement. As the technology of producing dispersion strengthened alloys improves, the problem of stability becomes more important (1). In particular, the effect of alloying additions on the stability of oxide dispersions must be understood more clearly as pure metal matrices are being replaced increasingly by alloyed ones, to achieve higher strength and better oxidation resistance.

Basically, the coarsening of oxide dispersions is a case of the general phenomenon known as Ostwald Ripening, i.e. the dissolution of the smallest particles with concomitant growth of the larger ones. The driving force of the process is the decrease of phase-boundary free energy as the surface-to-volume ratio of the dispersed phase is reduced.

A fairly complete theoretical understanding of the kinetics of the process has envolved from the work of a number of authors (2 - 7), notably C. WAGNER (5). Most of the experimental studies made on Ostwald Ripening in various metallic and nonmetallic systems are mentioned in ref. (8). Data on oxide dispersions in metal matrices have been reviewed by SEYBOLT (9).

The coarsening process proceeds by the diffusion of individual atoms of the constituents of the dispersed phase through the matrix. For oxide dispersions in metals, the rate-controlling step is normally the migration of the slowest species of the oxide, the

speed of which is governed by the diffusivity and the concentration
of this species in the matrix. According to Wagner, the rate law
of diffusion-controlled particle coarsening is

$$r^3 = r_0^3 + \frac{8\gamma Dc\ V^2}{9nRT} \cdot t \tag{1}$$

where r = a characteristic dimension of the (stationary) particle
       size distribution, e.g., the mean particle radius
    γ = the free energy per unit area of the particle/matrix
       phase boundary
    D = the diffusivity of the slowest migrating species
    c = the concentration of the slowest migrating species
    V = the molar volume of the oxide
    n = a parameter related to the numbers of cations and anions
       in a formula unit of the oxide *(5)*
while the remaining symbols have their usual meaning.

WORN and MARTIN *(10)* first pointed out the effect of impurities on
the growth of thoria particles in a nickel matrix. Small amounts
of both iron and chromium accelerated the coarsening in atmospheres
containing traces of water vapor. SIMS *(11)* foresaw the possibility
of an adverse effect of chromium additions to TD nickel, but found
no perceptible growth of thoria particles in either pure nickel or
Ni20Cr *(12)*.

SERGEENKOVA and BEREZUTSKIJ *(13)* found that the growth of alumina
particles in nickel is decreased by additions of titanium to the
matrix, in accordance with the well-established ability of this
element to lower the interface energy between alumina and nickel
in the molten state *(14, 15, 16)*. A similar effect of cobalt addition
to nickel was reported by HANCOCK, DILLAMORE and SMALLMAN *(17)*.

In principle, the coarsening of a given oxide could be affected by
impurities via the interfacial energy γ, the solubility c and the
diffusivity D of the rate-determining species. SEYBOLT discussed
some of these effects in his review *(9)*, but concluded that available
data did not allow the relative roles of these factors to be assessed.
No systematic attempt to improve the structural stability and the
service life of dispersion strengthened alloys by manipulation of
γ, D and c has been reported in the literature.

EXPERIMENTAL METHOD AND RESULTS

Three series of alloys were studied with nickel, iron, and iron with
40 w/o Ni as base. All contained 1,6 w/o ($\approx$ 3,5 v/o) of alumina.
Various additions of Cr, Mo, Mn, Ti and Co were made to the matrices.
The first four additives are used industrially to promote bonding
in metal-to-alumina seals; cobalt was included because of its importance in nickel- and iron-based high-temperature alloys. The
alloy series were designed to show, among other things, the effect
of a given additive in a face-centered cubic and in a body-centered
cubic matrix.

Starting materials were carbonyl nickel (International Nickel Ltd.,
type 123) and carbonyl iron (BASF, type CP) of Fisher particle
size 4 - 7 µm and 8 µm respectively, and Linde type B alumina powder
of nominally 0,05 µm size. Titanium was added as hydride powder,
other additives being elemental powders. The powders were mixed
and milled in alcohol in vibratory ball mills for 6 to 12 hrs.
After isostatic compaction and sintering in hydrogen at 1100°C
(1 hr for Ni-base alloys, 2 hrs for iron-base alloys) compacts of
~ 80% theoretical density were obtained. These were sealed in thick-
walled mild steel cans with titanium getter powder, heated to 1200°C
and extruded (with glass lubrication) at a ratio of 1:20. The
extruded can was machined off.

The coarsening of the oxide was followed by measuring the mean free
path between particles after heat treatment of the extruded specimens in hydrogen at various temperatures and times. Fig. 1 shows
an example of the structures developed after long heating where
the large oxide particles are beginning to develop crystallographic
habit shapes.

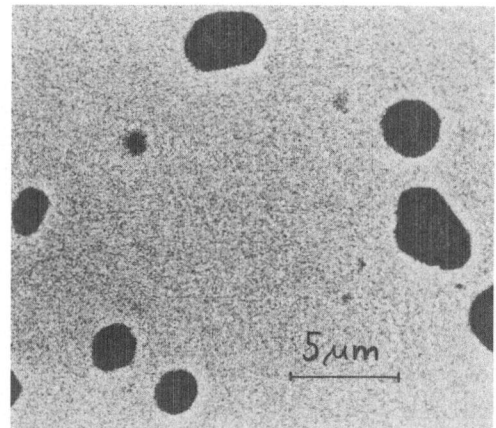

Fig. 1: Fe 15 Cr alloy after 100 hrs at
1450°C (scanning electron micrograph)

A light microscope with a linear scanning stage (18) and with a
160 x objective was used for the mean free path measurements. For
each measurement, between 500 and 1000 particles were traversed.
Attempts to measure the size distributions of the oxide particles
with a "Quantimet" (TV-type) image analyser (19) were successful
only for the most advanced stages of coarsening owing to resolution
limitations. It is clear, too, that the mean free path measurements
suffered from the same limitation albeit to a smaller degree since
direct visual observation allows full exploitation of the resolution
available. For these reasons, no mean free path measurements were
made on as-extruded specimens.

If the size distribution of the particles does not change its shape during coarsening, the mean free path is directly proportional to the particle size. For a system of monosized spheres of radius $\bar{r}$, the following relation holds between particle radius and mean free path:

$$\bar{r} = \frac{3f}{4} \cdot \bar{L} \tag{2}$$

where f is the volume fraction of the particle phase. According to WAGNER'S treatment of coarsening kinetics, the particle size distributions are expected to be stationary, and their shape is skewed so that the most frequent size $\hat{r}$ is encountered for fairly large particles ($\hat{r} = 1/2\ r_{max}$, the largest radius occurring in the population). The loss of the smallest particles in scanning with limited resolution is more tolerable with such a distribution than with a symmetrical one. Since the measured value of $\bar{L}$ is too large if some particles are lost, the particle size will be given approximately by

$$\bar{r} = const.\ \bar{L} \tag{3}$$

where the constant is smaller than 3f/4.

Fig. 2 shows the size distribution curves of the oxide particles in an alloy of Fe 40 Ni-5 Mn after various times of heating. The fine size fractions are cut off by resolution limitations. As far as the measurements go, they show that the distribution shape is stationary, justifying the use of eq. (3).

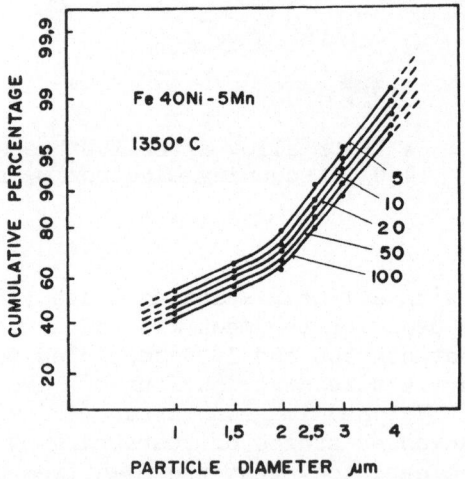

Fig. 2: Distribution curves of alumina particles

Fig. 3 shows the results of mean free path measurements on an alloy with fairly rapid growth; the matrix composition was Fe + 5% Cr and the temperature 1450°C. The slope of the line is 0,315. Within the experimental accuracy this agrees with the 1/3 power time dependence of the mean free path and the particle size predicted by eq. (1) for a case where r_0 << r.

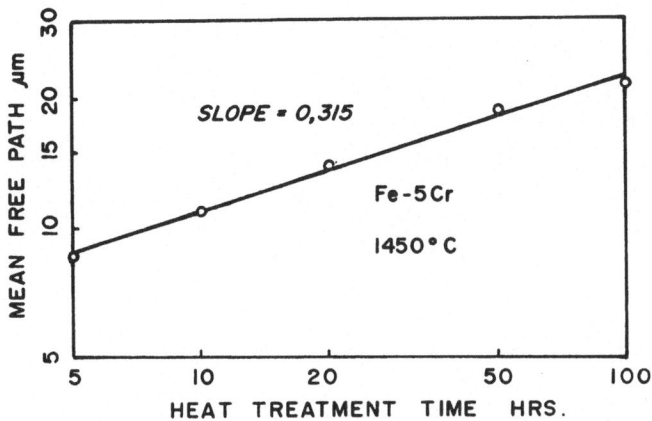

Fig. 3: Mean free path vs. heat treatment time

In case of less rapid growth, the influence of the incipient particle size (r_0 in eq. 1) makes itself felt in a deviation from the slope 1/3 in a plot like Fig. 3. Assuming the cubic rate law to hold, it is possible to extrapolate $\overline{L}_0$ from the measurements. The values so obtained range from 4,3 to 9,4 µm. For the specimen with a pure iron matrix the particle size in the as-extruded state was estimated from scanning electron micrographs, an example of which is shown in Fig. 4. The estimated particle radius was 0,13 µm, corresponding to a mean free path of 4,3 µm (by the monosize approximation, eq. (2)). The value $\overline{L}_0$ extrapolated from the lineal analysis results for this specimen was 4,6 µm. While the exactitude of the agreement in this case may be fortuitous, we feel justified in concluding that the extrapolated $\overline{L}_0$ values are of the right order of magnitude. Their variation from alloy to alloy is thought to be due to the coarsening during the heating of the billet and in the extruded material while it is still hot; large values of L_0 are found in alloys which show rapid particle growth.

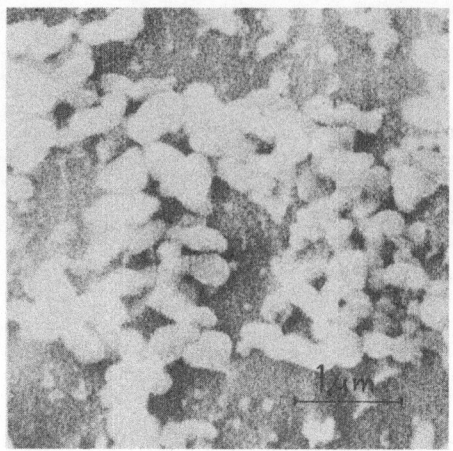

Fig. 4: Fe-Al$_2$O$_3$ as extruded
        Scanning electron
        micrograph.

Heavy etching was required,
as a result of which
particles from a thick zone
of material are accumulated
on the surface.

On the basis of the above, it appears justified to evaluate the
results in terms of rate constants defined as

$$K = \frac{\bar{L}_2^3 - \bar{L}_1^3}{t_2 - t_1} \qquad \left[\frac{\mu m^3}{hr.}\right] \qquad (4)$$

which, according to eqns. 1 and 3, are related to the rate determining quantities by

$$K = const. \frac{D \ c \ \gamma \ V^2}{n \ RT} \qquad (5)$$

The value of the constant depends on the shape of the particle
distribution; it will not be discussed further.

Table 1 shows the rate constants determined for all the alloys,
with matrix compositions and temperatures as stated, from the mean
free paths measured after t_1 = 10 hrs and t_2 = 100 hrs.

It is clear that coarsening is very much affected by the nature
of the matrix additives. Both retardation and acceleration of growth
occurs. This indicates basically different modes of interaction
between the alloying additive and the oxide. A special study of
this point was therefore made.

Table 1: Rate constants of particle coarsening

Alloy	Temperature $^\circ$C	Rate constant (μm^3/hr)
Nickel Base Alloys		
Ni 10 Cr	1350	4,7
Ni 5 Mo	1350	5,1
Ni 5 Cr	1350	7,0
Ni (pure)	*1350*	*16,0*
Ni 10 Mn	1250	180
Austenitic Iron Base Alloys		
Fe 40 Ni 10 Cr	1350	25
Fe 40 Ni 5 Mo	1350	30
Fe 5 Cr	1350	32
Fe (pure)	*1350*	*37*
Fe 40 Ni (pure)	*1350*	*43*
Fe 10 Co	1350	50
Fe 40 Ni 5 Mn	1350	128
Ferritic Iron Base Alloys		
Fe 5 Cr	1450	104
Fe (pure)	*1450*	*116*
Fe 5 Mo	1450	150
Fe 1 Ti	1450	360
Fe 15 Cr	1450	780

MATRIX - OXIDE - REACTIONS

By dissolving the matrix completely in hot hydrochloric acid, the
oxide particles were extracted from specimens of all alloys after
heat treatment for 100 hrs at 1350 or 1450°C. In most cases, the
extracted oxide powder was white, but for some alloys colouring
indicated a reaction between the oxide phase and the additive:
Ni-Cr and Fe-Cr alloys gave light pink and pink extracts, darker
at higher Cr contents, and the titanium-alloyed Fe- and Fe40Ni-
base specimens gave grey and black oxide powders.

Qualitative X-ray fluorescence analysis proved the presence of
traces of the alloying additive in the extracts from matrices of
Fe-10 Co, Fe 40 Ni-5 Mo, Ni-5 Mo, and of large quantities of
titanium in the Fe- and Fe 40 Ni-base specimens containing that
additive.

X-ray powder diffraction patterns were recorded of all oxide
residues with a Guinier type focusing camera. Measurable line shifts
were observed only for alumina powder extracted from an Fe-15 Cr
alloy, as given in the following table:

	a_{hex}	c_{hex}
Pure Al_2O_3	4,759 Å	12,991 Å
Extract from Fe-15 Cr	4,788 Å	13,077 Å
Pure Cr_2O_3	4,938 Å	13,581 Å

Alumina and chromia form a continuous series of solid solutions.
The lattice constants of the extracted oxide indicate an approximate
composition of $(Cr_{0.1}Al_{0.9})_2O_3$. As the colour of the oxide powders
indicates, some solid solution formation has occurred also in the
other Cr-bearing alloys (Fe5Cr, Ni5Cr, Ni10Cr, Fe40Ni10Cr) but in
such limited degree that the lattice constants were not affected.

The oxide from the Ni-Mn alloy gave an entirely different pattern
corresponding to the spinel phase galaxite, $MnO.Al_2O_3$, with no
remaining lines of alumina. Electron microprobe analysis confirmed
the composition of galaxite both in the central and peripheral
regions of the particles.

In the alloy of Fe 40 Ni with Mn, the oxide powder consisted partly
of alumina and partly of galaxite, indicating slower spinel
formation.

Extracts from titanium-alloyed specimens gave the normal diffraction
pattern of alumina without change in lattice dimensions. The strong
colouring and the X-ray fluorescence results indicate that Ti^{+3}
ions may have been taken into solid solution in the alumina lattice.

DISCUSSION

Let us first consider the nickel alloys. A calculation based on the
known solubility product of Al_2O_3 in Ni [20] and the diffusivities
in Ni of oxygen [21] and of aluminium (estimated on the basis of
HILLERT'S correlation [22] from the diffusivities of related elements
in Ni), and taking into account the oxygen partial pressure
(ca. 10^{-8} atm) at which our specimens were heat treated, indicates
that the product c.D has the same order of magnitude for Al and O,
so that the reaction rate is probably determined jointly by both
species.

It is not likely that the diffusivity of either Al or O is much affected by the alloying additives. LAZARUS'S theory of impurity effects on diffusion in dilute alloys (23) indicates that D_o and Q are affected simultaneously in such a way as to keep the diffusivity in the temperature range of technical interest almost constant.

If the alloying element is adsorbed at the phase boundary between oxide and metal, the phase boundary energy is lowered, and the driving force of the coarsening process is reduced.

The retardation of particle growth observed in most of our nickel alloys is thought to be due to this effect, possibly coupled with a decrease in the concentration of the rate determining species.

Manganese, on the other hand, produces a spectacular increase in the rate of coarsening. We have seen that a new oxide phase is formed. Manganese cannot by itself reduce Al_2O_3 to acquire the oxygen which it needs to form the spinel, $MnO.Al_2O_3$. However, some MnO is always present owing to the difficulty of excluding oxygen during fabrication. Compared to a mixture of MnO and Al_2O_3, the spinel phase has a lower free energy per mole of oxygen, yet its free energy is likely to be above that of pure Al_2O_3. The substitution of Al_2O_3 by an oxide phase of lower stability will increase the concentration of Al and O in the matrix, and this will lead to accelerated growth. According to eq. (1), a further increase is brought about by the larger molar volume of the new oxide phase; $V^2_{galaxite} = 2,73\ V^2_{alumina}$
It is also likely that the phase boundary energy is affected by the total change of the atomic arrangement at the interface, but this effect cannot be assessed as yet.

Turning now to the austenitic iron-base alloys, we note again a stabilizing effect of Mo and Cr additions, probably by the same mechanism as in Ni-base alloys. Manganese additions again produce a strong increase in growth rate, though not so spectacular as in pure Ni. It is worth remembering here that the formation of galaxite is less complete in FeNi matrix alloys than in Ni-based ones.

Finally, with the ferritic iron-based alloys we find essentially faster growth than for the austenic group. This difference persists in ferritic alloys heat treated at $1350^{\circ}C$; it may be related to the greater ease of diffusion in the bcc modification of iron.

Pronounced acceleration of particle coarsening again occurs in those cases where the oxide phase reacts with the alloying element. Cr has been shown to enter into solid solution in the alumina particles when present in high concentrations in the matrix. At $1450^{\circ}C$, the free energies of formation of the oxides are - 177 kcal/mole O_2 for Al_2O_3 and - 108 kcal/mole O_2 for Cr_2O_3. Thus the free energy of solid solutions with high Cr content will be

appreciably higher than that of pure Al_2O_3 (although of course
lower than the free energy of a mixture of alumina and chromia),
giving rise to an increase in the concentration of the migrating
species and thus to accelerated coarsening.   Low concentrations
of Cr, on the other hand, may actually bring the free energy below
the value of pure Al_2O_3 since the free energy curves of all solid
solutions start with a negative tangent at c = 0 and pass through
a minimum before rising at higher concentrations.

As concerns titanium, its solubility in alumina is so limited that
the solid solution in equilibrium with a matrix content of 1 per
cent may well be past its free energy minimum, in a region where
the free energy curve rises steeply. This would explain the distinct
acceleration in growth observed at a low level of titanium in the
matrix.

SUMMARY AND CONCLUSIONS

Our experiments show that the structural stability of alumina
dispersions in metal matrices can be strongly affected by the
addition of alloying elements, both in the positive and in the
negative direction. This opens up interesting possibilities for
the design of stable dispersion hardened alloys, which are not
limited to materials with alumina dispersions only.

The observed effects can be accounted for in a qualitative manner
by the change in oxide stability and phase boundary energy that
must be expected from the interaction of the alloying element with
the oxide. To proceed to a quantitative understanding of these
phenomena, more exact data on the solubility products of pure and
contamined oxide phases in alloyed matrices would be required.

ACKNOWLEDGEMENTS

We wish to thank Fagersta Bruks AB, Nyby Bruks AB and Höganäs AB
for assistance with various steps of alloy fabrication and for
the use of a Quantimet image analyzer. The work has been supported
financially by the Swedish Board of Technical Development.

REFERENCES

1. N.J. Grant, H.J. Siegel, R.W. Hall: Oxide Dispersion Strengthened
   Alloys, NASA SP-143, 1967.

2. G.W. Greenwood: Acta Met. 4 (1956) 243.

3. V.I. Psarev and I.V. Salli: Phys. Metals Metallogr. 5, 2
   (1957) 243.

4. J.M. Lifschitz and V.V. Slyozov: Phys. Chem. Solids 19 (1961) 35.

5. C. Wagner: Z. Elektrochemie 65 (1961) 581.

6. H.E. Exner and H. Fischmeister: Z. Metallkunde 57 (1966) 187

7. G.W. Greenwood: Institute of Metals Monograph no. 33 (1969), 103.

8. H.E. Exner and H. Fischmeister: Z. Metallkunde 6 (1970) 218

9. A.U. Seybolt: Metallurgical Society Conferences 47 (1968), 469.

10. D.K. Worn, S.F. Marton: Powder Metallurgy, Interscience, N.Y. 1961, 309.

11. C.T. Sims: Trans. **Met. Soc.** AIME 227, 6 (1963), 1455.

12. C.T. Sims: Metallurgical Society Conferences 47 (1968), 489.

13. V.M. Sergeenkova, V.V. Berezutskij: Sovjet Powder Metallurgy, No. 9, (1967), 706.

14. V.N. Jeremenko, V.I. Nizhenko: Zhurnal fiz. chimii, 35, 6 (1961), 1301.

15. B.C. Allen, W. D. Kingery: Trans. Met. Soc. AIME 215 (1959) 30.

16. W.H. Sutton: Report R64SD44, Missile and Space Div. General Electric, 1964.

17. J.W. Hancock, I.L. Dillamore, R.E. Smallman, 6th Plansee Sem. Preprint No. 30, Reutte 1968.

18. H. Fischmeister: Prakt. Metallographie, 2 (1966) 251.

19. C.F. Fisher, M. Cole: The Microscope, 16, (1968), 81.

20. O. Kubaschewski, E.Ll. Evans, C.B. Alcock: Metallurgical Thermo-chemistry, 4th Ed., Pergamon Press 1967, 257.

21. S. Goto, K. Nomaki, S. Koda: J. Japan. Inst. Met. 31, 4 (1967) 600.

22. J. Fridberg, L. Törndahl, M. Hillert: Diffusion i järn, Jernkontorets Annaler, 153 (1969) 263.

23. D. Lazarus: Solid State Phys. (Ed. Seits und Turnbull) 10, (1960), 71.

HARDENING WITH A DISPERSION OF SMALL HOLES

Michel A. EUDIER

Professor Ecole Centrale A. et M.

Consultant of METAFRAM

INTRODUCTION

Dispersion hardening is a broad subject and powder metallurgists have already played an important role in it. Examples are the manufacture of the TD Nickel and the theoretical work of Lenel and Ansell (1).

Although much research has been devoted to it, the phenomena are not well understood. The reason is that several factors are involved in the hardening which are not easy to evaluate. Among them are the total volume of the dispersed phase, the size of the particles, their shear resistance, and the coherency between the lattices of the particles and of the metal in which they are embedded.

The coherency can be perfect and in that case, there is a perfect continuity of the lattices so that there are only stresses at the interface between the particles and the matrix. When the coherency is imperfect, there are a certain number of defects at the interface. At the limit, when there is no coherency, there is no more binding at the interface and the matrix contains holes which are filled with the particle.

There are examples of almost perfect coherency in which the particles are soft and which results in a comparatively important hardening. One example, which is known to powder metallurgists, is the case of iron plus 2.5% copper. This alloy can be quenched from 920°C and then annealed for one hour at 575°C. The tensile strength is then almost doubled by this treatment, although the precipitated phase is nearly pure copper.

Because of that, we have thought that the tensions due to surface energy in the vicinity of a hole might have an influence on the tensile strength of sintered metals. It is well known that these tensions are important when the holes are very small in diameter. They are determined by the equation $P = 2A/R$ where P is the pressure at the surface of the hole, A is the surface tension of the metal and R is the radius of the hole. For nickel, which we shall considered further, $A = 1800$ dynes/cm and for $R = 100$ Å $(0,01\mu)$, the pressure P is roughly equal to 360 N or 25 Tsi. For $R = 20$ Å, $P = 125$ Tsi.

Yet it is astonishing to think that fine powder, able to give such small holes, has been studied previously at least to make barriers for the isotope separation of uranium and that very few remarks have been made on special properties of parts produced with these powders. The only paper we know is the one by Goetzel and Sheinberg (2).

THE SINTERING OF A FINE POWDER

We think that the reason why the dispersions of small holes in a metal have not been studied is that the sintering activity of the fine powders is very high and sinters very soon to the theoretical density. This is easy to understand because of the large surface per volume of these powders which results in a large energy to cause the sintering. We have previously mentioned the case of a powder, the grain diameter of which was around 100 Å. Such a powder, immediately after pressing, shrinks so quickly that it is possible to see the length of the sample diminishing at room temperature.

With a coarser powder, having a grain diameter of 300 Å, the sintering for 5 hours at 200°C into hydrogen purified through a palladium silver tube (Engelhardt apparatus) gives a perfectly dense metal. Figure 1 is a transmission electron mi - crograph which shows two remaining holes of an imperfectly sin- tered sample. The quality of the crystals is astonishing. No dis - location has ever been observed on 200 similar micrographs what- ever the remaining porosity was.

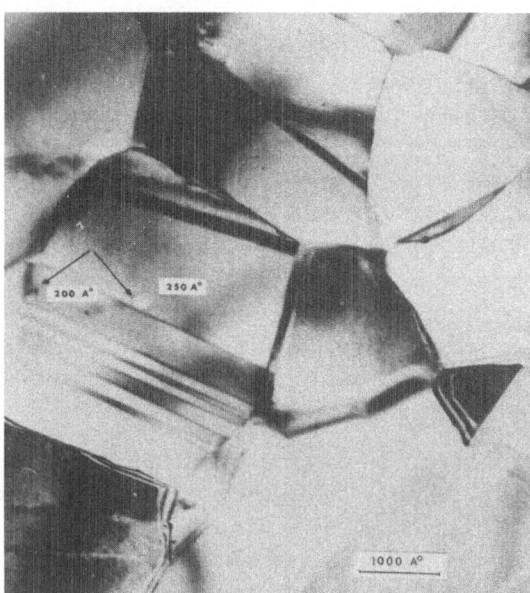

Fig. 1

It must be said also that, when such a powder is sintered at a higher temperature, big holes, a few microns in diameter, are formed, which cannot be easily eliminated afterwards.
The creation of these holes comes from uneven sintering (3) and
also from hydrogen diffusion as it has been explained by Rhines
and Birchenall (4). The theory of these authors is that hydrogen
is expelled from smaller holes, where the sintering pressure is
higher, towards the largest where P = 2 A/R is lower.

In order to maintain the small holes we have thought
of filling them with argon which does not diffuse through metals.
Rhines, Birchenal, and Hughes also studied the sintering under an
atmosphere of argon and it seems that an equilibrium is reached
for big pores, when they shrink, between the argon pressure and
P. For small pores we have observed that this equilibrium is
not reached. If it were, the final density would be very near the
theoretical one and this was not observed in our experiments. The
only explanation for this phenomenon is that the production of vacancies may be more difficult because there is a high probability that they would be filled with an argon atom and then cannot
diffuse.

Because of this phenomenon, internal stresses exist
closely around the pores and they change the physical and mechanical properties of the matrix.

SAMPLES PREPARATION

A variety of powders have been utilized for this study. Large grain size powders (1-50μ) are ex-carbonyle powders selected by elutriation of a regular Mond powder. Smaller grain size powders are produced by decomposing a nickel formate in hydrogen below 250°C in a fluidized bed apparatus. These powders have to be purified carefully because they contain oxygen and carbon compounds. They are treated for that in a fluidized bed apparatus with pure fluorine and again hydrogen at 200°C. Afterwards the handling of the powders, pressing and sintering occurs within glove boxes. By varying the decomposition temperature the specific surface of the powder is changed.

The samples made of fine powders are pressed at a low pressure (15 Tsi) so that they are permeable, and hydrogen can penetrate into the sample for a final reduction at 150°C. Then argon is admitted in the sintering tube after a vacuum has been made and the sample is heated in 5 hours up to 400°C and stays for 15 more hours at this temperature. The final analysis shows that there is no metal above 20 p.p.m. and no metalloid above 10 p.p.m.

Coarser powders are sintered at a higher temperature.

DETERMINATION OF THE MEAN PORE RADIUS

This determination is made by using several methods since none of them is good for all the range of radii and porosity. The electron microscope is utilized by transmission through a foil of the material and through replicas made on fractures. X rays and permeability of a gas through samples are also used. For this last method we assumed that, after the closure of the pores, their number has not varied. Their radius is then given by measurement of density.

It is surprising that all the methods give the same values within ± 15%. For one method the probable error is below 2% and the discrepancy between the methods can be explained by the fact that they do not measure exactly the same thing. The transmission method is the best but we have not succeeded to utilize it when the total porosity is over 3% due to the difficulty of making the foils. The permeability method is the easiest when porosity is high.

RESULTS OF EXPERIMENTS ON NICKEL

Due to the high cost of the preparation of the powders and samples we have only measured the electrical resistivity, hardness and tensile strength.

Resistivity

Figure 2 gives the results of the resistivity measurements as a function of the pore radius for a total porosity of 30%. Since the density of the samples varies from 30% to 3%, corrections are made using the formula

$$X \ / \ X_0 \ = \ 1 \ - \ k \mathcal{E} \quad (1)$$

where X is the electrical conductivity of the sample, X_0 that of the full density nickel, k a constant for a given radius and $\mathcal{E}$ is the relative porosity (total volume of the pores divided by the sample volume) - $k = 1.5$ for large pores. Relation (1) has been verified for three pore radii.

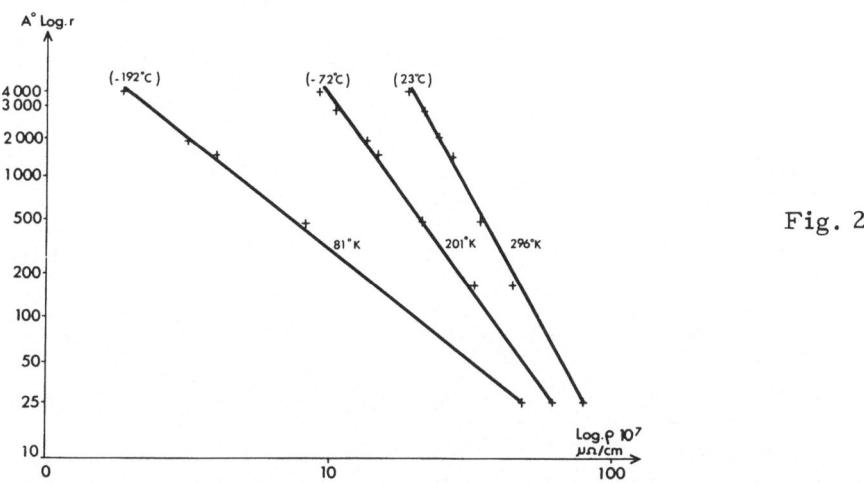

Fig. 2

In Figure 2 the results are given for three tempera - tures using logarithmic coordinates. The experimental points are located on straight lines. For larger diameters the resistivity is a constant. The large influence of the small pores is obvious. The two lines for the higher temperatures are meeting for a pore radius which is of the order of magnitude of a vacancy radius and the resistivity increase is then 120 $\mu\Omega$/cm. It gives a measure - ment of the influence of the vacancies on the resistivity. The ex - perimental value is then 4uΩ/cm for 1% vacancies which is equal to the value measured by Schmacher, Schule and Seeger (5).

Hardness

Fig. 3 gives the results obtained by measuring the Vickers hardness as a function of the relative porosity for two values of the pore radius. The laws are identical for both radii, when the relative hardness is plotted in ordinates, and the experimental values fit well with the curves.

$$H \ / H_o = 1 - 1.21 \varepsilon^{2/3} \qquad (2)$$

which is the same that fits for the tensile strength of porous metals (6.7). There is no value for $\varepsilon < 0.03$ and the curve for R = 200 Å suggests that for $\varepsilon = 0$ the hardness is high and this cannot be true. There must be a maximum below $\varepsilon = 0.03$.

The values of the hardness as a function of the pore radius are given by the curve at right in figure 4 in logarithmic coordinates (values are those which are extrapolated from curves similar to figure 3 for $\varepsilon = 0$)

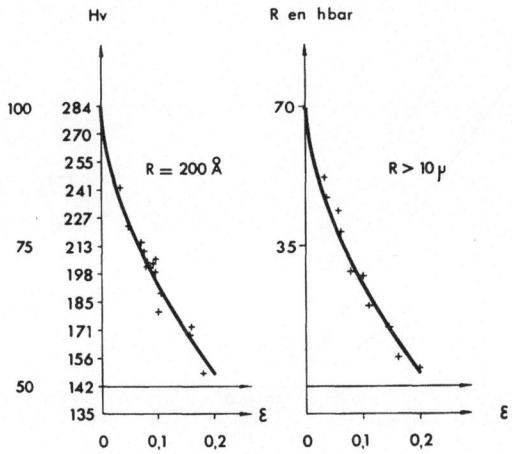

Fig. 3

The curve at left represents the hardness increase i.e. the difference between the hardness of the samples and the hardness of the ordinary nickel. It is a straight line the slope of which is 1.9 which gives a law very near the following

$$\Delta H_{\varepsilon = 0} = k' \sqrt{R} \qquad (3)$$

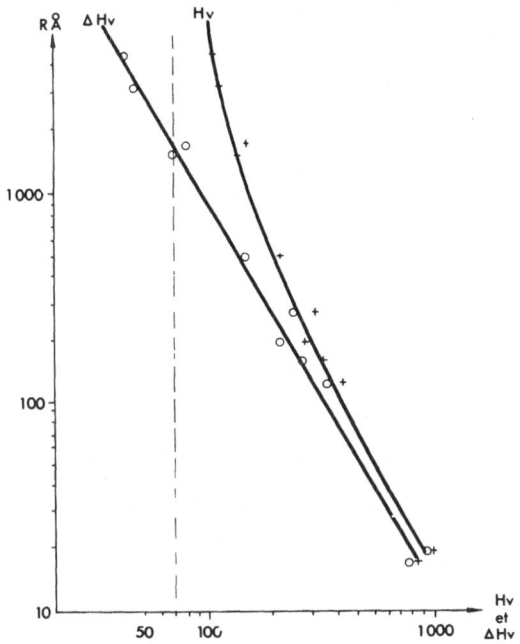

Fig. 4

Tensile strength

        After sintering, the tensile strength of the specimens, whatever the pore radius, does not exceed 45 Tsi.   The dispersion of the results is very high and there is no elongation before rupture when the pores are small.

        We have thought that this was due to the lack of deformation around the pores which are thus creating stress concentra - tions equal to the value which can be calculated by the theory of elasticity. This lack of deformation must come from an insuffi - cient production of dislocations. To create them, we workharden the samples by rolling. With 3% reduction in thickness, the ten- sile strength becomes equal to the value predicted by the hardness after the law which is valid for ordinary steels. An elongation be - fore rupture is then observed.

        The curve at left of figure 5 gives the tensile strength as a function of the pore radius for a porosity of 10% after 3% reduction in thickness by rolling.

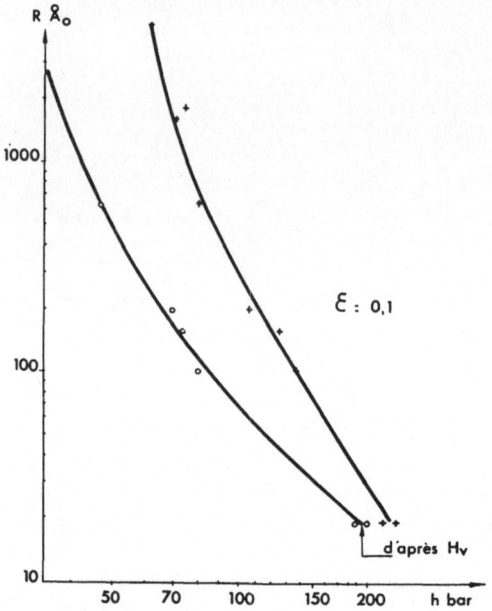

Fig. 5

The curve at right is the tensile strength of the same type of specimen after the maximum cold rolling reduction which is permissible without cracks. The reduction of thickness defined by $\rightleftharpoons = (e_0 - e) / e_0$ (where e_0 and e are the initial and final thickness) is 0.7 for radii over 10μ, 0.2 for a radius of 100 Å and 0.05 for the smallest radius.

The tensile strength for R $\cong$ 20 Å is unreliable, results are between 70 and 220 hbars (1 hbar = 0.7 Tsi).

For R = 100 Å figures 6 and 7 give the tensile strength of the samples as a function of log e_0/e and the value of the tensile strength increase after rolling as a function of $\log e_0/e$. This latter curve is a straight line. Its slope is 0.5 and the tensile strength values are extrapolated to $\mathcal{E} = 0$.

After rolling, the properties are not equal in all the directions of the sheet and figure 8 gives the tensile strength after the angle of the drawing direction with the rolling one in polar coordinates. The radius is 100 Å and four values of $\rightleftharpoons$ are utilized. For $\rightleftharpoons = 10\%$ the properties are near to being isotropic. Above, lower values are found which can be due to cracks which were not observable.

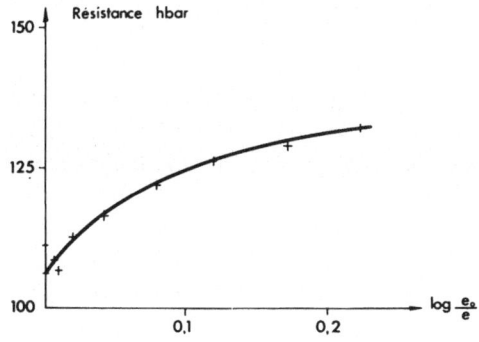

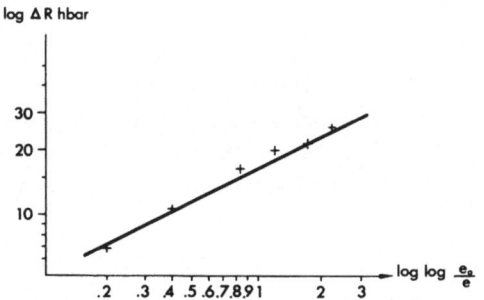

Fig. 6 and 7

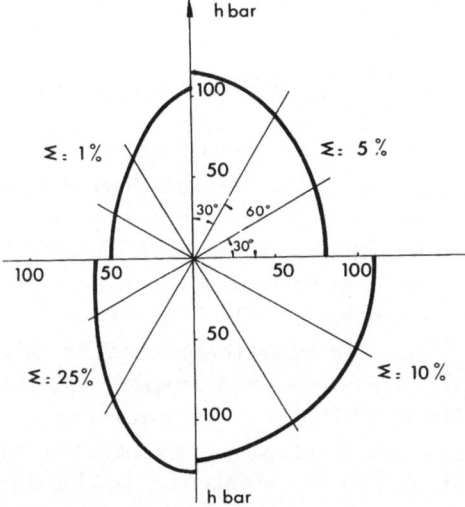

Fig. 8

RESULTS OF EXPERIMENTS FOR OTHER METALS

Because of the difficulties which are encountered in the preparation and purification of the powders, we have only made a few tests on other powders. Pure iron gives similar results. Solid solutions like iron-copper alloys do not give a much higher resistance. On the contrary, the compositions which can give ordered structures have a high tensile strength. The best we have tried is the iron-cobalt (50-50). With a mean radius of 120 $\overset{\circ}{A}$ a hardness of 600 Hv is reached. It is twice   of   what is obtained with the pure nickel.

THEORETICAL INTERPRETATION

There is not much hope in finding any practical application for such metals and alloys but their study is, in our opinion, one of the many steps which can help to understand the behaviour of dispersed phases and to make progresses to produce better metals.

Many conclusions can be drawn from these tests and the most important seems to be the fact that there is a perfect continuity of the properties as a function of the pore diameter. Hence, what we know about large pores must be applied at least partially to small pores or small particles before making any theory involving dislocations.

TENSILE PROPERTIES OF ORDINARY SINTERED METALS

When a load is applied to a sintered test bar, the deformation is initiated in the vicinity of the holes when the elasticity limit is reached in these regions. It occurs for less than one half of the elasticity limit of the full density metal. A very slight bending of the tensile curve is then observed and, when the porosity is below 10%, the deformation is locally important but negligeable for the total sample. When the elasticity limit is passed over in all the metal, then the deformation is important, and this gives an apparent elasticity limit which is very near the elasticity limit of a fully dense metal. As a first approximation the elasticity limit of a porous metal is constant, whatever is the density. It is only reduced when the rupture occurs in the vicinity of the hole before the elasticity limit has been passed over in the regions which are far from the holes. Then there is practically no elongation before rupture. The metal seems to be brittle.

Using this idea and with a few assumptions it is then

possible to determine mathematically the form of the tensile curve
of a porous metal and also the elongation of this metal as a function of the total porosity when the properties of the fully dense
metal are known. The theoretical curves cannot fit exactly with
the experimental results but the results are comparatively good.
Several authors have made such mathematical treatments and
we wish to mention especially the work of Butcher and Cope (8)

With a cubical model we have obtained the curves of figure 9 for
iron and nickel and the experimental points are not too far from
them.

        Using the same idea it is also possible to preview the
maximum tensile strength of all low alloy steels as a function of
porosity as we have shown previously.

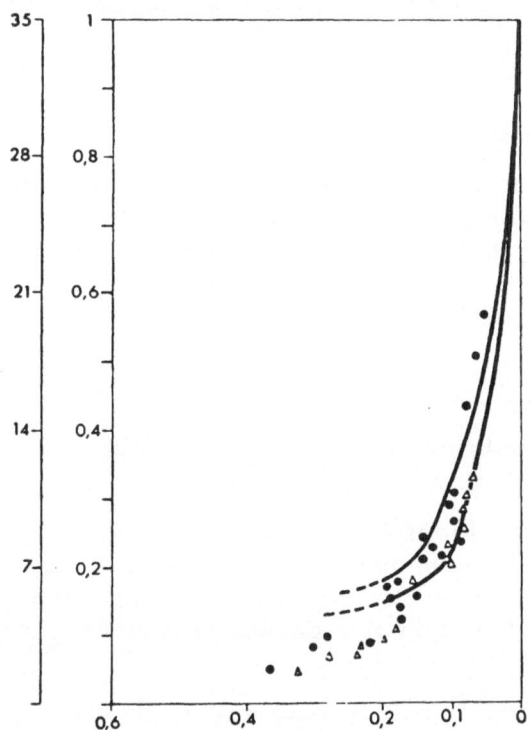

Fig. 9

TENSILE PROPERTIES OF A DISPERSION OF SMALL HOLES

        When a metal containing large hole is deformed the
holes tend to a shape which is near an ellipsoid. When the holes
are small there is a force which acts against this deformation

since, if this deformation exists, there is an increase of the surface of the hole when the volume of the hole keeps a constant value.
If we suppose that the hole is not deformed and that the metal is
shearing around it, there must be a certain number of defects
which compensate this lack of deformation of the hole. Ashby (9)
makes the assumption, in the case of dispersed particles, that piles
of dislocation loops are then produced between the particles. We
have applied his idea to the small holes.

When a deformation of the entire sample occurs after
the apparent elasticity limit has been passed, the dislocations are
meeting these loops produced by the local deformation and it results in an increase of the stress which is necessary to deform
the metal.

It is obvious that a fixed local stress is necessary to
produce the first loop and it explains why the influence of the radius begins at a rather precise value which is in our experiments
0.4μ. For a radius above this value the pores behave like large
pores. Below, they produce dislocation loops.

Another fact is that the loops interfere with each other.
This fact seems to have been neglected by Ashby. The loops are
repulsed from each other so that between two holes a stress is
produced which acts against the creation of new dislocation loops.
If not, in the case of holes, the hardness increase would vary after the law

$$H_v = k \, / \, R \qquad (4)$$

Because of the repulsion, the law which can be calculated has the form of equation 3 obtained experimentally.

The existence of the loops has been controlled (10) and
a drawing after an electron microscope photograph is given in
figure 10. The mathematical approach would be too long here, but
it is rather simple and very few unknown parameters have to be
used.

One important item difficult to determine, is the
relation between the total deformation of the sample and the local
deformation which creates the loops. This relation is a function
of the work hardening law of the metal which, up to now, has not
been considered by theoreticians. Unhappily, the mathematical
treatment is then very difficult, but at the end of the phenomenon,
when the holes are very small, we can suppose that for a small
percentage of total porosity the material will have no elongation

before rupture. Then the holes remain spherical and the rupture occurs when the stress is equal to the sum of the local tension due to the surface energy and of the elastic limit of the full density metals. It corresponds to our results.

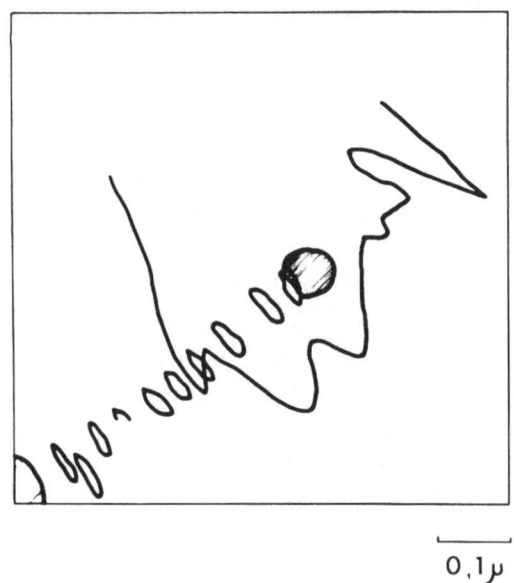

Fig. 10

0,1μ

Another fact coming from the results on resistivity is that a vacancy behaves like a small hole. Hence its energy of formation can be calculated starting from the surface tension. Such a calculation has been made by Brooks (11) but with a more precise calculation one obtains the value of 1.38 e V measured by Nakamura (12).

EXAMPLE OF A PRACTICAL APPLICATION

Although the preceeding paragraphs are mainly theoretical, it is possible to find practical aspects that define tendencies for research. One example is the manufacture of valve seats for high temperature. The task was to find a comparatively soft material so that the valve can fit perfectly to the seat, but which is corrosion resistant and which, after a few hours of working, is ve-hard. This specification gave us the idea of using a precipitation hardenable steel. Such an alloy must first work harden very quickly after a small deformation, and this is obtained with a ferritic material since the work hardening curve is favourable for a high content of the precipitated phase. Then this material must be embedded in an austenetic steel resistant to corrosion. We have started from a steel containing chromium 13%, nickel 7%, carbon 1.5%. The sintering occurred above the eutectic temperature so that a liquid phase, rich in chramium and nickel, is created around the nodules of an alloy which is poorer in chromium and nickel. After quenching and aging at 750°C, we obtain the desired structure. After a few tests, in which we have either a too brittle or a too soft structure, it is possible to get a final hardness of 50 R_C . This metal can be deformed easily by cold hammering, and, after this hardening, the value of the hardness is above 58 R_C The precipitated phase, mainly a chromium carbide, is stable up to 700°C. Like the metals containing a dispersion of holes, such a material has a poor tensile strength when it is not work hardened.

The question then arises as to whether or not, a small amount of a phase which is not hard has an influence on the tensile strength of a dispersed phase. Returning to nickel we have prepared samples containing large nickel grains mixed with a fine powder which gives holes having a radius of 100 Å.

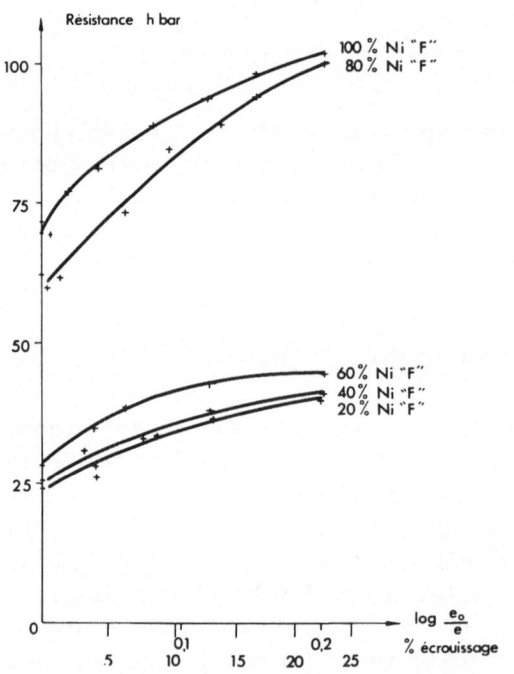

Fig. 11

Figure 11 shows the results. When the volume of the soft regions is comparatively low, and after work hardening, the difference is small. We think that this is a proof that the soft re - gions are consolidated by the fact that dislocations are stopped by the hardened regions.

CONCLUSIONS

The object of this paper was to show the analogy which exists between dispersed holes and dispersed particles. It has not been possible to give many details on how the comparison of holes and particles can help to impose the existing theories of dispersion hardening but we have given a few ideas and we hope that these results will be useful to others.

REFERENCES

1 - G. S. Ansell and F. V. Lenel-Acta Metallurgica 8, 614, (1960)
2 - C. G. Goetzel and M. A. Steinberg - Modern Devlpts in P. M. 1,
                    194, Plenum Press (1966)
3 - M. Eudier - Powder Met. G. B. 6 12 , 17, (1963)
4 - F. N. Rhines, C. E. Birchenall and L. A. Hughes TAIMME 188,
                    378, (1950)
5 - D. Schmarer, W. Schule and A. Seeger - Z. Naturforsch 17a,
                    228, (1962)
6 - L. H. Cope - Metallurgia 72   432   165   (1965)
7 - M. Eudier - Planseeber. Pulvermet. 14, 29, (1966)
8 - B. R. Butcher and L. H. Cope - J. of P. M. 4, 4 , 49, (1968)
9 - M. F. Ashby - Philos. Mag. 14 , 1157 , (1966)
10 - Humphrey and Martins - Phil. Mag. 16, 927 (1968)
11 - H. Brooks - A. S. M. Impurities and Imperfections, 1 (1955)
12 - Y. Nakamura - J. Phys. Soc. Japan 16, 2167, (1961)

DISPERSION-HARDENED MATERIALS WITH COPPER AS A BASE METAL

H. Schreiner and H. Ohmann

Siemens-AG., Nürnberg, BR Deutschland

1. Introduction

Materials having high strength at elevated temperatures, and good electrical conductivity are of particular importance for conducting springs, contacts, parts in electrical machinery, welding electrodes, as well as for components in electronic and vacuum technology.

Dispersion-hardened sintered materials, having a copper matrix with mechanically and thermally stable oxides of Al, Ti, Be, Th,etc.as the dispersed phases, are superior to the precipitation-hardened copper alloys of the types Cu-Cr, Cu-Be, Cu-Zr, when comparing the structural stability at elevated temperatures and the electrical conductivity.

For technical and economical reasons these dispersion-hardened copper-base materials are best produced by the method of powder metallurgy i.e. by compacting and sintering a mixture of copper powder and suitable oxide powders, and then subsequent extrusion. The final shape may be obtained by machining or cold-forming.

The most important step in the powder metallurgy process is the preparation of the mixture of copper and metal-oxide powders, since this largely determines the geometry of the dispersion, expressed by

the particle size and the particle spacing of the
dispersed phase. The dispersion-hardening effect
depends partially on the properties of the dispersed
phase and its interaction with the copper matrix,
but above all on the quality of dispersion. The
mean particle size should be less than 1 micron,
and the mean particle spacing of the dispersed phase
should be adjusted for the required purpose by
suitable choice of the percentage volume of the
dispersed phase.

The dispersion-hardening effect can be evaluated from
the following parameters:

The strain-hardening with cold-forming:
The particles of the dispersed phase prevent motion
of the dislocations and cause increased blocking of
the dislocations. For this reason, dispersion-hardened   materials harden even more than those hardened
by mixed crystal formation.

Softening behaviour:
The mobility of dislocations is limited by the deposited particles to such an extent, that even at
temperatures which are 90 ^{o}/o of the absolute melting temperature, no pronounced recrystallization
and only a small decrease of the hardening occurs.
Only above this temperature do the strength and
hardness decrease considerably. The increased
strength obtained by cold-forming is therefore very
well maintained up to high temperatures.

Creep behaviour at elevated temperatures over long
periods of time:
The creep rate is decreased by the thermally stable
particles of the dispersed phase, such that at temperatures of over 600^oC, the dispersion-hardened
copper metal-oxide alloy materials should even be
superior to high-grade steels.

The coarsening of the deposited particles at high
temperatures:
Unlike precipitation-hardened materials, the deposited particles of thermally stable oxides are
insoluble in the copper matrix, so that even for
long periods at high temperatures, there is no
coagulation of these particles, and therefore the
dispersion-hardened effect is also maintained.

2. Experimental Aspects

2.1 Mechanical mixing of the powdered components

This method is unquestionably the most versatile on account of the wide choice of type and quantity of the dispersed phase, and is also simple and economical. The quality of the dispersion is determined however by the fineness of the initial powders, and the efficiency of mixing. Coagulation and segregation of the dispersed phase are unavoidable.

Dendritic electrolytic copper powder (mean particle size 30 microns) and spherical Al_2O_3 (mean particle size 20 microns) can be used as the initial powders. In preliminary tests the most favourable mixing conditions were determined: the mixers which could be used were the ball mill, the tumble mixer and the paddle mixer. Four different mixtures were placed in each mixer in turn, and mixed for 4 hrs. These mixtures were the dry powder alone, and then the powder mixed with stearic acid, amyl acetate or softened water. The best method was found to be mixing with soft water in the paddle mixer. By this method, composite powders were prepared containing 1,25, 2,5, 5 and 10 %-volume of Al_2O_3, TiO_2 and SiO_2.

2.2 Microfine precipitation by chemical methods

A substantially finer distribution of the deposited oxides is expected by preparing the composite powders using microfine chemical precipitation of their components. Above all, by this method, segregation and coagulation during further processing of the powder is avoided. Two different methods are available: the finest stable oxide in normal commercial usage, is suspended in an aqueous solution, and the copper component precipitated upon it from a suitable medium. In this way the quality of the dispersion is still linked with the fineness of the oxide powder, but segregation due to subsequent processing is avoided.

In order to obtain an even finer distribution of the oxides in the copper, both components of the composite powder are precipitated together.

The further processing of these composite powders is such that first, the inert metal is
transferred into the stable oxide, and in a
following annealing process the copper component
is reduced.

By these two methods just described, composite
powders are prepared containing Al_2O_3, TiO_2 and
SiO_2 in the proportions 1,25, 2,5, 5 and 10 %-
volume.

2.3 Internal oxidation of copper-alloy powders

By pressure-atomizing copper alloys containing
small quantities of Al, Ti and Si, powder was
produced with particle sizes of less than 0,3 mm.
Nevertheless it is important to avoid oxidation
of the inert metal during melting. It is necessary for particularly careful metallurgical
handling of the copper melt, and also an inert
gas is used as the pressure medium. The internal
oxidation of powder particles occurs quickly,
since the diffusion lengths are short, and the
time for total oxidation of spherical particles
is considerably less than that required for
plates of the same thickness. As a result of
these short diffusion lengths the effect of
particle coarsening seldom appears, so that the
oxide dispersion can be maintained uniform and
fine over the whole cross-section.

Maintaining the optimum annealing atmosphere for
internal oxidation gives, of course, special
difficulties with the internal oxidation of
powder particles. The method possible for
compacts, namely the use of the "Rhines-Packing"
method in cuprous oxide is out of the question
for powder particles. In order to perform this
there are two suitable methods:

The copper alloy powder is annealed in an
atmosphere with a low oxygen partial pressure,
corresponding to the pressure of dissociation
of Cu_2O. These conditions can be better explained by the schematic drawing of the two-
chamber method in Fig. 1:

In this case, the powder to be internally oxidized is placed in a chamber 1, at the optimum
temperature for internal oxidation.

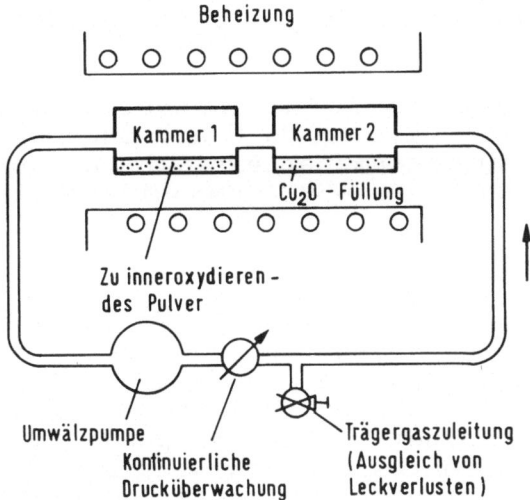

Fig. 1:

Two-chamber method for the internal oxidation
of copper alloy powder

The chamber can be in the form of a rotating
drum or a flow-bed. The cuprous oxide-filled
chamber 2 is connected to chamber 1. Chamber 2
is held at the same temperature as that of
chamber 1. By means of an inert carrier-gas in
the closed circulatory system, the nascent
oxygen formed by dissociation of the Cu_2O, with
the appropriate partial pressure, is led into
chamber 1. By this method, the most favourable
oxygen partial pressure adjusts itself, depending on the selected temperature of the internal
oxidation, and is simpler and more accurately
maintained than by injecting O_2 into a carrier-
gas.

Poniatowski and Clasing[8] suggested oxidizing
the powder first of all in air. By doing this,
the necessary oxygen for oxidation of the inert
metal in these alloy powder particles, is
obtained first from scale comprising Cu_2O and
CuO on the powder particles. A second annealing
in an inert atmosphere then produces the
diffusion of the oxygen into the inside of the

powder particles, and the oxidation of the inert
metal.

Finally, in a third annealing operation, the
excess scale of Cu_2O and CuO must be removed in
a reducing atmosphere. The authors named above
produced a Cu-BeO composite material using this
rather expensive method; the price per kilogram
was about DM 100,--.

2.4 Further processing of the composite powders

From the composite powders prepared by the
methods described earlier, cylinders were compacted and sintered. These compacts were welded
into copper mantles to prevent oxidation during
extrusion. The compacts prepared in this way
were extruded horizontally into rods of 12 mm
diameter. The pressure ratio $a_1:a_2$ was 17:1. The
necessary force for the extrusion depends on the
type and quantity of oxide in the compact, and
lies between 5 and 10 Mp/cm^2.

With these compacts in the extruded state, the
properties like tensile strength, hardness and
electrical conductivity were tested and
determined.

To test the hardening and softening behaviour,
the samples were first annealed for four hours
under vacuum at 1.000^oC, in order to remove the
rest of the stored energy produced by the
extrusion process, and to produce a definite,
soft state. Then the samples were cold-formed by
rolling, and the hardening behaviour examined.

The samples which are cold-formed about 75 % by
cold-rolling or round-hammering, were then each
annealed under vacuum for 1 hr. at temperatures
between 100° and 1.000^oC, in order to examine
the softening behaviour.

3. Experimental Results

3.1 Structural Tests

The structure of the composite materials, prepared as in 2. was examined metallographically,
both in the pre-sintered as well as in the extruded state, with the aid of an optical

microscope and to some extent electron optics.

The preliminary tests, for determining the best
method of mixing, showed that the most favourable dispersion was achieved using the paddle
mixer with softened water (Fig. 2). Although
the distribution of the oxide particles is not
uniform using this method, coagulation of the
particles occurred many times. Therefore in
places there are oxide deposits with a total
particle size up to 10 microns.

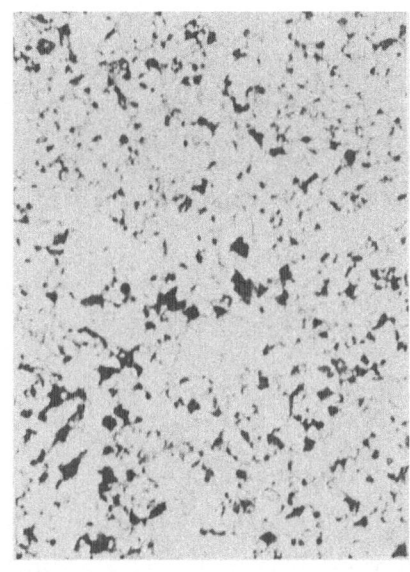 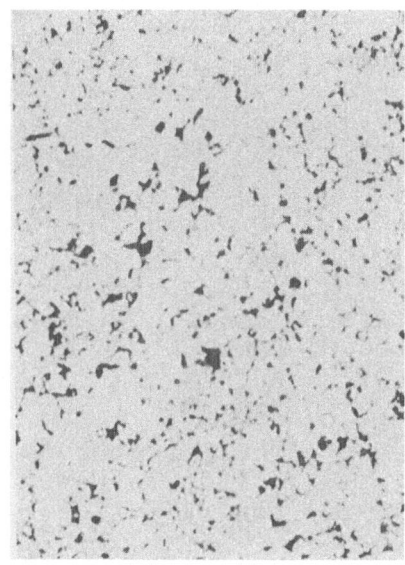

500:1
Dry powder
mixed alone

500:1
Wet-mixed in
softened water

Fig. 2:

Structures of mechanically mixed Cu-Al$_2$O$_3$
(3 %-Vol.) Composite Materials

Fig. 3 shows the structures of Cu-Al$_2$O$_3$-Com-
posite Materials, obtained by microfine precipitation. The simultaneous precipitation of both
components leads to a finer and more uniform
oxide dispersion than the precipitation of Cu
upon oxide particles in suspension. With the
former method the particle sizes of the oxide
deposits are a few microns.

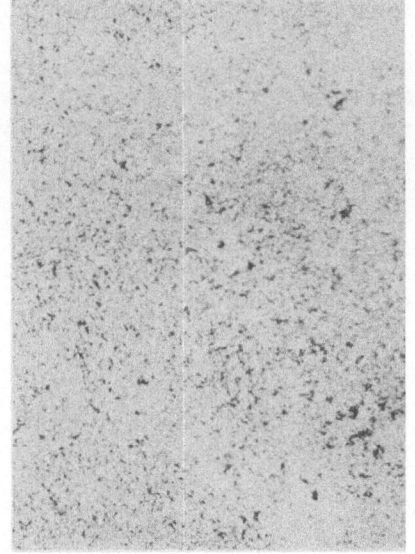 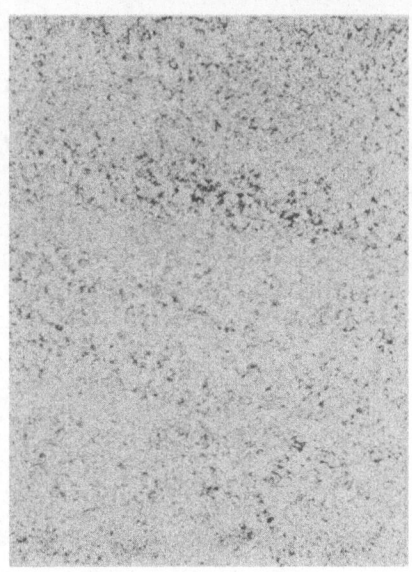

500:1   500:1
Precipitation of Cu   Combined precipitation
upon Al_2O_3 in sus-   of Cu and Al
pension

Fig. 3:

The structure of $Cu-Al_2O_3$ (1,5 %-Vol.)
Composite Materials, produced by microfine
precipitation

Fig. 4 shows electron-optically made photographs
of the structure of these two composite materials obtained by chemical means, which confirm
the findings discovered with an optical microscope.

3.2 Electrical Conductivity

Since, by diffusion hardening, the copper matrix remains unchanged to a large extent, many
physical properties change only proportionally
to the volume fraction of the dispersed phase.
Since this fraction is only 10 % at the most,
the changes, e.g. the electrical conductivity,
are relatively small when comparing the values
with those of pure copper, as can be seen in
Table 1.

500:1   500:1   15000:1
Precipitation   Combined precipitation
of Cu upon Al_2O_3 of Cu and Al
in suspension

Fig. 4:

The structure of $Cu-Al_2O_3$ (1,5 %-Vol.)
Composite Materials

Table 1:

Electrical conductivity in $m/\Omega\,mm^2$ of extruded
Cu-Oxide-Composite Materials

Type	SiO_2	TiO_2	Al_2O_3
Mechanically mixed %			
0	58,6	58,6	58,6
1,25	56,7	56,3	55,6
2,5	55,3	55,3	53,8
5,0	53,9	53,7	52,3
10,0	52,0	51,9	49,2
Microfine precipitated %			
1,25	56,1	55,8	54,9
2,5	54,8	54,4	53,1
5,0	52,6	52,1	51,8
10,0	50,9	49,9	48,8

3.3 Strain-Hardening

The curves shown in Fig. 5 of the strain-hardening of different materials, represented by
increase in hardness against degree of deformation, show the considerably greater hardness
of the dispersionhardened materials compared
with pure copper. Also Al_2O_3 has a far greater
effect than SiO_2.

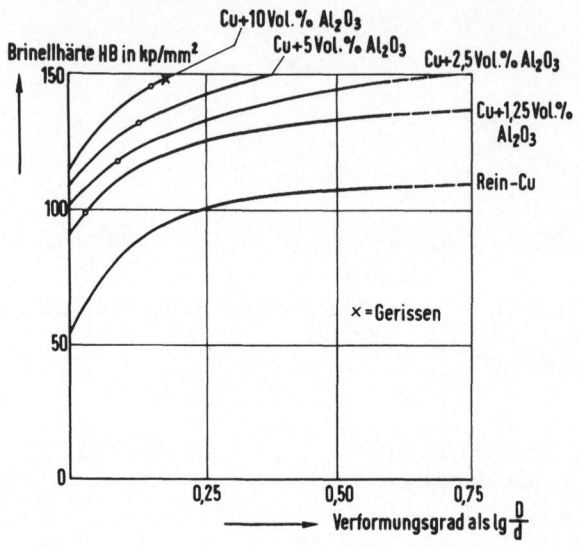

Fig. 5:

Hardening curves of Cu and Cu-Al_2O_3-Composite
Materials

3.4 Softening_Behaviour

Some of the curves shown in Fig. 6 of the
softening process of composite materials,
represented by the decrease in hardness against
the annealing temperature, show the strong
influence of the geometry of the dispersion on
this deciding criterion. The composite materials
containing Al_2O_3 come nearest to the typical
softening behaviour of a dispersion-hardened
material. With the composite material containing
SiO_2, a sharp decrease in hardness occurs at
temperatures above $200^\circ C$, because the SiO_2
has a relatively small heat of formation
of 101 to 103 kcal/g-Atom Oxygen, and it tends

to coagulate.

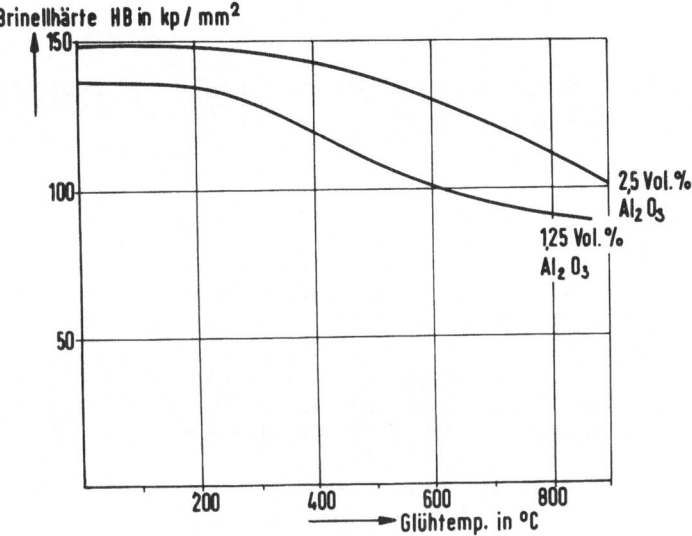

Fig. 6:

Softening curves of Cu and Cu-Al$_2$O$_3$-Composite
materials

4. Conclusions

The powder metallurgical production of composite
materials out of copper with finest deposited
particles of mechanically and thermally stable
oxides, is a very suitable way of producing a
material with high elevated-temperature strength
and good electrical conductivity. The examination
of the possible methods for producing such composite
materials, showed that the main properties, which
are criteria of the quality of the dispersion, such
as strain-hardening and the softening behaviour,
depend chiefly on the geometry of the dispersion,
in addition to the type of oxide used. The choice
of a suitable method for the production of a new
commercial material is also affected by economical
considerations.

Mechanical mixing of components in powder form is
versatile and necessitates the least manufacturing

expenditure. Although the fineness and uniformity
of the dispersed phase, even in ideal conditions
regarding the type of mixer and the mixing medium,
do not correspond to the theoretical ideas, materials so produced are already superior to
precipitation-hardened copper alloys, when comparing
the softening behaviour. By annealing at 900°C they
only lose about 25 % of the strain-hardening caused
by cold-forming,

Microfine precipitation undoubtedly requires higher
manufacturing costs than mechanical mixing. The
average particle size of the oxide deposits is,
however, an order of magnitude smaller, and so the
corresponding softening behaviour is better. Using
appropriate chemicals and the best technological
form of the precipitation process, a high-grade
material can be economically produced.

The method of internal oxidation of alloy-powder
particles is even more expensive. Only in good
conditions, particularly with a small concentration
of the inert metal, and therefore also of the
dispersed oxide (1,5 %-Vol.), does this method give
finer and more uniform oxide dispersion than the
microfine precipitation. With larger proportions,
oxide precipitations occur mainly at the grain
boundaries, or else in the form of clusters or
streaks.

Literature:

1) N.I.Grant u. O.Preston: Trans. AIME 209 (1957)
   S. 349
2) K.M.Zwilsky u. N.I.Grant: Trans.AIME 221 (1961)
   S. 371
3) K.M.Zwilsky u. N.I.Grant: Trans.AIME 209 (1957)
   S. 1197
4) K.M.Zwilsky u. N.I.Grant: Metal Progr. 80 (1961)
   S. 108
5) M.Adachi u. N.I.Grant: Trans. AIME 218 (1960)
   S. 881
6) P.Predecki u. N.I.Grant: Proc. ASTM 62 (1962)
   S. 639
7) O. Loebich u. I.Vock: Z.Metall 19 (1965) H. 11
   S. 1178
8) M.Poniotowski u. M.Clasing: Z.Metallkunde 59
   (1969), H. 3, S. 165

THE HIGH-TEMPERATURE STRENGTH OF OXIDE-STRENGTHENED NICKEL ALLOYS

R. F. Cheney and W. Scheithauer, Jr.

Sylvania Electric Products Inc., Chemical & Metallurgical Div.

Towanda, Pennsylvania   18848

INTRODUCTION

The role of dispersed inert oxide particles in strengthening metals at room temperature and below has been evaluated often. Much less has been done to measure the affects of an oxide dispersion on high-temperature properties. For that reason, we prepared $Ni-2ThO_2$ and $Ni-15Mo-ThO_2$ sheet with varying volume fractions and sizes of ThO_2 and investigated the metallurgical conditions affecting the strengths at 1090°C. Matrices of Ni and Ni-15Mo were chosen because of their relative simplicity, our previous experience in their manufacture, and their potential as useful alloys for high-temperature applications.

SUMMARY

Dispersion-strengthened $Ni-2ThO_2$ and $Ni-15Mo-ThO_2$ alloys were prepared by P/M methods and conditions affecting their tensile strengths at 1093°C (2000°F) were evaluated. Small ThO_2 and strongly developed crystallographic textures were both necessary for good strengths at 1093°C. $Ni-15Mo-ThO_2$ alloys having volume fractions, f, of 0, 0.0003, 0.003, 0.03, and 0.10 were made and evaluated. The 0.2% yield strengths at 1093°C increased in proportion to $f^{1/3}$, but only for sheet with mean ThO_2 diameters of about 150Å and strong textures. The yield strengths of sheets without textures and with mean ThO_2 of 400-500Å showed little dependence on volume fraction. $Ni-2ThO_2$ sheet with about 200Å ThO_2 had strong texture and good strength at 1093°C while sheet with about 400Å ThO_2 had little texture and poor strength. The thermomechanical processing of the alloys greatly influenced their strengths because of the affects on texture, grain structure, and the spatial distribution of the ThO_2.

MATERIALS AND PROCEDURES

Ni-15Mo-ThO$_2$ Alloys

The powders were made by spray drying and selective reduction[1,2]. The reduced powders were isostatically pressed into 2.5" dia x 5" long billets, sintered, canned in mild steel, evacuated and then extruded to flat bars. After removing the cladding, the extrusions were rolled to sheet, all by the same deformation-anneal schedule.

The ThO$_2$ size and loading were varied. Volume loadings of 0, 0.03, 0.3, 3.0, and 10.0% were used. The average ThO$_2$ diameter was controlled by adjusting the sintering temperature as shown in the table below. The ThO$_2$ size was estimated by x-ray line broadening.

Sintering Schedule	Target ThO$_2$ Diameter, Å	Measured ThO$_2$ Diameter, Å
5 hrs at 750°C	50-150	150
2 hrs at 950°C	200-300	150
2 hrs at 1150°C	400-500	400
2 hrs at 1300°C	>500	400

Densification of the sintered billets was completed by extruding to flat bar at 870°C. Normal densification during sintering is impeded by the ThO$_2$ dispersions in proportion to the ThO$_2$ size and volume loading. Thus the sintered densities ranged from the pressed density of 60% up to 99%, depending on sintering temperature and composition. The extrusion ratios ranged from 7.1 to 10.4, the lowest density billet being extruded at the highest ratio to equalize the work energy input to the matrix. Densities after extrusion were 98-100%.

After extrusion the unclad billets were rolled to sheet using the following procedure:

1) Heat-treat for 5 hours at 800°C plus 1 hour at 1200°C.
2) Warm roll at 760°C to a 40% thickness reduction.
3) Cold roll using a 10% reduction per pass with intermediate anneals for 1 hour at 1200°C.

Sheet was made successfully from all but the Ni-15Mo-10ThO$_2$ alloy.

Ni-2ThO$_2$ Alloys

As with the Ni-15Mo-ThO$_2$ alloys, the powders were made by spray drying and selective reduction. Pressing and sintering were also similar. However final densification was not done by extrusion. Instead, the billets were heated in hydrogen at 1050°C, press forged and then rolled according to the following schedule.

1) Warm roll to 65% reduction in thickness.
2) Cold roll to 70% reduction in thickness.
3) Recrystallize 1 hr at 1000°C in H_2.
4) Cold roll 30% reduction in thickness.
5) Anneal 1 hr at 1200°C in H_2.
6) Repeat 4 and 5.

RESULTS AND DISCUSSION

Powder Properties

Typical chemical and physical properties are shown in Tables I and II below:

TABLE I - CHEMISTRY

Element Range, ppm

1-10	5-50	10-100	50-500
Ba,Co,Cu Mn,Pb,Sr	S,P	C,Al,Ca,Mg Fe,O_2,W	Si

TABLE II - PHYSICAL PROPERTIES

Property	Composition			
	Ni-15Mo-10ThO$_2$	Ni-15Mo-0.3ThO$_2$	Ni-15Mo	Ni-2ThO$_2$
Fisher Sub-Sieve Sieve, μm	0.65	2.6	5.5	1.35
BET-Specific Surface, m^2/g	5.9	0.93	0.52	2.16
BET-Average Diameter, μm	0.11	0.68	1.28	0.312
Bulk Density, g/cc	1.30	2.19	2.65	1.72

Electron Metallography

Electron micrographs of ThO_2 extracted from the surface of polished sheet are shown in Figure 1 for representative alloys. The average ThO_2 diameter as estimated from x-ray line broadening is shown with each micrograph.

High-Temperature Strength

The mechanical properties were measured at $1093^{\circ}C$ ($2000^{\circ}F$) after annealing the samples at $1200^{\circ}C$ for 1 hr in H_2. Excellent strengths were developed for those sheets with less than 250Å ThO_2 even with as little as 0.3v/o ThO_2. The best 0.2% yield strengths (ksi) at $1093^{\circ}C$ were: Ni-$2ThO_2$, 17.0; Ni-15Mo, 13.1; Ni-15Mo-$0.03ThO_2$, 13.8; Ni-15Mo-$0.3ThO_2$, 21.3; Ni-15Mo-$3.0ThO_2$, 23.2. Ultimate strengths were usually within 2-4 ksi of the yield strengths and elongations were 2-6%.

Stress-rupture properties of Ni-15Mo-$0.3ThO_2$ and Ni-15Mo-$3ThO_2$ were investigated for samples taken from material which had given our best $1093^{\circ}C$ tensile properties. The samples were tested at $1093^{\circ}C$ in a vacuum of about 5×10^{-5} torr. The results are shown in Table III below:

TABLE III - STRESS-RUPTURE PROPERTIES

Material	Stress (ksi)	Rupture Time (Hrs)	Elongation Percent
Ni-15Mo-$0.3ThO_2$	16	0.05	5
Ni-15Mo-$0.3ThO_2$	12	0.30	6
Ni-15Mo-$0.3ThO_2$	9	11.7	3
Ni-15Mo-$0.3ThO_2$	7	128.8	4
Ni-15Mo-$3ThO_2$	12	467.9	–
Ni-15Mo-$3ThO_2$	14	58.6	–

ThO_2 Size and Loading

The relationship of ThO_2 size and volume loading to the high-temperature strength of the Ni-15Mo-ThO_2 alloys is shown dramatically in Figure 2. The strengths of sheet having ThO_2 of an average diameter less than 250Å are represented by the darkened circles, and are proportional to the cube root of the ThO_2 volume fraction. The sheets with coarser ThO_2 had much poorer strengths and showed little or no effect of volume loading.

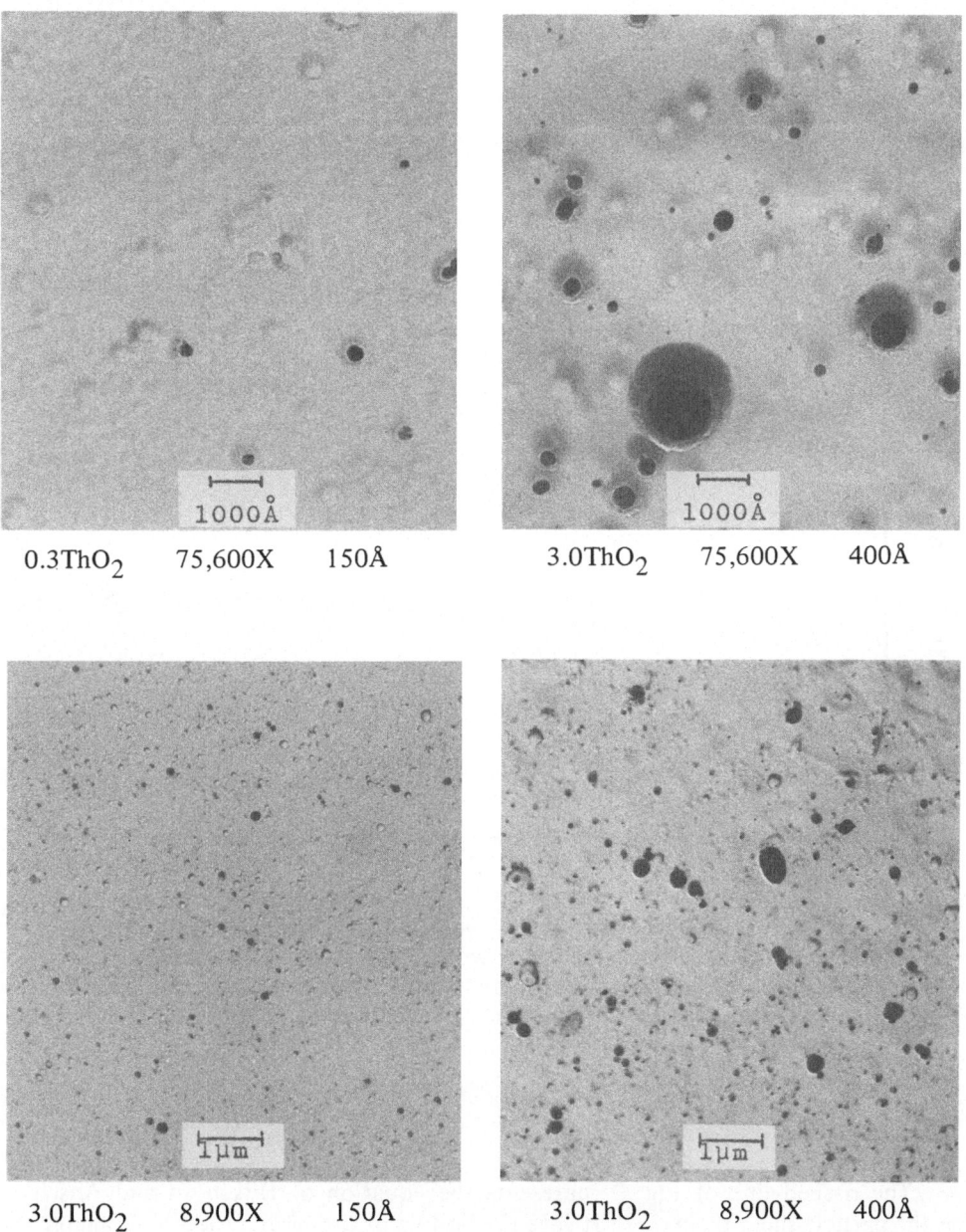

Figure 1.   Extraction electron micrographs showing ThO_2 dispersions in Ni-15Mo-ThO_2 sheet cold worked 80% and annealed 1 hour at 1200°C.

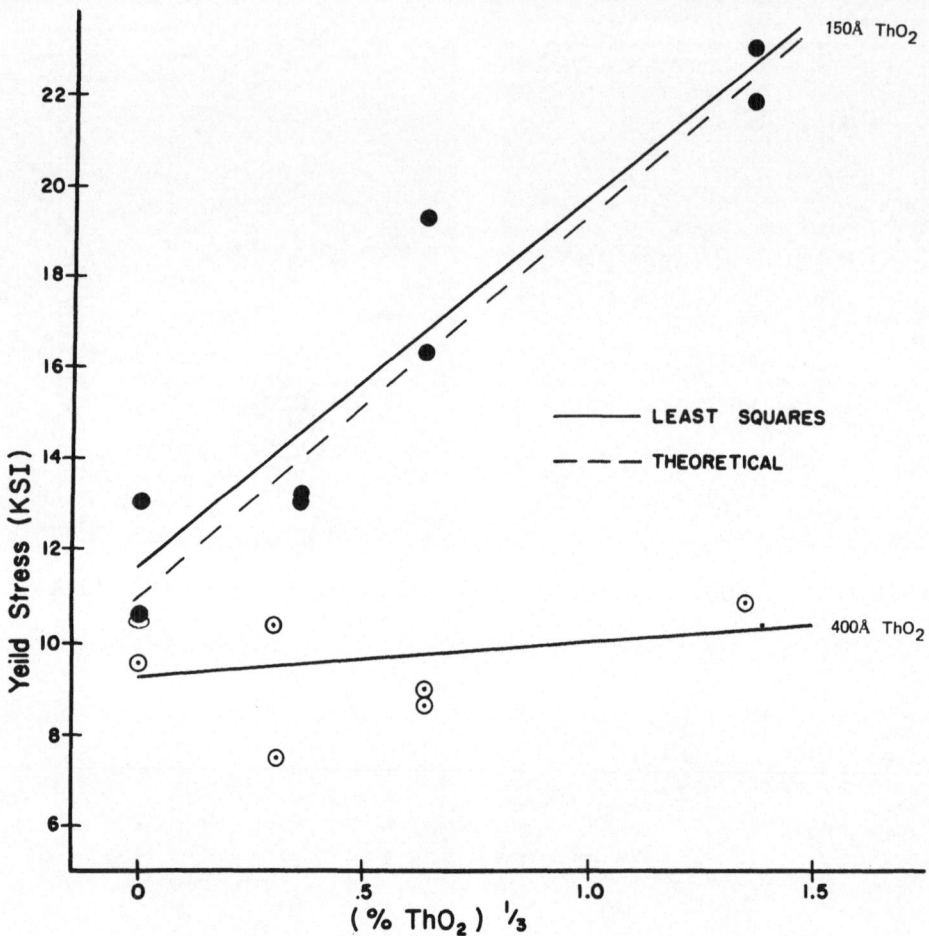

Figure 2.   Yield strength at 1093°C versus thoria volume loading for Ni-15Mo-ThO₂ sheet after 80% cold work.   The darkened circles represent sheet with strong texture and small thoria.   The open circles represent sheet with larger ThO₂ and weak texture.

The dashed line of Fig. 2 represents the equation of Hirschorn and Ansell for the particle shear theory[3]. It gave the best fit of the several theories available. Their equation is as follows:

$$T = T_m + (6/\pi)^{1/3} (T_p - T_m) f^{1/3}$$

T is the yield strength of the alloy containing the dispersion, T_m is the strength of the matrix, T_p is the shear strength of the particles and f is the volume fraction of the dispersoid. For our calculation, T_m was taken as 11.0 ksi which is the average Ni-15Mo yield strength at $1093^{\circ}C$. We used T_p = 435 ksi which is the compressive strength of ThO_2 at $1093^{\circ}C$[4].

Evaluation of the Ni-2ThO$_2$ sheet substantiated the importance of ThO_2 size to the development of good strength at $1093^{\circ}C$. The transmission electron micrographs of Fig. 3 show Ni-2ThO$_2$ with large average ThO_2 diameter and another sample with small average ThO_2 diameter. The ThO_2 diameters in those samples were measured from electron micrographs using a Zeiss TGZ-3 counter. The mean, median and standard deviation of the representative distributions are indicated with each micrograph. The yield strengths (ksi) are also shown, the stronger alloy having the smallest ThO_2, as expected.

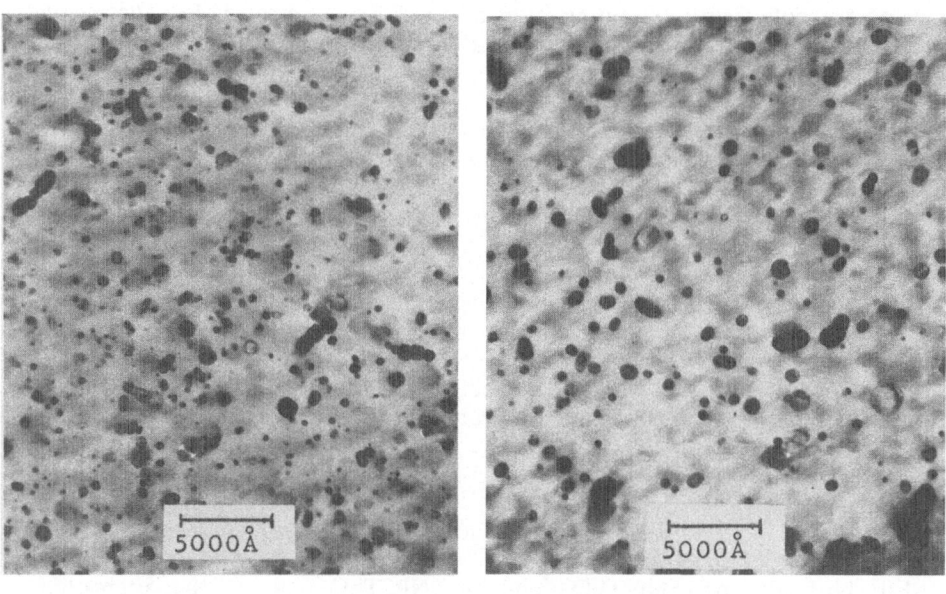

212 - Mean,Å	394 - Mean,Å
201 - Median,Å	321 - Median,Å
69 - Sigma,Å	233 - Sigma,Å
15.4 - Yield Str, ksi	8.3 - Yield Str, ksi

Figure 3.   Electron transmission micrographs of ThO_2 dispersions in Ni-2ThO$_2$ sheet.   Data on ThO_2 size distribution and yield strength at $1093^{\circ}C$ are also shown.

Residual Strain and Crystallographic Texture

The Ni-15Mo-ThO$_2$ alloys were examined for residual elastic strain by x-ray line broadening[5] and there was no correlation with strength. However, intense textures were noted in some sheets. Upon closer examination we found that the strong sheet containing less than 250Å ThO$_2$, as represented by the darkened circles of Figure 2, always had an intense texture. The sheets with coarser ThO$_2$ and poorer strengths had weak textures. Typical pole figures for strong and weak sheets of Ni-15Mo-0.3ThO$_2$ and of Ni-2ThO$_2$ are shown in Figure 4. ThO$_2$ sizes and yield strengths are also shown.

To better illustrate the affect of crystallographic texture it was convenient to quantitatively display the data. We did that by selecting a pole location which best represented the textures under investigation and then determining the average "times-random" intensity of (111) poles approaching that ideal location. We selected the (112) [11T] pole location and then created concentric reference circles on the pole figure extending from 5° to 20° around this pole. The average "times-random" intensity was then calculated for the (111) poles falling within the circles.

The texture data are shown in Figure 5 for Ni-15Mo-0.3ThO$_2$ sheet after 40, 60, and 80% reduction-in-thickness at room temperature. The sheet had a ThO$_2$ size of 150Å. The yield strengths are shown for each reduction. The stronger the texture, the stronger the sheet. The samples were heated at 1200°C for 1 hour prior to testing.

Figure 6 is similar to Figure 5 except the data are from four different Ni-15Mo-0.3ThO$_2$ sheets, all having the same metalworking schedule and all reduced 80% in area by rolling. Each sheet came from a billet sintered by a different schedule and therefore each sheet had a different intended ThO$_2$ size. The yield strengths at 1093°C are indicated for the respective intensity plots. The stronger the texture, the stronger the sheet and the smaller the ThO$_2$ size.

Deformation-Anneal Affects

For both the Ni-15Mo-ThO$_2$ and the Ni-2ThO$_2$ alloys, the appropriate deformation-anneal sequence produced large strength increases when applied to alloys with ThO$_2$ of small diameter. Exactly why is not clear. An understanding would seem to require a very thorough evaluation of the affects of ThO$_2$ deagglomeration, grain size/shape, substructure and texture.

One of the obvious affects of deformation was to break up clusters of ThO$_2$ particles thereby improving the effectiveness of the ThO$_2$ dispersion. Unfortunately we have not yet developed a quantitative measure of this affect.

The deformation-anneal sequence had a marked affect on the grain structure of the Ni-2ThO$_2$ but little affect on that of the Ni-15Mo-ThO$_2$ alloys. Both types of grain structure are shown in Figure 7. The highly textured long

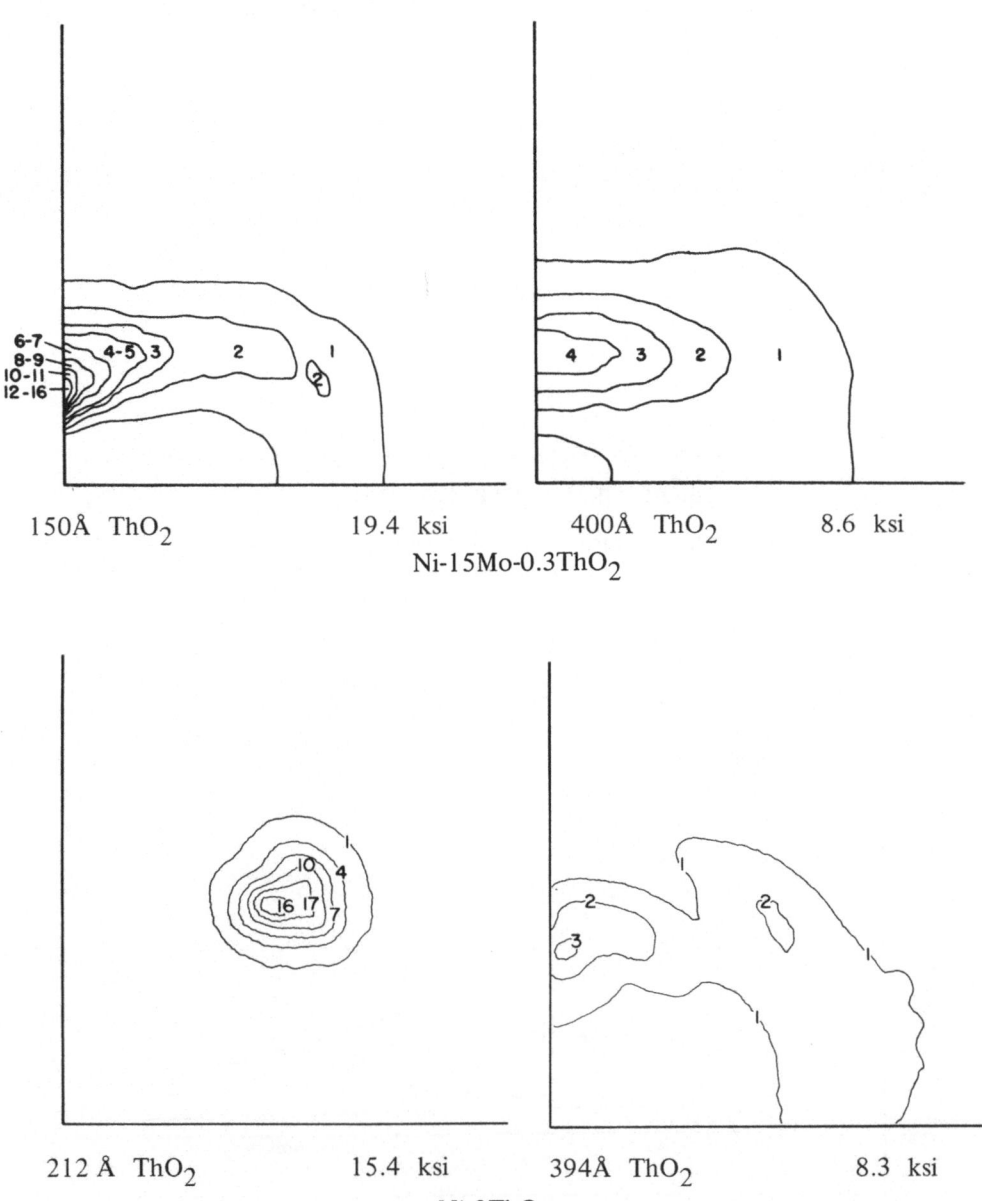

Figure 4. Typical (111) pole figures of strong and weak Ni-15Mo-0.3ThO$_2$ and Ni-2ThO$_2$ sheet. Mean ThO$_2$ size, yield strength at 1093°C and rolling direction (RD) are shown.

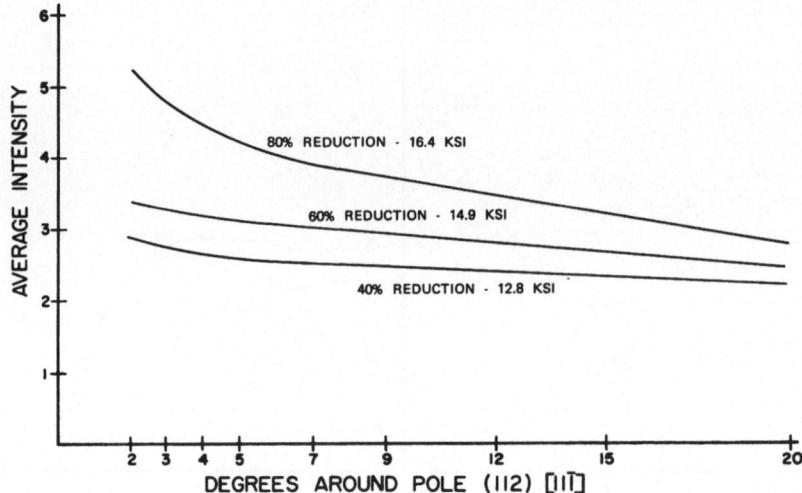

Figure 5. Average times-random intensity of (111) poles as a function of distance from the (112) [11T] pole position for Ni-15Mo-0.3ThO$_2$ sheets after 40, 60, and 80% cold work. Yield strengths at 1093°C are indicated for each sheet. ThO$_2$ size was constant at 150Å.

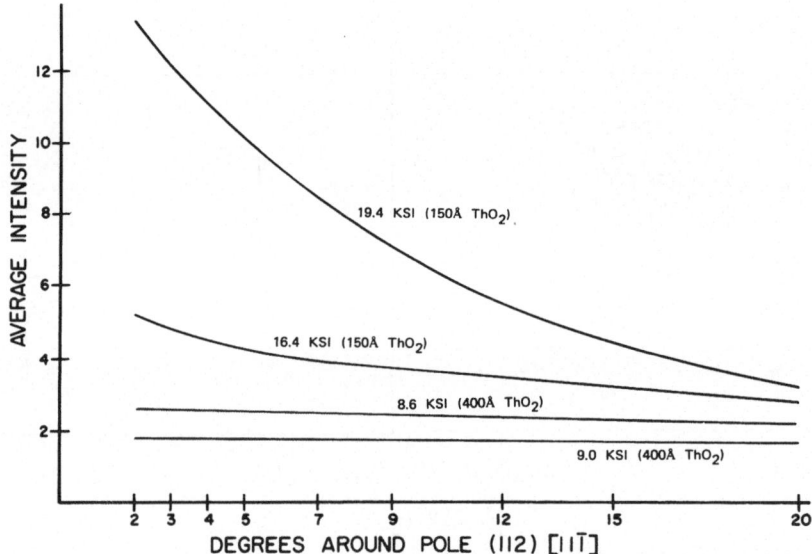

Figure 6. Average times-random intensity of (111) poles as a function of distance from the (112) [11T] pole position for Ni-15Mo-0.3ThO$_2$ sheets from billets having various thoria sizes and sintering temperatures. All samples were reduced 80% in cross-section by rolling. Yield strengths at 1093°C and ThO$_2$ size are indicated for each alloy.

interlocking grains of the Ni-2ThO$_2$ are formed during the recrystallization anneal after 70% cold work. The strength at 1093°C increased markedly, ∿4-5 ksi, after the recrystallization. We found no similar behavior in the Ni-15Mo-ThO$_2$, probably because of the necessarily different processing. The textures and grain structures of the strong sheets were stable up to 1400°C for both the Ni-15Mo-ThO$_2$ and Ni-2ThO$_2$ alloys.

The amount of deformation between anneals was important to the strengths of the two matrices but was very different for each one. For the Ni-15Mo-ThO$_2$ alloys frequent anneals were necessary for strong sheet. For the Ni-2ThO$_2$ sheet, frequent annealing would have prevented development of the beneficial texture and grain structure.

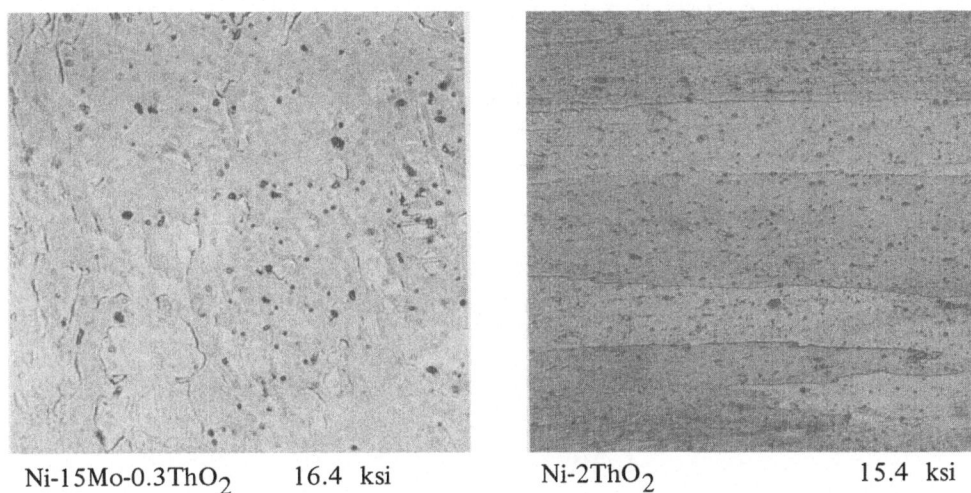

Ni-15Mo-0.3ThO$_2$ 16.4 ksi   Ni-2ThO$_2$ 15.4 ksi

Figure 7. Light micrographs (450X) showing grain structures of strong Ni-15Mo-0.3ThO$_2$ and Ni-2ThO$_2$. Typical yield strengths at 1093°C are indicated.

Too much total deformation, regardless of annealing frequency, weakened both the Ni-15Mo-ThO$_2$ alloys and the Ni-2ThO$_2$. After about 60-70% room-temperature deformation with anneals after each 10% reduction in thickness the strengths of the Ni-15Mo-ThO$_2$ sheet would decrease. The same was true of the Ni-2ThO$_2$ sheet after it had received the recrystallization anneal. Its strength was found to increase for up to two deformation-anneal cycles of 30% reduction each. Further working decreased the strengths. Interestingly, the texture decreased in intensity with increasing deformation beyond the point of optimum strength. That was true for both Ni-15Mo-ThO$_2$ and for Ni-2ThO$_2$.

ACKNOWLEDGMENTS

A portion of this work was sponsored by the Metals and Ceramics Division, Air Force Materials Laboratory, Directorate of Laboratories, Wright-Patterson Air Force Base, Dayton, Ohio under Contract F33616-67-C-1462, "Development of Dispersion-Strengthened Nickel-Base Superalloys". We also wish to acknowledge the efforts of Lester F. Barnes and James R. Spencer.

REFERENCES

(1) Scheithauer, Jr., W., "The Manufacture and Properties of High-Conductivity, High-Strength Cu-ThO$_2$", Proceedings of the International Powder Metallurgy Conference, July 1970.

(2) Scheithauer, Jr., W. and Cheney, R. F., "Development of Dispersion-Strengthened Nickel-Base Sheet Alloys", Final Report AFML-TR-69-285 to Contract F33615-67-C-1462, February 1970.

(3) Hirschorn, J. S. and Ansell, G. S., "Dispersion Strengthening Models", Acta Met., Vol. 13, p. 572, 1965.

(4) Refractory Ceramics for Aerospace, published by the American Ceramic Society, 1964, p. 265.

(5) Pines, B. Y., Dokl. Akad. Nauk SSSR, Vol. 103, p. 601, 1953.

THE MANUFACTURE AND PROPERTIES OF HIGH-CONDUCTIVITY HIGH-STRENGTH Cu-ThO$_2$

W. Scheithauer, Jr., R. F. Cheney, and N. E. Kopatz

Sylvania Electric Products Inc., Chemical & Metallurgical Div.

Towanda, Pennsylvania   18848

SUMMARY

Methods are described for producing plate, sheet, rod and wire of high-purity copper with dispersions of thorium oxide particles. Powders are first produced by spray drying, calcining and selective hydrogen reduction. The reduced powders are isostatically compacted. Densification which cannot be accomplished by conventional sintering techniques is done either by hot forging, hot rolling or hot extrusion. After the thermomechanical densification step, the material can be further worked into wrought forms by somewhat conventional techniques.

The primary problems encountered in the powder metallurgy approach were in obtaining low-oxygen consolidated forms and in obtaining dispersions with good spatial distributions. Low-oxygen consolidated forms were obtained by modifying the hot consolidation technique, the critical feature being that the porous billets be heated only in a hydrogen atmosphere until the resultant forms are 100% dense. The problem of improving the thoria spatial distribution was not solved and will require additional investigation.

High purity permits attainment of conductivities in excess of 95% IACS. With our best ThO$_2$ dispersions, 600°C yield strengths to 20,000 psi are obtained with Cu-1ThO$_2$. Better dispersions should yield higher strengths. After annealing at temperatures as high as 1000°C, the room-temperature 0.2% yield strengths are still as high as 20,000-25,000 psi.

INTRODUCTION

There is currently a need for a material with high electrical and thermal conductivity and with good mechanical properties at elevated temperature and at room temperature after a high temperature exposure. The material being used to meet the conductivity requirement is OFHC copper, but it is inadequate with respect to the mechanical properties. Iron-, nickel-, cobalt-, and some copper-base alloys have the necessary mechanical properties, but are completely inadequate with respect to thermal and electrical conductivity.

The industry which has the primary need for such material is the microwave tube industry. Because high electrical and thermal conductivity are essential, OFHC copper is presently used almost exclusively. During tube manufacture the OFHC parts are exposed to temperatures of 0.85 to 0.95 Tm, and consequently recrystallize, undergo grain growth, and even distort severely. The recrystallized parts are weak at room temperature. During tube use the parts are generally operated at elevated temperature and are inherently weak and thus must be thick or well supported.

Of the mechanisms known and used for strengthening metals, oxide-dispersion strengthening is, when properly applied, the only strengthening mechanism that does not seriously degrade the thermal and electrical conductivity. The process of oxide strengthening has been known and used for years. Much of the past work in the field of oxide strengthening has been comprehensively reviewed by Bunshah and Goetzel[1], Hansen[2], Rice[3], and Ansell[4]. It is well known that oxide strengthening will work and will produce the desired end results. It is not well known how the mechanism works or how to produce oxide-strengthened (OS) alloys economically and in many cases technically.

Many of the methods that have been attempted for manufacturing OS copper have given poor materials from the viewpoint of purity and of quality of the dispersion. Other methods have produced relatively pure materials with good dispersions and good mechanical properties, but are limited in size of material. Methods such as internal oxidation produce good OS copper, but because of diffusional considerations the pieces so produced are limited to thin cross-sections, generally 1/8".

The purpose of this effort was to demonstrate the feasibility and reliability of a pilot powder metallurgy process for the manufacture of OS copper in bar and plate form.

PROCESS DETAILS AND DISCUSSION

General

OS copper forms were prepared in an eight-step process involving solution preparation, spray drying, calcining, selective hydrogen reduction, isostatic compaction, sintering, hot consolidation, and primary and secondary metalworking. The overall process is depicted in the flow-chart in Figure 1. The details of the processing steps are described in the sections that follow.

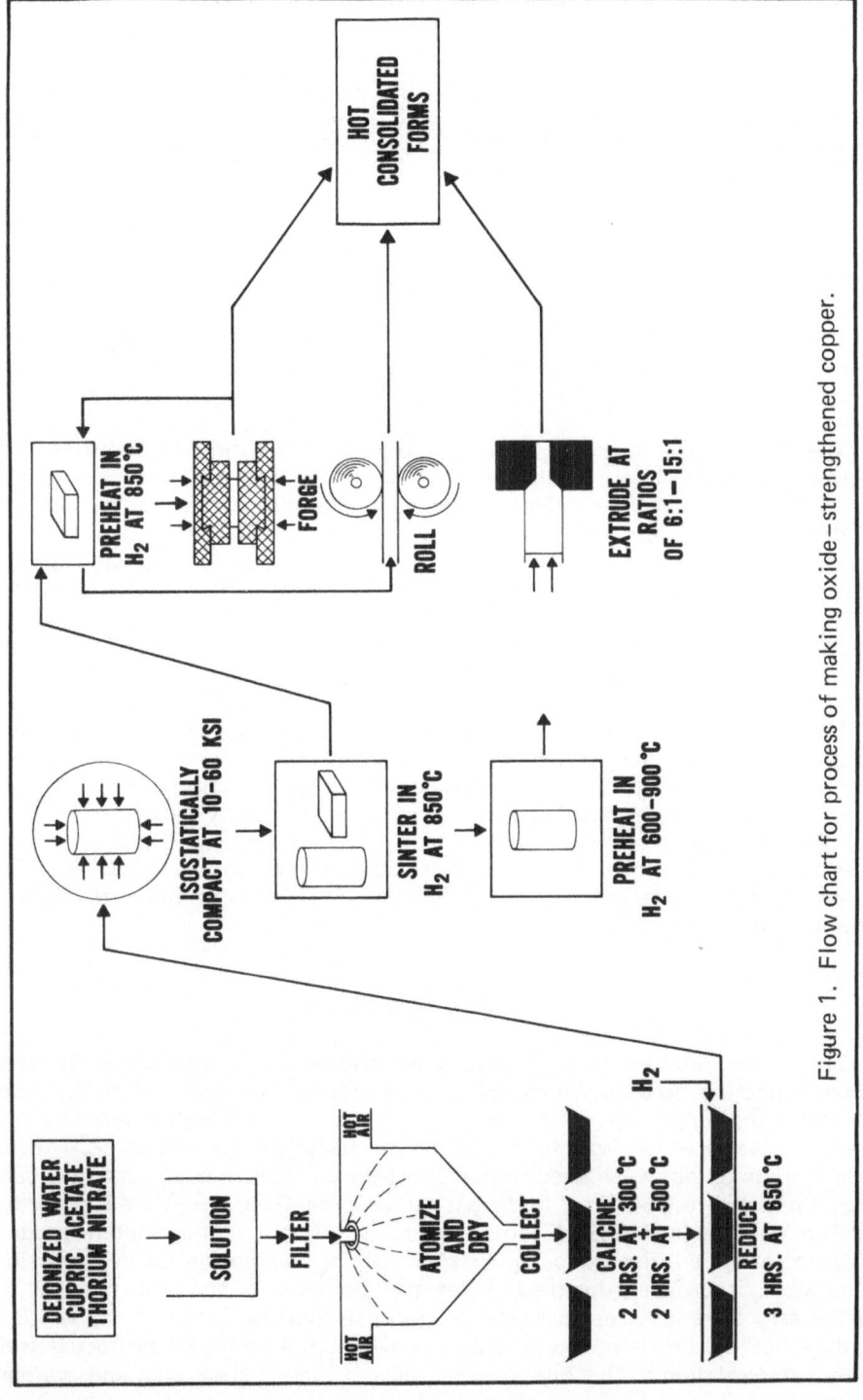

Figure 1. Flow chart for process of making oxide–strengthened copper.

Solution Preparation and Spray Drying

The first step in our process is to prepare a stable solution containing the desired alloy components. The prerequisites for the solution are that it must be pure and economical to make and that it must be capable of drying to a crystalline solid in the range of 150-450°C. Generally, metallic salts which are soluble in water are used.

We selected reagent-grade cupric acetate and thorium nitrate as the starting raw materials. Both are soluble and form a stable solution in deionized water at room temperature. The purpose of starting with a solution is that an ionic mixture is obtained. This type of mixture gives the most random distribution of copper and thorium ions that is possible. Throughout our process we strive to maintain this random distribution of copper and thoria (initially copper and thorium).

The next step is to spray-dry the solution to form a fine, dry powder. The principle, procedure, and equipment for spray drying has been adequately described[5]. Basically it involves atomizing the solution to a very fine droplet size, drying the droplets instantaneously by impinging a hot gas stream on them, and collecting the dried particles. The dried particles are an intimate mixture of cupric acetate and some thorium compound which we have not identified.

Calcining

The next operation is to convert the spray-dried product to an oxide mixture. This is done by calcining in air. It serves several purposes: to decompose the cupric acetate to cupric oxide, to decompose the thorium compound to thorium oxide, and to lower the volatile impurity content, primarily carbon and sulfur. The calcining is usually done in two steps. The first step is to fire in air for 2 hours at 300°C, and the second is to fire in air for 2 hours at 500°C. The product of this operation is cupric oxide containing a very fine dispersion of thorium oxide.

Selective Hydrogen Reduction

The oxide mixture is next selectively reduced in hydrogen. It is called selective reduction because the cupric oxide is reduced to copper while the much more stable thorium oxide remains as an oxide dispersed throughout each powder particle. We use a boat-tube reduction technique whereby 200 g of cupric-thorium-oxide is charged into copper-lined Inconel boats and manually stoked through a tube with a hydrogen atmosphere. The stoke rate is such that the residence time in the 650°C hot zone is about 3 hours. The resultant product of this operation is a fine copper powder containing the thorium oxide dispersion. The powder, if properly handled, is, at this point, pure and well reduced. It is rather easy to reduce cupric oxide to copper in hydrogen even at temperatures less than 650°C; however, the problem is to keep the powder protected from subsequent oxidation. The fine powders have a high surface area and are very susceptible to surficial oxidation upon exposure to the atmosphere. After reduction, the powders are mixed and blended, screened -200 mesh, and stored in a nitrogen atmosphere.

Compaction and Sintering

The powders are next isostatically compacted into either cylindrical or rectangular billets depending on the hot-consolidation technique to be used. The density of the billet is dependent on the pressing pressure and can be varied between about 50% with a pressure of 10 ksi and 85% with a pressure of 60 ksi. The compacted billets are sintered in hydrogen to effect an increase in their green strength and to remove the final traces of free oxygen. Sintering is usually done with an anisothermal cycle involving a maximum temperature of 850°C. Sintering does not cause an increase in density, and only at temperatures approaching 1000°C, where serious degradation of the dispersion occurs first, does one observe a density increase.

Hot Consolidation

Since densification or consolidation cannot be achieved by conventional sintering, we rely on other techniques. The basic ones that we have used are hot extrusion, hot forging and/or hot rolling. By these techniques, the pressed and sintered billets can be consolidated to fully dense forms which are suitable for additional metalworking.

Initially all billets were consolidated by a hot extrusion. The sintered billets were placed in steel cans, evacuated to a pressure of 1 micron and sealed. The can-billet assembly was then extruded by conventional extrusion whereby the assembly was heated to 850°C, lubricated, and extruded to a round or rectangular cross section at extrusion ratios of 6:1 up to 15:1.

One of the biggest problems we encountered in our work was in obtaining consolidated forms which were very low in oxygen. Regardless of the care exercised in evacuating and sealing the can-billet assembly, few satisfactory extrusions were obtained. Our means of evaluating the extrusions was to heat the extrusions at 950°C in a hydrogen atmosphere and then measure the expansion and check for porosity formation using visual, macroscopic, and microscopic techniques. Usually expansions of from 0.5 to 7.5% were measured. Sometimes expansions of as high as 50% were observed. This problem is classically referred to as hydrogen embrittlement.

To eliminate the problem of oxygen contamination, we used a technique which was developed in relationship to our work on OS nickel. It involved heating the sintered slab (rectangular billet) in hydrogen and quickly densifying it by hot drop-forging, hot press-forging, hot rolling, or a combination of these operations. This generally yielded a form which was 100% dense and which exhibited no expansion when heated in hydrogen.

Metalworking

After 100% dense forms which are free from H_2 embrittlement are obtained, additional metalworking can be carried out more or less conventionally. However, we know, based on our work with OS nickel[6], that the metalworking temperature, the amount of deformation, the direction of rolling, the annealing conditions, and the frequency of anneals are all critical to the development of strength. The physical metallurgy of the material as influenced by metalworking was not thoroughly investigated and is beyond the scope of this paper.

PROPERTIES AND DISCUSSION

In this section we will discuss primarily two main properties, electrical conductivity and mechanical, and they will be discussed with reference to the two main variables in our process, purity and quality of dispersion. Other variables affecting these two main properties and other properties will be discussed briefly.

Purity

The purity of the finished product is controlled by two factors: The purity of the raw materials and the control of in-process contamination. Typical purity of the finished product is shown in Table I.

TABLE I

TYPICAL CHEMICAL ANALYSES OF FINISHED PRODUCT

Element	PPM	Element	PPM	Element	PPM	Element	%
Ag	1-10	Mn	1	C	10	ThO_2	1.51
Al	1-10	Mo	50-500	O_2	25		
B	1-10	Ni	5-50	P	2		
Ba	1-10	Pb	1-10	S	13		
Ca	1-10	Si	10-100	Si	240		
Co	1-10	Sn	5-50				
Cr	1-10	Sr	1-10				
Fe	10-100	W	50-500				
Mg	1-10	Zn	1				

Elements that are somewhat high are Fe, Mo, Si, and W. These are introduced as contaminants from our processing areas and equipment. More stringent control would result in less in-process contamination.

Electrical Conductivity

The electrical conductivity of $Cu-ThO_2$ is primarily a function of the impurity level, as is the case with pure copper. It is affected by oxygen, and by metallic impurities to varying degrees, as shown in the work of Pawlek and Reichel[7]. Oxygen primarily affects the conductivity after the oxygen-bearing material is heated in hydrogen and porosity results.

Since ThO_2 is a dielectric material, we expect that it lowers the conductivity in proportion to its reduction of the current carrying cross-sectional area. Thus for a volume loading of two per cent, we would expect a maximum conductivity of 97% IACS. It is possible that the thoria particles may further reduce the conductivity if the particles are small enough and if the volume fraction is such that the interparticle spacing is less than the mean free path of the electron in copper.

The conductivities that we measured on various forms of OS copper are shown in Table II.

TABLE II

TYPICAL CONDUCTIVITY VALUES

Composition	Condition	Conductivity,%IACS
$Cu-0.03ThO_2$	As Hot Consolidated	96.5
$Cu-0.3ThO_2$	As Hot Consolidated	97.0
$Cu-1.0ThO_2$	As Hot Consolidated	93.8
$Cu-2.0ThO_2$	As Hot Consolidated	92.1
$Cu-3.0ThO_2$	As Hot Consolidated	88.5
$Cu-0.03ThO_2$	Annealed one hour at $800^{\circ}C$ in H_2	94.6
$Cu-0.3ThO_2$	Annealed one hour at $800^{\circ}C$ in H_2	95.1
$Cu-1.0ThO_2$	Annealed one hour at $800^{\circ}C$ in H_2	92.6
$Cu-2.0ThO_2$	Annealed one hour at $800^{\circ}C$ in H_2	92.1
$Cu-3.0ThO_2$	Annealed one hour at $800^{\circ}C$ in H_2	88.9
$Cu-1.0ThO_2$	As Hot Consolidated	96.2
$Cu-1.0ThO_2$	Annealed one hour at $800^{\circ}C$ in H_2	96.2
$Cu-1.0ThO_2$	Annealed one hour at $1050^{\circ}C$ in H_2	96.7

Dispersion

The quality of the dispersion in OS copper is critical to the mechanical properties. The critical features are the size and size distribution and the spatial distribution of the dispersoid particles. Generally, we had no problem in obtaining small particle size and narrow size distributions; however, the spatial distribution has in many cases been poor.

Examples of some of the spatial distributions that we have obtained are shown in Figure 2. A poor distribution is shown in Figure 2a and 2b, while

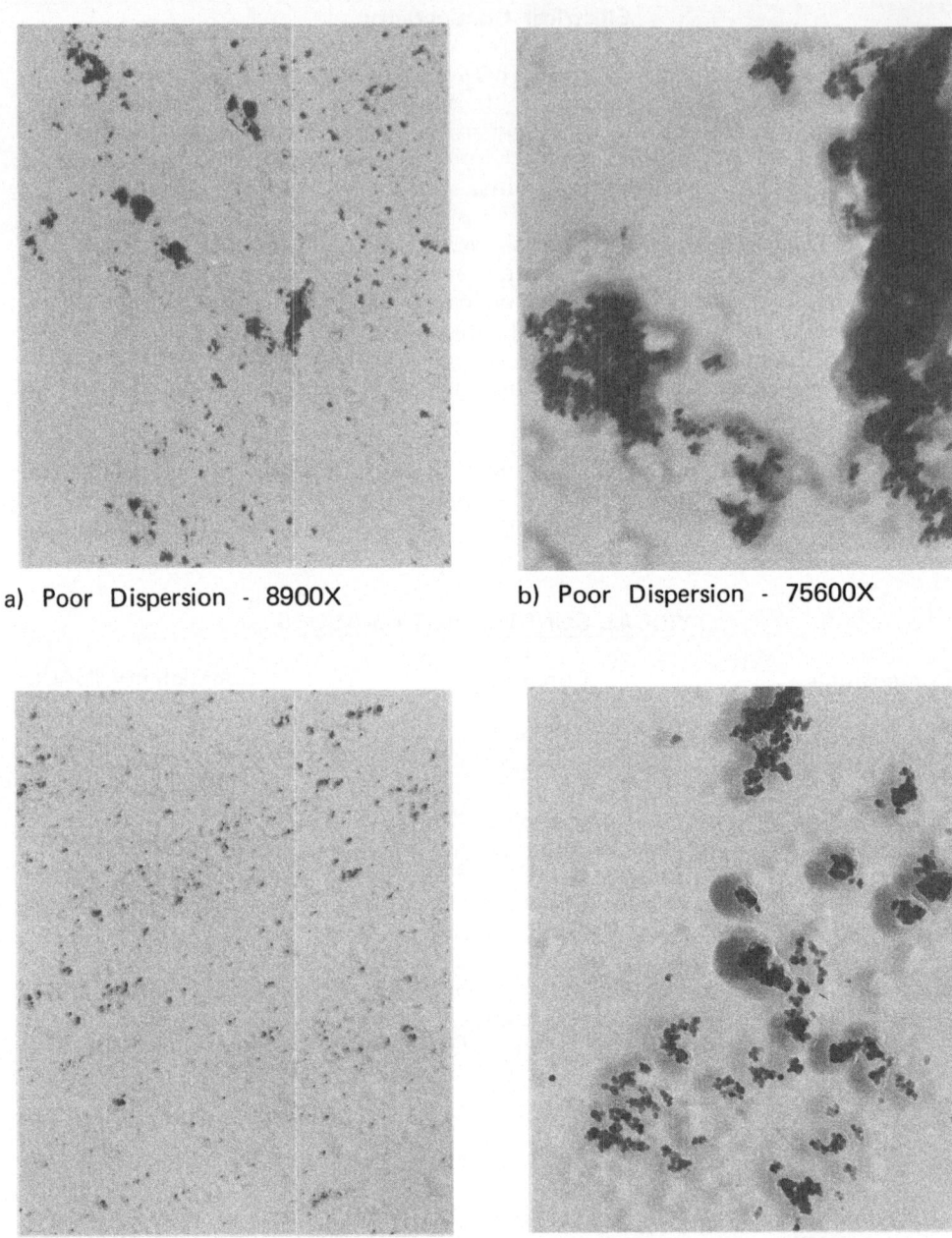

a) Poor Dispersion - 8900X   b) Poor Dispersion - 75600X

c) Best Dispersion - 8900X   d) Best Dispersion - 75600X

Figure 2.   Extraction electron micrographs showing a poor dispersion and our best dispersion of one volume   per cent ThO_2 in copper.

our best distribution is shown in 2c and 2d.

Mechanical Properties

Tensile properties at both room temperature and at 600°C were measured on various forms of wrought OS copper. The properties are listed in Table III. Some values for OFHC copper are shown for comparative purposes. The properties vary as a function of the form, composition, quality of the dispersion, and the metalworking schedule. However, these are merely presented to show some representative values.

TABLE III

MECHANICAL PROPERTIES

Room Temperature Tensile Properties

Composition	Form	Condition	UTS,ksi	0.2%YS,ksi	Elong.,%
OFHC	1/4" Plate	As-Processed	32.2	10.2	51.0
OFHC	1/4" Plate	Annealed[1]	29.5	3.6	65.0
Cu-1ThO$_2$	1/4" Plate	As-Processed	41.6	30.6	30.0
Cu-1ThO$_2$	1/4" Plate	Annealed[1]	40.6	27.9	32.0
Cu-1ThO$_2$	1/4" D Rod	As-Processed	39.9	29.2	21.0
Cu-1ThO$_2$	1/4" D Rod	Annealed[1]	39.2	27.9	25.8
Cu-2ThO$_2$	1/4" D Rod	As-Processed	60.5	57.1	7.5
Cu-2ThO$_2$	1/4" D Rod	Annealed[1]	41.2	31.4	21.2
Cu-2ThO$_2$	60-Mil Wire	As-Processed	58.4	48.8	2.3
Cu-2ThO$_2$	60-Mil Wire	Annealed[1]	42.3	31.4	18.2
Cu-2ThO$_2$	60-Mil Wire	Annealed[2]	35.3	24.7	36.8

600°C Tensile Properties

Composition	Form	Condition	UTS,ksi	0.2%YS,ksi	Elong.,%
OFHC	1/4" Plate	As-Processed	5.6	2.4	42.0
OFHC	1/4" Plate	Annealed[3]	4.9	2.4	56.0
Cu-1ThO$_2$	1/4" Plate	As-Processed	7.5	6.3	8.0
Cu-1ThO$_2$	1/4" Plate	Annealed[3]	7.5	6.5	7.2
Cu-1ThO$_2$	30-Mil Sheet	Annealed[1]	19.3	17.4	9.3
Cu-1ThO$_2$	1/4" D Rod	As-Processed	9.5	8.4	9.4
Cu-1ThO$_2$	1/4" D Rod	Annealed[1]	9.3	7.9	8.0
Cu-2ThO$_2$	1/4" D Rod	As-Processed	14.8	14.4	6.3
Cu-2ThO$_2$	1/4" D Rod	Annealed[1]	15.0	14.6	6.4
Cu-2ThO$_2$	60-Mil Wire	As-Processed	29.7	25.7	2.7
Cu-2ThO$_2$	60-Mil Wire	Annealed[1]	24.4	19.6	7.3

(1) One hour at 600°C in H$_2$

(2) One hour at 1000°C in H$_2$

(3) 500 hours at 600° in vacuum

Brazing, Outgassing, and Machinability

The brazing, outgassing, and machining characteristics were compared with those of OFHC copper. We made the following observations. OS copper can be brazed similarly to OFHC copper with some minor modifications in technique. Its outgassing characteristics are about the same as those of OFHC copper. OS copper, while harder and stronger than OFHC copper, is generally considered to have superior machining characteristics.

ACKNOWLEDGMENT

This work was primarily sponsored by the Electronics Branch, Manufacturing Technology Division, Air Force Materials Laboratory, Air Force Systems Command, Wright-Patterson Air Force Base, Ohio under Contract No. F33615-67-C-1965, "Manufacturing Methods For Dispersion-Strengthened Copper". We would also like to acknowledge the efforts of Glenn A. Shaffer.

REFERENCES

(1) Bunshah, R. F., and Goetzel, C. G., "A Survey of Strengthening of Metals and Alloys", WADC Technical Report 59-414, March 1960.

(2) Hansen, N., and Lilholt, H., Bibliography on Dispersion-Strengthened Materials, Research Establishment Riso #48, Sept. 1962; Hanse, Liholt, H., and Hahnkam, 1962-1963, #48 Supplement #1, July 1964, Danish Atomic Energy Commission, Riso Roskilde, Denmark.

(3) Rice, L. P., "Metallurgy and Properties of Thoria-Strengthened Nickel", DMIC Memorandum 210, October, 1965.

(4) Ansell, G. S., Cooper, T. D., Lenel, F. V., "Oxide Dispersion Strengthening", Proceedings of the Second Bolton Landing Conference, June, 1966.

(5) Scheithauer, Jr., W., "Manufacturing Methods for Dispersion Strengthened Copper"; First Interim Progress Report, Contract No. F33615-67-C-1695, November, 1967.

(6) Cheney, R. F. and Scheithauer, Jr., W., "The High-Temperature Strength of Oxide-Strengthened Nickel Alloys", Proceedings of the International Powder Metallurgy Conference, July 1970.

(7) Pawlek, F. and Reichel, K., "The Effect of Impurities on the Electrical Conductivity of Copper. I. The Electrical Conductivity of Pure Copper, Its Maximum Value and Its Control by Impurities", Z. Metallkunde, Vol. 47, 1956, p. 347-56.

DISPERSION STRENGTHENED FERRITIC HEAT-RESISTING STEEL

Teishiro Oda and Takashi Daikoku

Nagasaki Technical Institute

Mitsubishi Heavy Industries, Ltd.

I - INTRODUCTION

Dispersion strengthened alloy has been attracting much attention as a new type of heat resisting material. An outstanding feature of this type of alloy is that it maintains superior strength over a long period at elevated temperatures, since neither the coagulation of the dispersed particles nor the solution of those particles into the matrix occurs, as was found in SAP developed by Irmann.

For this reason, the investigation into this new alloy has been conducted very actively by many researchers for the purpose of developing new material or making clear the strenghtening mechanism theoretically. It was also reported that TD Nickel has been practically used for the construction of the jet engine parts and gas turbine nozzle guide vanes. Nevertheless, this type of alloys is still in an early stage of development and its utilization is limited to some special purposes.

In the study described in this paper, an attempt has been made to develop the dispersion strengthened 13% Cr steel which is, in the elevated temperature strength, well comparable to the austenitic heat resisting steels.

The reason for trying the strengthening of 13% Cr steels is as follows; if the ferritic steel can be strengthened in creep, it will no doubt give a large advantage in the field of the heat resisting material for turbine blades and bolting, because the ferritic material is lower in price and has lower thermal expansion coefficient and higher material damping, compared with the austenitic one.

As the first step, the authors investigated into the improvement of the high temperature strength of the sintered 13% Cr
steels by the method of dispersing fine   aluminium oxide particles but with no marked success. Then, the authors tried to
achieve the improvement by employing the above dispersion
strengthening method, combined with the technique of the liquid
phase sintering and the subsequent hot-working.

II – EXPERIMENTAL PROCEDURE

For the production of the dispersion strengthened alloys,
various techniques have hitherto been proposed. In this experiment, the mechanical mixing method was employed. As substance to
form the matrix, prealloyed 13CrMoTi and AISI 406 steel powders
(-150 mesh) of the chemical compositions as shown in Table 1 were
used. As the dispersoid, a very fine γ -Al$_2$O$_3$ with mean particle
size of 20 mμ (DEGUSSA, Type P110C-1) was selected.

For further improvement, small fraction of Fe-Ti alloy
powder (26.61 wt.% Ti, -325 mesh) was admixed, which resulted in a
liquid phase at the sintering temperatures for the matrix powders.

Mixing of powders was carried out in the horizontal ball
mill by the wet process using carbon tetrachloride. The mixtures
were dried at about 100°C and then annealed at 800°C in the purified hydrogen to improve the compressibility. As a lubricant,
0.5 to 1.0% of zinc stearate or liquid paraffin was added to the
annealed mixes, prior to compacting. In compacting the powders,
either a die with a rectangle-shaped hole, 18 mm in width and 100
mm in length, or a die having a cylindrical hole, 45 mm in diameter was used, according to the experimental purpose. Compacting pressure was 8 tons/cm^2 and the height of green compact to be
obtained was adjusted by the weight of powders.

A preliminary survey was made to find the optimum sintering
conditions for each alloy steel powder employed. As a result, it
was decided as follows:

1350°C x 5 hr in vacuum for 13CrMoTi steel
1400°C x 5 hr in hydrogen atmosphere for AISI 406 steel

Moreover, the effect of hot working on the mechanical properties of the sintered compacts was investigated by hot forging
them, either in as-sintered state or sintered and canned in mild
steel. In the following. the latter will be referred to as
sheath forging to be distingushed from the former.

Table 1. Chemical composition of alloy steel powders

Powder	C	Si	Mn	S	P	Cr	Al	Mo	Ti
AISI 406 steel	0.03	0.64	0.69	0.017	0.013	12.91	2.86	–	–
13CrMoTi steel	0.04	1.35	0.15	0.02	0.022	12.91	0.14	1.91	0.41

Table 2. Effect of γ –Al_2O_3 addition on mechanical
properties of sintered 13CrMoTi steel

γ –Al_2O_3 addition (%)	Sintered density (gr/cm^3)	0.2% Proof stress (kg/mm^2)	Ultimate tensile strength (kg/mm^2)	Elongation (%)	Vickers hardness (10 kg load)
0	7.65	30.0	45.2	23.2	181
0.5	7.52	34.5	50.7	14.0	184
1.0	7.41	34.2	48.6	11.4	178

III – EXPERIMENTAL RESULTS

Effect of Addition of γ –Al_2O_3 on Mechanical Properties
of Sintered Compact

In the beginning, the effect of addition of γ –Al_2O_3, mixed
with 13CrMoTi steel powder for various times of 3 to 72 hours,
on the room temperature tensile properties of the sintered
compacts was examined to obtain the result that the sintered 13-
CrMoTi steel with dispersed 0.5% of γ –Al_2O_3 is stronger than the
sintered compact without addition of γ –Al_2O_3, due to the
strengthening effect of dispersed γ –Al_2O_3. Regarding the effect
of the mixing time, the strength of the sintered compacts was
gradually increased as mixing time was prolonged. Tensile elongation, on the contrary, showed a slight decrease with the increase of mixing time. The tensile properties of the sintered
AISI 406 steels as a function of the amount of dispersed γ –Al_2O_3
was also investigated. As a result, it was confirmed that the
strengthening effect appeared even when only 0.1% of γ –Al_2O_3 was
added and there was no remarkable change with increased addition
of γ –Al_2O_3 up to 0.5%.

The authors investigated, moreover, the strengthening effect
in the case of further addition of γ –Al_2O_3 in the sintered compact. Table 2 shows an example of the results obtained, indicating
that the tensile strength and elongation of the sintered 13CrMoTi
steels decreased slightly when the amount of γ –Al_2O_3 added was

increased from 0.5% to 1.0%. With regard to the tensile properties of the sintered AISI 406 steel, the same tendency was also
observed. The strength of the sintered compacts decreased to a
further extent if γ -Al$_2$O$_3$ was added over 1%. On the contrary,
increasing the amount of γ -Al$_2$O$_3$ was necessary for improvement
of the elevated temperature strength, particularly creep rupture
strength, of the sintered compact, as mentioned in the following.

Improvement of Strength of Sintered Compact

**Improvement of Sintering Characteristics by Liquid Phase
Sintering.** It has been well known that a progress of sintering
can be remarkably accelerated by introducing the technique of the
liquid phase sintering, because of the greater mobility of metal
atoms in a liquid phase.

As the solubility of titanium in α -iron is anticipated to
be high, the authors attempted to improve the sintering characteristics of the concerned alloy steels with or without dispersed
γ -Al$_2$O$_3$, by addition of Fe-Ti alloy as the liquid phase constituent during the sintering process.

Fig. 1 shows the effect of addition of Fe-Ti alloy on the
tensile properties of 13CrMoTi steel without dispered γ -Al$_2$O$_3$.
The amount of Fe-Ti alloy added is referred to as Ti %. It is
clear from this figure that the addition of Fe-Ti alloy has a remarkable effect to improve the strength of the sintered compact
and more than 1% Ti should be added to achieve a sufficient in-

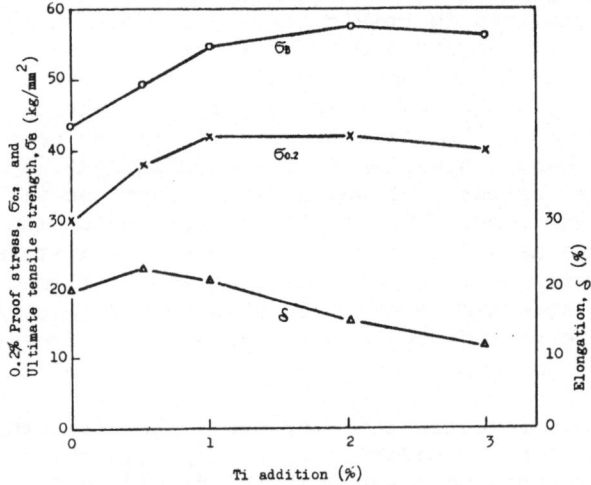

Fig. 1 Effect of addition of Fe-Ti alloy
on tensile properties of sintered
13CrMoTi steels

crease in strength. However, addition of more than 2% of Ti
causes   slight decrease in the strength. According to the
microscopic observation, a densification of the sintered compact
was promoted and the porosity decreased considerably by addition
of only 0.5% of Ti. By the addition of 2% Ti, a further improvement in the sintering characteristics of these alloys was attained and the existence of pores in the product was sometimes
observed.

The relationships between the tensile properties of the
sintered AISI 406 steel with 0.5% of γ -Al₂O₃ and the amount of
Fe-Ti alloy added are shown in Fig. 2. It is observed, in this
case too, that addition of Fe-Ti alloy not exceeding 2% Ti is effective for improvement of the strength of the sintered compact,
while it brings about only slight decrease in the tensile elongation. Addition of Fe-Ti alloy showed the same effect for the
improvement of the sintering characteristics of 13CrMoTi steel
with dispersed γ -Al₂O₃.

<u>Improvement of Strength of Sintered Compact by Hot Working</u>.
As mentioned above, the tensile strength of the alumina-disper-
sion strengthened steels was improved by means of the liquid
phase sintering. However, some pores were still retained in the
sintered compacts obtained in such a way. Therefore, the effect
of further densification of the sintered specimens by hot
working on their mechanical properties was investigated.

The test results on the sintered AISI 406 steels with 0.5%
of γ -Al₂O₃ are shown in Table 3. In this case, the sintered

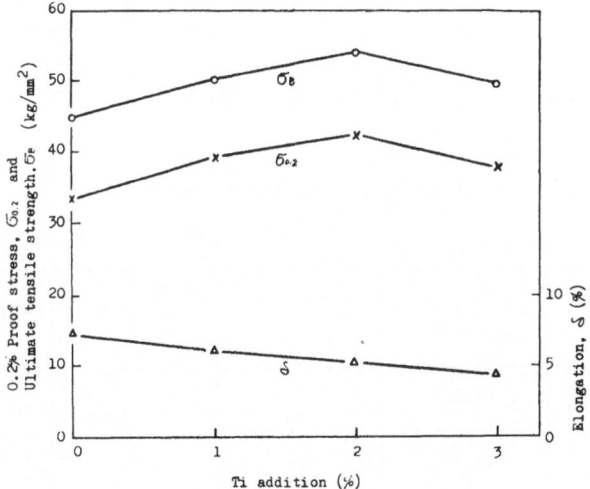

Fig. 2 Effect of addition of Fe-Ti alloy on
the tensile properties of sintered
AISI 406 steels with dispersed
γ -Al₂O₃ particles

Table 3. Effect of addition of Fe-Ti alloy and hot forging on tensile properties
of sintered AISI 406 steel with dispersed 0.5% of γ-Al$_2$O$_3$

Ti addition (%)	Treatment	0.2% Proof stress (kg/mm^2)	Ultimate tensile stress (kg/mm^2)	Elongation (%)	Reduction of area (%)
0	As sintered	33.3	45.2	7.2	8.7
	Hot forging	33.7	48.2	13.6	15.3
1	As sintered	39.9	50.8	6.0	4.0
	Hot forging	49.7	63.3	8.0	9.2
2	As sintered	43.5	53.7	5.2	3.3
	Hot forging	59.6	71.2	4.8	4.4
3	As sintered	38.5	49.5	4.6	2.8
	Hot forging	53.8	62.8	4.0	5.7

Table 4. Tensile properties of sintered AISI 406 steel
at elevated temperature (650°C)
(hot forging after sintering)

γ-Al$_2$O$_3$ addition (%)	Ti addition (%)	0.2% Proof stress (kg/mm^2)	Ultimate tensile strength (kg/mm^2)	Elongation (%)	Reduction of area (%)
0	0	10.5	14.7	27.2	20.6
0.5	0	10.2	13.4	6.0	6.0
0.5	1.0	14.2	21.4	46.0	42.5
0.5	2.0	19.4	28.3	35.8	30.4

Table 5. Effect of forging ratio on tensile properties
of sintered AISI 406 steels
(γ-Al$_2$O$_3$:1%, Ti addition:2%)

0.2% Proof stress (kg/mm^2)	Ultimate tensile strength (kg/mm^2)	Elongation (%)	Reduction of area (%)	Forging ratio
39.7	45.5	2.2	1.4	—— (as-sintered)
48.0	59.8	4.2	4.1	1.5 : 1 (hot forging)
53.2	67.1	7.2	7.8	6.5 : 1 (sheath forging)

specimens were hot forged at temperatures of 1100°C to 900°C
with a forging ratio of about 1.5:1. It can be seen from data in
this Table, the hot forging has little effect on the improvement
of the tensile strength of the sintered specimen with no addition
of Fe-Ti alloy. On the other hand, the sintered compact with
Fe-Ti alloy added, which had high strength even in as-
sintered state, was further markedly improved by hot forging.
From these results, it can be said that a remarkable improvement
of the strength of the dispersion strengthened steels concerned
is attained by the combination of liquid phase sintering and hot
working.

Table 4 shows the tensile properties at elevated temperature
of various kinds of the sintered AISI 406 steels forged after
sintering. Strengthening due to the dispersed γ -Al$_2$O$_3$ particles
was scarcely observed, when γ -Al$_2$O$_3$ was added by alone. In
contrast, when Fe-Ti alloy was added together with γ -Al$_2$O$_3$, the
elevatea temperature strength of the sintered compact was greatly
improved.

The improvement of hot workability of the sintered compact
is another effect of addition of Fe-Ti alloy. This could be
certified by the fact that the sintered specimens with 0.5% of
γ -Al$_2$O$_3$ were sometimes failed to forge, but when Fe-Ti alloy was
added, the sintered compact containing more than 1% of γ -Al$_2$O$_3$
could be satisfactorily forged.

The effect of a high forging ratio of about 6.5:1 on the
mechanical properties of the sintered compact was examined. For
this purpose, the sintered specimens, about 45 mm in diameter and
35 mm in height, sheathed in mild steel case of which thickness
was 15 mm, were forged.

As shown in Table 5, the tensile properties of the sintered
AISI 406 steel can be further improved with increased forging
ratio. With respect to the tensile properties of the sintered
13CrMoTi steel, the same tendency was also observed.

Such an improvement of the tensile properties of the sintered compact by sheath forging is partly due to the work hardening
by higher degree of hot working, but is considered to be largely
attributable to the better densification of the compact.

Table 6 shows the elevated temperature tensile properties of
the sintered specimens with 1% of γ -Al$_2$O$_3$ and 2% Ti, sheath forged after sintering. Compared with the results shown in Table 4,
it is understood that the increased addition of γ -Al$_2$O$_3$ from
0.5 to 1.0% is more favorable for improvement of the elevated
temperature properties, and regarding the strength and ductility
at elevated temperatures, the sintered 13CrMoTi steel is superior

Table 6. Example of tensile properties at elevated
temperature (650°C) of dispersion strengthened steels
(γ-Al$_2$O$_3$:1%, Ti addition:2%, sheath forged)

Matrix	0.2% Proof stress (kg/mm^2)	Ultimate tensile strength (kg/mm^2)	Elongation (%)	Reduction of area (%)
AISI 406 steel	22.5	32.2	14.4	18.4
13CrMoTi steel	24.2	34.2	33.8	57.9

Table 7. Comparison of creep rupture time
(Test temperature:650°C, Stress applied:10 kg/cm^2)

Matrix	γ-Al$_2$O$_3$ (%)	Ti addition (%)	Creep rupture time (hr)	Remarks
AISI 406 steel	-	-	0.9	AISI 406 wrought steel
	0.5	-	1.98	Hot forging
	1.0	-	4.4	Hot forging
	1.5	-	2.8	Hot forging
	-	1.0	1.3	Hot forging
	-	2.0	1.5	Hot forging
	0.5	1.0	77.8	Hot forging
	0.5	2.0	181.4	Hot forging
	0.5	3.0	203.7	Hot forging
	1.0	2.0	284.6	Hot forging
	1.0	3.0	244.0	Hot forging
	1.0	2.0	1177.1	Sheath forging
	2.0	2.0	148.5	Hot forging
	2.0	2.0	423.4	Sheath forging
	2.0	2.5	184.5	Hot forging
	2.0	3.0	377.2	Hot forging
	2.0	3.0	606.9	Sheath forging
13CrMoTi steel	-	-	3.5	Sheath forging
	0.5	-	8.8	Sheath forging
	1.0	-	19.9	Sheath forging
	0.5	1.0	141.9	Sheath forging
	1.0	2.0	3973.2	Sheath forging
	1.5	3.0	342.5	Sheath forging

to the sintered 406 steel.

Creep Rupture Test Results

In order to evaluate this type of alloys as the heat resisting material, the creep rupture tests were carried out on smooth bar test pieces, using a multiple type tester. Under a constant load, testing stress was adjusted by changing the diameter of the parallel part (gauge length; 25 mm) of the test specimens.

The creep rupture strength of various sintered specimens are shown in Table 7 and, for the sake of comparison, the data of AISI 406 wrought steel are also given in the table. The test temperature and stress applied were 650°C and 10 kg/mm^2, respectively. As can be seen in the table, the addition of γ -Al$_2$O$_3$ brings about only a slight improvement for the creep rupture properties. On the other hand, if Fe-Ti alloy is added together with γ -Al$_2$O$_3$, the creep rupture is greatly delayed. Considering that creep rupture times were not prolonged by separate addition of Fe-Ti alloy, this improvement can not be attributed to the effect of Fe-Ti alloy itself. The test results, moreover, show that the sheath forging of the sintered compacts improves their creep rupture strength to a large extent. By the sintered 13CrMoTi steels, better results in rupture strength were obtained, compared with the sintered AISI 406 steels.

In Fig. 3, the creep rupture properties at 650°C of the present two sintered steels are shown. For comparison, the rupture strength of wrought steel of nearly the same chemical composition as AISI 406 steel powder and of Type 422 and 321 wrought steel as typical examples of martensitic and austenitic heat

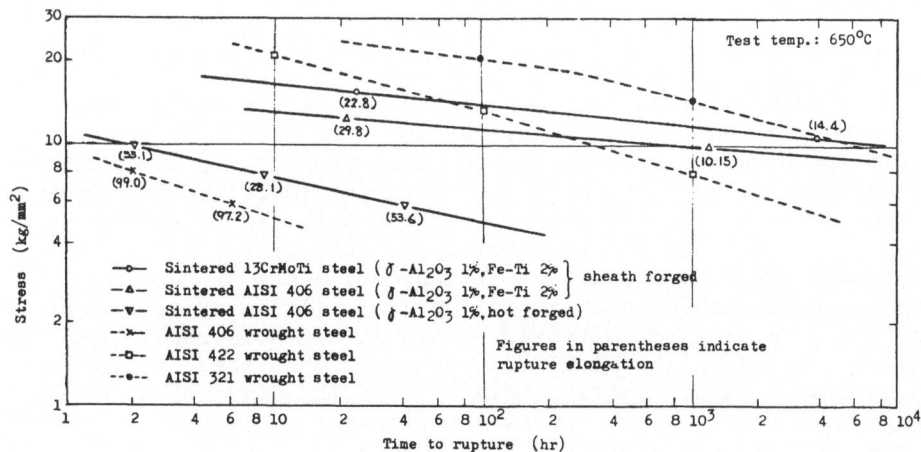

Fig. 3 Results of creep rupture test

resisting steels were also shown in the same figure. The creep
rupture strength of AISI 406 wrought steel is very low and it
decreases rapidly with the lapse of time. The sintered and hot
forged AISI 406 steel with dispersed 1% of γ -Al$_2$O$_3$ is slightly
superior in creep rupture strength than AISI 406 wrought steel,
while the rupture strength of the sintered AISI 406 steels with
dispersed γ -Al$_2$O$_3$ and Fe-Ti alloy, further processed by sheath
forging, was very high and their rupture times at the same stress
levels were much longer than that of the sintered specimens with
addition of γ -Al$_2$O$_3$ only.

        The rupture strength of the sintered 13CrMoTi steels is
higher than that of the sintered AISI 406 steels, all the other
conditions being the same. No convincing explanation of this
superiority in the creep rupture strength of the sintered
13CrMoTi steel over the sintered AISI 406 steel can be given,
but it is likely to be related to the precipitation hardening.
In fact, a great number of fine precipitates, confirmed to be
intermetallic compound (Fe.Si)$_2$Mo as a result of the electron diffraction analysis, was found in the matrix of the sintered
13CrMoTi steels, but such a microstructural change during creep
rupture test was not observed in the sintered AISI 406 steel.

        The short-time creep rupture strength of the dispersion
strengthened 13CrMoTi steel is lower than that of Type 422 and
321 wrought heat resisting steel. However, the long-time creep
rupture strength is superior to that of Type 422 steel, and it
becomes higher than that of Type 321 steel after 5,000 hr,
because of the small rate of lowering of the rupture strength as
shown in Fig. 3. The sintered AISI 406 steel also shows comparable creep rupture strength to that of Type 321 austenitic steel
after a certain long time.

                      IV - CONCLUSION

        Dispersion strengthening is one of the most interesting
methods for increasing the creep rupture strength of alloys. In
order to improve the elevated temperature properties, in particular, creep rupture strength of the ferritic steel, the fundamental studies on 13% Cr steels were carried out. As a result, the
dispersion strengthened type of alloys of superior elevated
temperature properties could be obtained by means of the liquid
phase sintering and subsequent hot working. As an example, the
long-time creep rupture strength of the sintered 13CrMoTi steel
with dispersed γ -Al$_2$O$_3$ at 650°C was higher than that of Type 321
wrought steel. Consequently, this type of the dispersion
strengthened steels are considered to be very promising as the
new heat resisting materials.

P/M REFRACTORY MATERIALS
AND SPECIALTY PRODUCTS

RELATIONS BETWEEN STEREOMETRIC MICROSTRUCTURE AND

PROPERTIES OF CERMETS AND POROUS MATERIALS

S. Nazaré, G. Ondracek,[+] and F. Thümmler

Institut für Material- und Festkörper-

forschung, Kernforschungszentrum Karlsruhe

Germany

Abstract

The characteristic parameters of a dispersed phase in a matrix
of another phase may strongly influence the properties of a
two-phase material. This is especially true for cermets, which
consist of a phase having often very different properties, because one phase is a ceramic, the other is a metal. In this sense,
porous specimens, as achieved generally with powder metallurgical
methods, are a special case of cermets. This is why the effects
of several parameters of a dispersed phase in a matrix on some
properties shall be considered here in a more general manner.

At first theoretical considerations are discussed for the concentration function of the electrical conductivity of cermets.
Some other properties like permeability and thermal conductivity
are also mentioned. The limiting cases of the equations enable
us to consider the influence of pores in single phase material
as well as particle shape, distribution and size of the dispersed
phase on cermet properties. These considerations from a theoretical point of view are then compared with experimental results
available. Conclusions are drawn about the relationship between
the stereometric microstructure of cermets and porous materials
and the properties mentioned.

[+]Speaker. Work partly perfomed at the University of Surrey,
Dept. of Metallurgy, U.K., during a stay as a visiting
scientist.

1. Introduction

In determining the properties of materials prepared by powder
metallurgical methods, the question of porosity and its influence on the properties becomes relevant. Basically a porous
material is a special case of a two phase material, in which
the properties of the two phases are extremely different. Such
materials are called cermets and can be defined as combinations
with at least one phase with predominantly metallic and at least
another with predominantly nonmetallic bonding [1]. The theoretical considerations for the concentration functions of cermets
must therefore also be valid for porosity and furthermore in
simplified form. The electrical conductivity and other similar
properties of cermets will be considered here in relation to the
concentration. The expressions shown here will be used for
porous materials.

Whilst considering cermets, it is useful to differentiate between
the structure as related to the materials involved, and the
structure as given by the stereometric arrangement of the phases
in the composite. The structure as related to the materials
describes the interaction between the constituents and the
bonding at the interfaces, and is dependent e.g. on the concentration and thermodynamic properties of the phases involved.
Without considering this point any further, it should be noted
that the following discussion is valid for stable cermet combinations. Unlike unstable cermets the components of the former
do not react with one another [1]. The stereometric structure
describes the geometry and the arrangement of the phases in the
cermet. Its parameters are therefore independent of the material
[2], and we shall consider its influence on the properties.

2. Stereometric Structure of Cermets

The density of UO_2-Cu cermets as a function of the concentration is shown in fig. 1. It can be seen that within a certain interval,

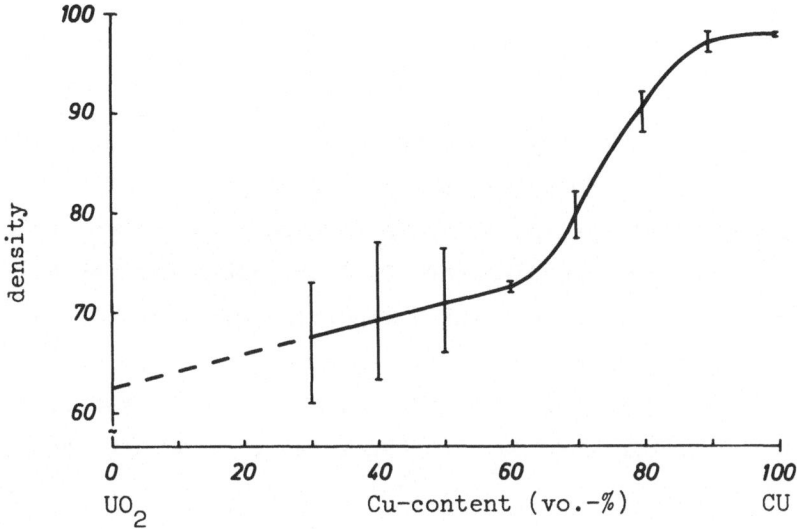

Fig. 1:   Density of UO_2-Cu cermets as a function
          of concentration [2]

a change in the slope of the curve takes place. A similar relationship has been found for the variation of density and other
properties for various cermet combinations [3-5]. It should be
mentioned, however, that in general the concentration function
of the density must not show such a slope, because of their
dependance of the production technique. But, nevertheless, it
shows one of the first characteristics of the stereometric
structure of cermets - the change in the matrix. In the case of
a two phase cermet combination, there exists a region with a

metallic and a region with a ceramic matrix, i.e. the particles
of one phase are separated by a continuum of the other phase, in
which they are dispersed (matrix structure). With porous materials
it would correspond to the case of non-interconnected porosity. The
two regions with a matrix structure are separated by a region where
the matrix change occurs. Here both phases are continuous. The
corresponding case for porosity would be where the pores are interconnected.

The region where these two types of structures exist depends on the
preparation method, on the particle size and shape of the starting
powders. The particle size can be characterised as a stereometric
factor by the mean particle diameter. The particle shape can be
defined by using a form factor, which generally gives the deviation
from a spherical shape. Particles with an irregular shape are
roughly characterised through the three axis of the ellipsoid. Other
stereometric factors for the matrix structure are shown in fig.2

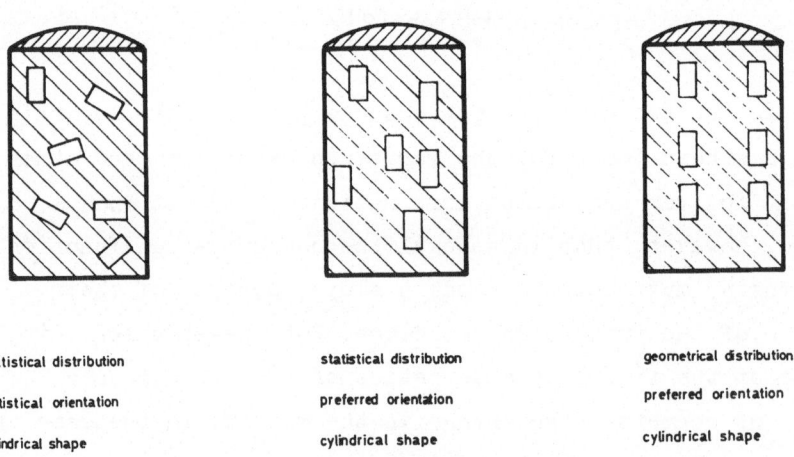

statistical distribution   statistical distribution   geometrical distribution

statistical orientation   preferred orientation   preferred orientation

cylindrical shape   cylindrical shape   cylindrical shape

Fig.2: Stereometric factors in a cermet with a
        matrix structure [8]

and comprise the orientation and distribution of the dispersed
phase. The orientation can be defined by means of texture parameter and the distribution through its degree [2]. When various
preferred orientations exist, the distribution function of the
texture parameters must be considered.

The various parameter of the stereometric structure of cermets
e.g. particle shape and size, orientation and distribution can
also be adapted to porous materials. Instead of the particles of
the second phase we have the pores. We shall now consider in which
way the stereometry influences the concentration function of the
electrical resistivity of cermets and porous materials.

3. The Concentration Functions of Cermet Properties

We shall at the outset deal with variation of the electrical resistivity with concentration. A study of various existing equations
for cermets [8] and porous materials [14-30, 51] shows that the
following two relations are most general and have a minimum of
restrictions [9]:

$$\left[(1-V_K)\left(\frac{1}{\rho_m}-\frac{1}{\rho_c}\right)\sum_{i=x}^{z}\frac{\cos^2\alpha_{im}}{\frac{1}{\rho_c}+\left(\frac{1}{\rho_m}-\frac{1}{\rho_c}\right)F_{im}}\right]+\left[V_K\left(\frac{1}{\rho_K}-\frac{1}{\rho_c}\right)\sum_{i=x}^{z}\frac{\cos^2\alpha_{iK}}{\frac{1}{\rho_c}+\left(\frac{1}{\rho_K}-\frac{1}{\rho_c}\right)F_{iK}}\right]=0 \quad (1)$$

$$\int_{0}^{V_D}\frac{dv}{1-v}=-\int_{\rho_M}^{\rho_c}\frac{d\rho}{\rho\left(\frac{1}{\rho_D}-\frac{1}{\rho}\right)\sum_{i=x}^{z}\frac{\cos^2\alpha_{iD}}{\frac{1}{\rho}-\left(\frac{1}{\rho_D}-\frac{1}{\rho}\right)F_{iD}}} \quad (2)$$

The suffixes mean:

K = ceramic phase ; m = metallic phase ; D = dispersed phase ;
M = matrix phase ; C = cermet ; i = x,y,z = ellipsoid axis ;
ϱ = electrical resistivity ; V = concentration of the phases
in vol.-% ; α = angle between the axes of the ellipsoid and the
direction of the field ; F = form factor $(F_x + F_y + F_z = 1)$

The influence of the stereometry can be seen by the fact that
equation 1 is valid when both phases are continuous, whereas
equation 2 holds for the case where one phase is dispersed in a
continuous matrix of the other.

The texture factor $(\cos^2 \alpha_i)$ takes into account the orientation
of the phases to the field and the form factor (F) the particle
shape. A factor for the size of the particles does not occur in
these equations. Experimental data show [10,11] that particle
size does not influence the resistivity. Whereas particle size is
considered in the derivation of the equations, but cancels out in
the final result, a factor for the distribution is not included
because statistical distribution is assumed.

Since our comparison with experimental data will deal with cermets
having a matrix structure and porous materials with non-interconnected porosity, we shall restrict our discussion to equation
2 only. This equation was originally derived for the dielectric
constant [9] and was later adapted for the electrical resistivity [8].

In the case of rotational ellipsoid (spheroids x = y = z) it
reduces to the simpler form:

$$\int_{0}^{V_D} \frac{dv}{1-v} = - \int_{\rho_M}^{\rho_c} \frac{d\rho}{\rho\left(\frac{1}{\rho_D} - \frac{1}{\rho}\right)\left[\frac{\cos^2\alpha_{xD} + \cos^2\alpha_{yD}}{\frac{1}{\rho} + \left(\frac{1}{\rho_D} - \frac{1}{\rho}\right)F} + \frac{\cos^2\alpha_{zD}}{\frac{1}{\rho} + \left(\frac{1}{\rho_D} - \frac{1}{\rho}\right)(1-2F)}\right]} \tag{3}$$

The above equations (1,2 and 3) are derived by considering that in
an initially homogeneous field, a stray field is created when a
very dissimilar substance is introduced. The stray field is dependent on the shape of the material, which is related to the
deelectrization factor in the field equations. In our case it becomes the form factor (F). It can be calculated for the rotational
ellipsoids with the following equation [6]:

$$F = \frac{x^2 z}{2} \int_0^{\infty} \frac{du}{|x^2+u|^2 \sqrt{|z^2+u|}} \qquad (4)$$

and is shown in fig.3 for various ratios of the three axes. The
mean axial ratio of the dispersed phase or pores can be determined

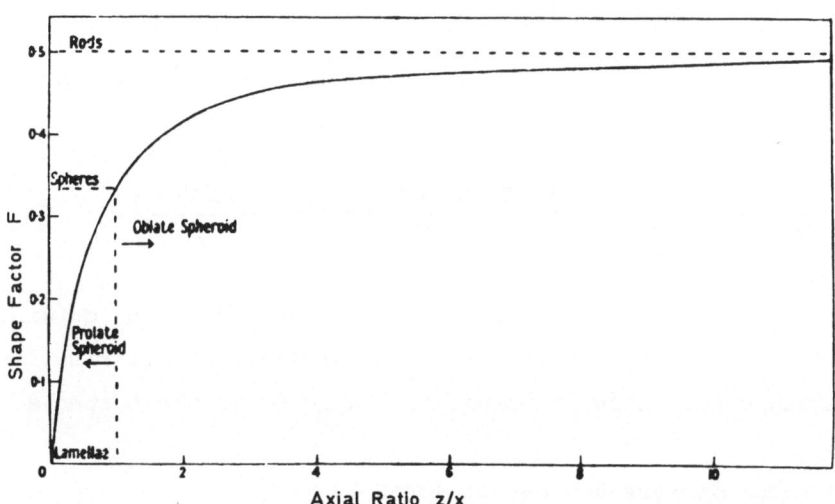

Fig. 3: Variation ot the form factor for various
         ratios of the ellipsoid axes [6]

by metallographic procedures, and the form factor (F) can then be
directly obtained from fig. 3.

The use of equation 2 which is implicit and has to be integrated
is somewhat cumbersome for practical purposes. An approximate
explicit solution of this equation is however available for
texture-free material. It is valid for the cases [12] where:

$$\frac{\rho_M}{\rho_D} >> 1 \quad \text{and} \quad \frac{\rho_M}{\rho_D} << 1$$

If the texture factor is introduced in these equations we obtain
in the case of the electrical resistivity the following approximate
equations for a spheroidal dispersed phase.

When $\dfrac{\rho_M}{\rho_D} \quad >> 1$ we have:

$$\rho_c = \rho_M \left(1 - v_D\right)^{\dfrac{\cos^2 \alpha_{xD} + \cos^2 \alpha_{yD}}{F} + \dfrac{\cos^2 \alpha_{zD}}{1-2F}} \quad (5)$$

and when $\dfrac{\rho_M}{\rho_D} << 1$ and $\dfrac{\rho_c}{\rho_D} << 1$ we obtain:

$$\rho_c = \rho_M \left(1 - v_D\right)^{\dfrac{\cos^2 \alpha_{xD} + \cos^2 \alpha_{yD}}{F-1} - \dfrac{\cos^2 \alpha_{zD}}{2F}} \quad (6)$$

Exact solutions of the general equation (eq. 3) can be obtained
for the case of texture-free material (statistical distribution
of the dispersed phase or pores; $\cos^2 \alpha_{iD} = \frac{1}{3}$) in three special
cases:

a) spherical dispersed phase or pores [9, 13]

$$x = y = z; \quad F = \frac{1}{3}; \quad \cos^2 \alpha_{xD} = \cos^2 \alpha_{zD} = \frac{1}{3}$$

$$1 - v_D = \frac{\rho_c - \rho_D}{\rho_M - \rho_D} \sqrt[3]{\left(\frac{\rho_M}{\rho_c}\right)^2} \quad (7)$$

b) lamellar dispersed phase or pores [9, 13]

$$x = y \gg z; \quad F = 0; \quad \cos^2 \alpha_{xD} = \cos^2 \alpha_{zD} = \frac{1}{3}$$

$$\rho_c = \rho_M \frac{3 - v_D \left(1 - \dfrac{\rho_D}{\rho_M}\right)}{3 - 2v_D \left(1 - \dfrac{\rho_M}{\rho_D}\right)} \tag{8}$$

c) cylindrical dispersed phase or pores [9]

$$x = y \ll z; \quad F = \frac{1}{2}; \quad \cos^2 \alpha_{xD} = \cos^2 \alpha_{zD} = \frac{1}{3}$$

$$1 - v_D = \frac{\rho_M}{\rho_c} \frac{|\rho_c - \rho_D|}{|\rho_M - \rho_D|} \left[\frac{\rho_c}{\rho_M} \frac{|\rho_M + 5\rho_D|}{|\rho_c + 5\rho_D|} \right]^{\frac{2}{5}} \tag{9}$$

The implicit solutions (7 and 9) can be used in practice in the following way: The resistivity of the cermet or the porous material is calculated for various concentrations of the dispersed phase or pores by using the known values of the resistivity of the two phases.

In fig.4 the theoretical relation is compared with the experimental values for cermets. In this particular case a single theoretical curve is justified since the conductivities of the metals considered and the ceramic UO_2 differ very much. The plotted relation represents the electrical conductivity relative to the compact matrix material. The agreement is good, when the experimental accuracy is considered. The stereometric assumptions like e.g. spherical particle shape and as a result statistical orientation of the dispersed phase were approximately fulfilled in this case. These cermets were mostly prepared by using metal coated spherical UO_2-particles which were subsequently hot pressed isostatically. To check the validity of the stereometric effect on thermal conductivity, UO_2-Cu-Cermets (5 vol.-% Cu) were prepared. In one case the embedded phase consisted of platelets, in the other case of cylinders. The thermal conductivity was measured at three tem-

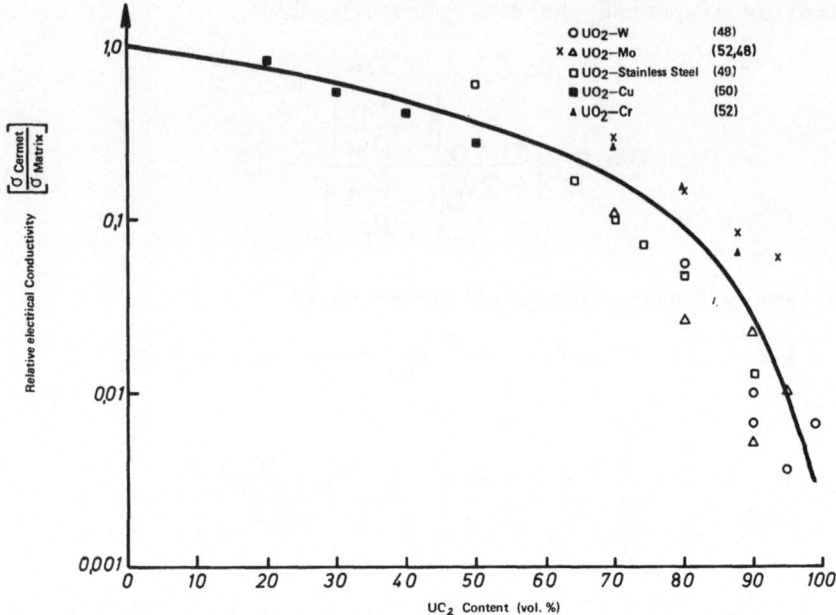

Fig. 4:   Relative electrical conductivity of cermets

peratures (40, 93.75, 163°C Colora apparatus). In fig.5 the preliminary results are given. Up to now no   test was done to confirm
their reproducibility. But they seem to indicate, that in agreement with the theoretical analysis the conductivity of cermets
strongly depends on the stereometric structure.

To show the validity of the equations discussed previously the
case of spherical geometry of the inclusions shall now be considered more in detail by using porous material as an example. The
theoretical relation in fig. 6 for porous materials is also calculated for spherical pores (statistical orientation).   The
experimental data show deviations from the theory because the

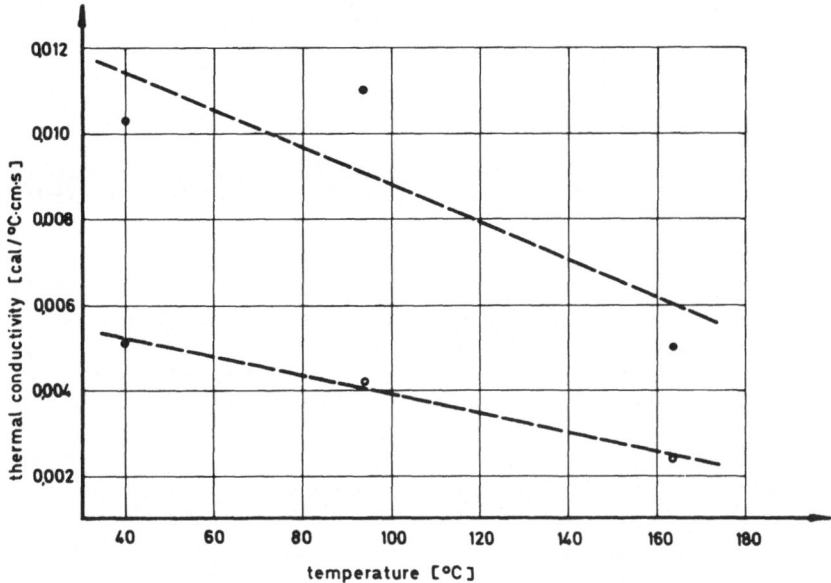

Fig.5: Preliminary results about the effect of dispersed
particle shape on the thermal conductivity of
UO_2-Cu cermet (5 vol.-% Cu)

• platelets
◦ cylinders

stereometric structure does not fully conform with the assumption
made.

Finally we shall deal in short with other cermet properties.As a
result of the mathematical analogy in the relations for the
electrostatic field:

$$\vec{D} = \varepsilon \cdot \ell \qquad (10)$$

the magnetostatic field:

$$\mathscr{B} = \mu \cdot \mathfrak{h} \qquad (11)$$

the electric current in the steady state:

$$\vec{\imath} = \sigma \cdot \ell = \frac{1}{\varrho}\ell \qquad\qquad (12)$$

the heat flow in the steady state:

$$\vec{q} = -\lambda \ \mathrm{grad} \ T \qquad\qquad (13)$$

(The symbols mean:

$\vec{D}$ = dielectric deviation; ε = dielectric constant; ℓ = electrical field strength; $\vec{B}$ = magnetic induction; μ = magnetic permeability; $\mathfrak{h}$ = magnetic field strength; $\vec{\imath}$ = electrical current density; σ = electrical conductivity; ϱ = electrical resistivity; λ = thermal conductivity; T = temperature)

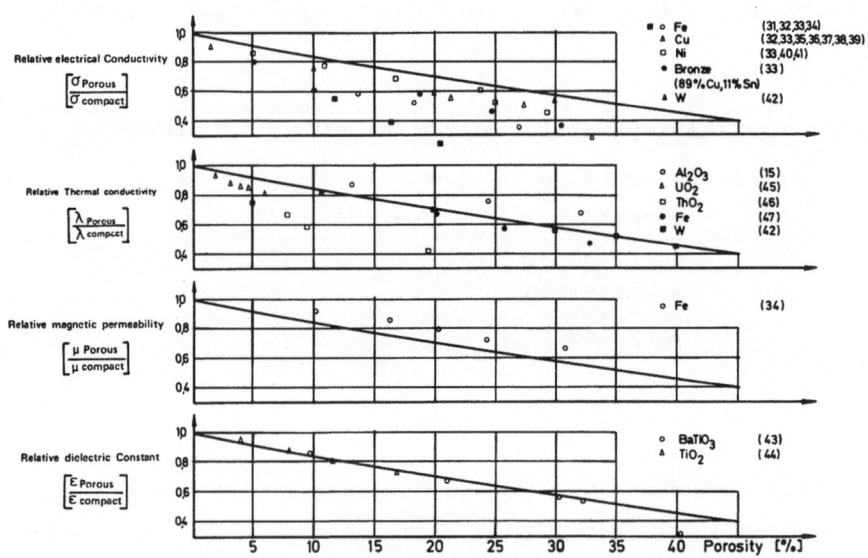

Fig.6:   Relative electrical and thermal conductivity,
         magnetic permeability and dielectric constant
         of porous materials
                ──────── theoretical curves

The relations which have been derived for the electrical resistivity can be adapted for: the thermal conductivity, the magnetic permeability, the dielectric constant.

The transformation of equation 6 for the thermal conductivity of porous materials (λ_P) yields:

$$\lambda_P = \frac{\lambda_o}{(1-P)^n} \tag{14}$$

With λ_o = thermal conductivity of the compact material

   P = porosity and

$$n = \frac{2 \cos^2 \alpha_{xD}}{F - 1} - \frac{\cos^2 \alpha_{zD}}{2 F}$$

For spherical porosity $(\cos^2 \alpha_{iD} = \frac{1}{3}; F = \frac{1}{3})$
we have:

$$\lambda_P = \lambda_o (1-P)^{1.5} \tag{15}$$

The above relation is similar to the simplified form of an equation derived for the thermal conductivity with a different approach [14, 15, 16]. The comparison with the available experimental data of porous materials yields again fair agreement considering the fact that the porosity in the specimens could possibly be neither spherical nor statistically oriented.

When the electrical resistivity in equation 7 is replaced by the reciprocal of the magnetic permeability we obtain the results shown also in fig.6. The same considerations are obviously valid in this case as well in the case of the dielectric constant.

The equations considered above are therefore well suited to explain experimental data, by taking into account stereometric factors like particle or pore shape and orientation. For this purpose the stereometric structure must be known as well as the properties of the pure phases. The validity of the equations is restricted to the case where the distribution is statistical. Furthermore they are only valid for stable cermet combinations and isotropic phases. They do not take into account the degree of bonding at the interfaces.

Acknowlegdement: The authors would like to thank Mr. E.von Staden for supplying some of the literature data. Thanks are also due to Miss B.Schulz for her cooperation.

References

[1]   G.Ondracek, K.Splichal, Proc. III.Int.Pulvermetallurgische Tagung, Karlsbad (1970)

[2]   A.Jesse, G.Ondracek in KFK 845, Kapitel 6 (1969)

[3]   E.Gebhardt, G.Ondracek, F.Thümmler, J.Nucl.Mat.13, Nr.2, S.210-219 und 229-241 (1969)

[4]   D.Schmidt, Diss.Universität Karlsruhe, DLR-FB-69-96 (1969)

[5]   G.Ondracek, E.Patrassi, B.Schulz, KFK 922 (1968)

[6]   Winkelmann, Handbuch der Physik, Bd.V, S.124 (1908)

[7]   K.Schröder, G.Reissig, R.Reissig, Mathematik für die Praxis, Bd.III, S.31, Verlag Hari Deubel, Frankfurt/M. und Zürich (1964)

[8]   G.Ondracek, B.Schulz, Ber.DKG (1970)

[9]   W.Niesel, Annalen der Physik, 6.Folge, Bd.10, S.336-348 (1952)

[10] F.Wachholtz, A.Franceson, Kolloid-Z. 92, S.75 und 1958 (1940)

[11] R.de la Rue, C.W.Tobias, J.Elektrochem.Soc. 106 (9) S.827-833 (1959)

[12] G.P.de Loor, Proefschrift Universität Leiden (1965)

[13] D.A.G.Bruggeman, Ann.Physik Bd.24, S.636; Bd.25, S.645 (1935/1936)

[14] A.L.Loeb, J.Am.Ceram.Soc., Vol.37, Nr.2, Pt.II, p. 96-99 (1954)

[15] J.Francl, W.O.Kingery, J.Am.Ceram.Soc., Vol.37, Nr.2, Pt. II, p. 99-107 (1954)

[16] W.D.Kingery, J.Francl. R.L.Coble, I.Vasilos, J.Am. Ceram.Soc., Vol.37, Nr.2, Pt.II, p.1o7-11o (1954)

[17] C.J.F.Böttcher, Rec.Trav.Chim. Pays-Bas 64, p.47 (1945)

[18] A.Biancheria, Transactions Am.Nucl.Soc.9, p.15 (1966)

[19] M.J.Brabers, Proc. 2nd Conf. 6, p.122 (1958)

[20] G.Breit, Suppl. Nr.46 des Comm.Leiden (1922)

[21] P.le Goff, C.Prost, Genie Chemique-Chimie et Industrie,
      Vol.95, Nr.1 (1966)

[22] R.B.Grehila, T.J.Tien, J.Am.Ceram.Soc., Vol.48, Nr.1 (1965)

[23] H.J.Juretschke, R.Landauer, J.A.Swanson, J.Appl.Phys.27,
      S.839 (1956)

[24] H.J.Juretschke, R.Steinitz, J.Phys.Chem.Solids 4, p.118(1958)

[25] W.D.Kingery, J.Am.Ceram.Soc.42, P.617 and Pregress in
      Ceramic Science, Vol.II, p.224 (1959)

[26] V.P.Maskovets, J.Appl.Chem. USSR 24, p.391 (engl.Translation)

[27] H.W.Russell, J.Am.Ceram.Soc., Bd.18, H.1, p.1-5 (1935)

[28] G.V.Samsonov, V.S.Neshpor, Dokl.Akad. Nauk SSSR. 104,
      p.405 (1956)

[29] V.V.Skorokhod, Powder Metall., p.180 (1963)

[30] A.Slavinski, J.de Chemie Physign 23/26, S.710/368
      (1926/1929)

[31] D.A.Oliver, Iron and Steel Institute, Special report 38
      (9), 63 (1947)

[32] F.Sauerwald et al, Z.Elektrochemie 38 (1932)

[33] P.Grootenhuis et al, Proc.Phy. Soc. 65 (1952)

[34] V.A.Danilenko, Por.Met. 2 (50) 44 (1967)

[35] K.Adlassnig et al, Radex Rundschau 79 (1950)

[36] C.G.Goetzel, J.Inst.of Metals 66 (1940)

[37] H.H.Hausner, Pow.Met.Bull 3 (1948)

[38] R.Kieffer, W.Hotop, Kolloidzschr. 104 (1943)

[39] W.Trzbiatowski, Z.Phys.Chem. A 169 (1934)

[40] G.Grube et al, Z. Elektrochemie 44 (1938)

[41] H.H.Hausner et al, The Physics of Pow.Met.Symposium Sylvania
      Electricalproducts Inc. (London: Mcgraw Hill) (1951)

[42] S.N.L'Vov et al, Por.Met. 5 89 (1966)

[43] D.F.Rushman et al, Proc.Phy.Soc. 59 (1947)

[44] A.R.Hippel, Report VII N.D.C; MIT (1944)

[45] T.R.MacEwan, J.Nucl.Mat. 24, 1o9 (1967)

[46] M.Murabayash et al, J.Nucl. Sci and Technology 6 (1969)

[47]   G.I.Aksenov, Por.Met. <u>54</u>, 39   (1967)

[48]   I.Amato et al, Fiat Sezione Energia Nucleare R.103
       (1965)

[49]   R.W.Dayton et al, BMI-1259 (1958)

[50]   A.Jesse, Dissertation Univ. Karlsruhe (1970)

[51]   Ribaud, Chaleur et Industrie <u>18</u> (1937)

[52]   P.Weimar, Dissertation Univ. Karlsruhe (1969)

Porous Stainless Steel - The Unique Filter Medium

N. Nicholaus and R. Ray

Pall Corporation, Glen Cove, New York

The "Unique Filter Medium"... an imposing title for a product made from a handful of tiny stainless steel granules. However, when the properties of porous stainless steel are examined it will be seen that this statement is not presumptuous.

An old saying claims that the ultimate filter medium would have a pore size of zero, no resistance to flow, unlimited strength and would cost about the same as tissue paper. Porous stainless steel is by no means the ultimate, but it often is the only material that could perform satisfactorily under the conditions imposed.

This paper will describe various manufacturing techniques for porous stainless steel, advantages and disadvantages of each, properties and applications of this material for filter and non-filter purposes.

What is porous stainless steel? In the broadest sense, it is a sintered metal product containing a predetermined useful pore structure. This is in complete contrast with the basic thought behind most powder metallurgy products - the achievement of 100% density at all costs. The starting material may be powder, fibers, wire, machined shavings or even wire mesh. By the proper application of powder metallurgy techniques these materials, singly or in combination, may be converted into porous stainless steel.

MANUFACTURE

We will limit ourselves to techniques using powder for the starting material.

In one case, a briquette is prepared by mixing a small amount of organic binder and lubricant with a stainless steel powder and compacting this mix at a pressure of approximately 50 tons per square inch. This briquette possesses enough green strength to be handled without damage.   It is then heated to a temperature sufficient to volatilize the binder (around 600° to 800°F). Sintering is then accomplished at temperatures around 2200°F in a reducing atmosphere.

The main advantage for this procedure lies in its adaptability to high production rates for simple shapes as flat discs, simple cylinders, cones, etc.

The main disadvantages are increased resistance to passage of fluids and the possibility of contamination due to uncontrolled decomposition of the organic binder with subsequent reduced corrosion resistance. Another disadvantage is the limitation in size of the parts that can be produced because of the limited availability and high cost of large compacting presses. This limitation can be partially overcome using isostatic pressing.

In another process, a mixture of powder and binder is extruded through dies to give (most commonly) circular or flat cross sections . Other shapes are also produced. The extrusions are then volatilized and sintered in the same manner as discussed previously.

Advantages of the extrusion process are the ability to produce seamless cylinders, fairly high production rates and lower flow resistance than compacting.

Disadvantages of the process are limited sizes and shapes and reduced resistance to corrosion due to the binder.

In a third process a tray or mold is loosely filled with powder (without any binder or lubricant present). This is then sintered at 2400°F or higher in a reducing atmosphere.

Advantages of this process are lowest resistance to flow of the three processes discussed, freedom from contamination from an organic binder and so the highest resistance to corrosion and the ability to produce larger flat sheets, up to 18" x 96".

Disadvantages are inability to produce seamless tubes and the process is not suitable for automated production of standard shapes.

PROPERTIES

The important and critical properties of porous stainless steel include thickness, average porosity (100% minus % density), the mean pore size, pore size distribution, pore shape and degree of interconnected porosity.

Porous stainless steel is available in thicknesses from .020" to 4" but most common thicknesses are 1/32, 1/16, 1/8 and 1/4". Normal tolerances are ± .010".

For filtration, porosities will range from a low of 20% to a high of 70%. Lower porosities have higher strength but higher flow resistance.

Mean pore sizes commonly will range from 1 micron to 165 microns. Manufacturers will offer a family of 5 to 6 pore sizes covering this range. This mean pore size bears a definite relationship to degree of filtration. Mean pore sizes are varied by suitable selection of particle size and shape in the starting powder and control of process variables.

Pore size distribution can be controlled to narrow limits by selecting a powder with a very precise screen analysis. The wider the spread the less efficient will be the filter medium. In a typical material, over 90% of the pores should be within ± 10% of the mean pore size.

Pore shapes in this type of filter medium are usually irregular. Whereas this induces greater flow resistance than do pores of circular cross section, the irregularity enhances particle capture.

For filtration to occur, flow passages must connect one face
of the medium to the other. For example, a porous material
with 50% voids could still be a barrier to flow if all the passages
were dead-ended. For media of equal porosity, the one with the
higher degree of interconnected pores would have a higher permeability. A comparison of flow-pressure drop characteristics
for similar competitive materials will provide a qualitative measure of the relative degree of interconnected porosity.

An important factor in the choice of a filter medium is the
filtration rating (a function of pore size and pore size distribution),
an arbitrary value indicating a given percentage removal of particles of a specific size using a standard test procedure; for example, 98% (by weight) removal of all particles larger than 10
microns.

The most common method of determining filtration ratings
is by actually filtering a slurry of micron sized glass beads and
microscopically measuring the size distribution of those which
have passed through the filter by collecting them on an ultrafine
filter paper or membrane. For convenience in quality control
of production material it is possible to correlate glass bead findings with a simple "bubble point" test. The porous material is
submerged in a suitable fluid of known surface tension and wetting properties while gradually increasing air pressure is applied
from the opposite face. The pressure at which the first bubble
occurs is a measure of the largest pore in the material and the
pressure at which uniform bubbling occurs is equivalent to the
mean pore size. Empirically the bubble point and pore size are
related by the following equation; which is based on stainless
steel and Solox 190 alcohol.

$$\text{Bubble Point (mm Hg.)} = \frac{395}{\text{Pore Size (Micron)}}$$

Table I shows the filtration ratings for some typical grades
of porous stainless steel.

These parameters (which are specifically a function of form
and not material) directly control other properties which must be
considered in the selection or design of porous stainless steel
media for use in both filtration and nonfiltration applications.
Among the most important of these properties are flow resis-

Grade	Mean Pore Size (Microns)	Removal Ratings, Microns			
		When Filtering Liquids		When Filtering Gases	
		Nominal (98%)	Absolute (100%)	Nominal (98%)	Absolute (100%)
C	165	55	160	45	110
D	65	22	55	8	20
E	35	12	35	4	11
F	20	7	25	1.3	3
G	10	3	15	0.7	1.8
H	5	2	12	0.4	1.0

TABLE I

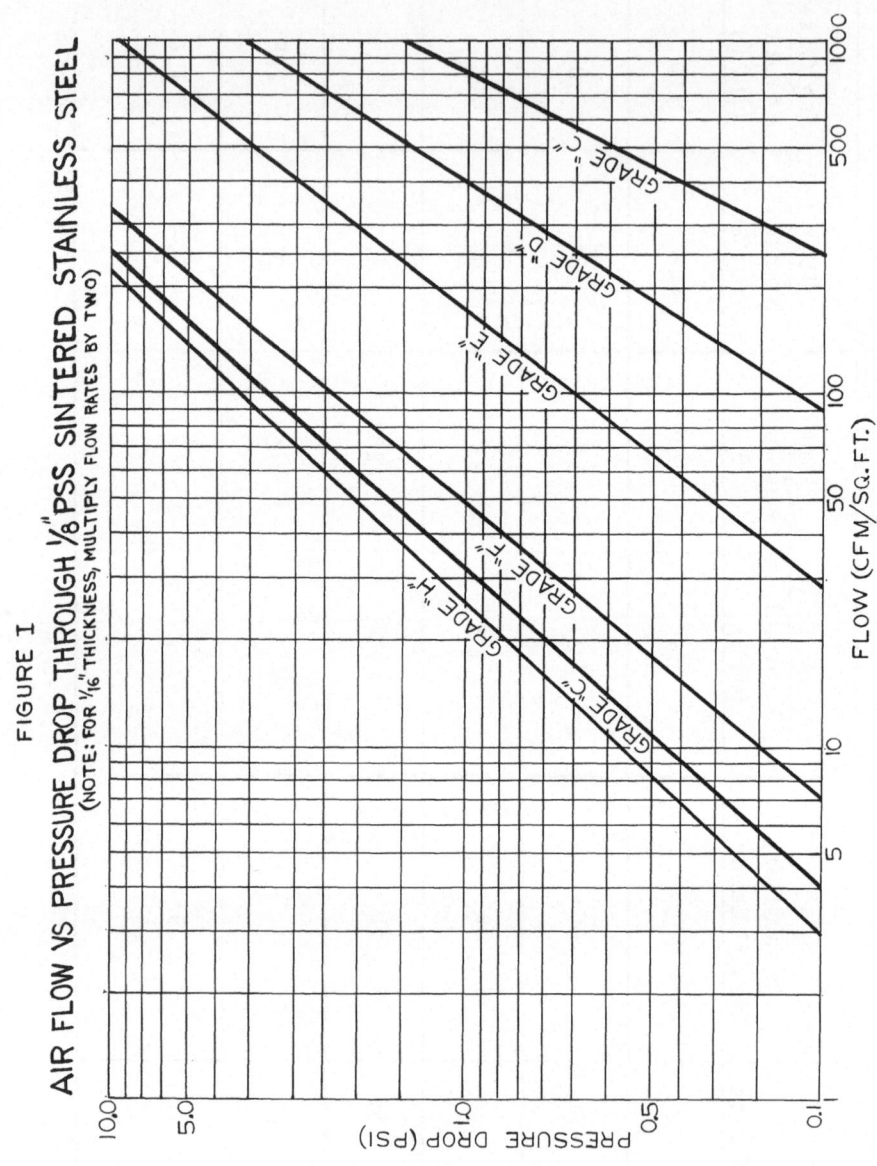

FIGURE I

AIR FLOW VS PRESSURE DROP THROUGH ⅛" PSS SINTERED STAINLESS STEEL
(NOTE: FOR 1/16" THICKNESS, MULTIPLY FLOW RATES BY TWO)

tance, dirt capacity, electrical and thermal conductivity (as related to the conductivity of the 100% dense solid), strength, ductility and corrosion resistance (as related to the 100% dense solid).
Other properties which enhance the application of porous stainless
steel as opposed to non-metallic porous materials are high specific strength, rigidity, resistance to temperature cycling over a
range of over 2000°F, cleanability, chemical stability and freedom from media migration.

In any application flow resistance is extremely important.
Generally flow resistance is inversely proportional to pore size
and directly proportional to thickness for materials of equal porosity. Figures 1 and 2 show the relation between flow and clean
pressure drop for some typical grades of porous stainless steel
for both air and water.

FIGURE 2

WATER FLOW VS PRESS. DROP THRU. ⅛"PSS SINTERED ST. STL.

(NOTE: FOR ¹⁄₁₆" THICKNESS, MULTIPLY FLOW RATES BY TWO)

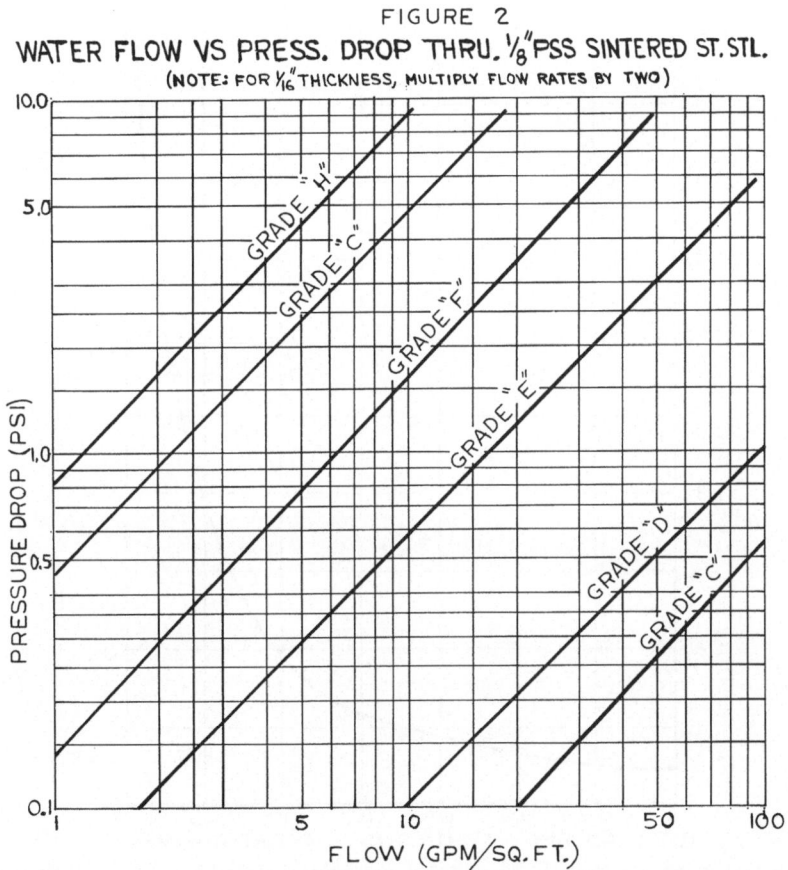

However, as the porous metal builds up a coating of contaminant the pressure drop continually increases until it ceases to pass any more fluid. This is why consideration of dirt capacity is so important. The greater the useful life of the filter before it requires cleaning, the lower the cost of operation. This parameter is dependent upon the same external factors as the filtration rating. Many standard tests have been devised which make use of either AC (air cleaner) fine or AC coarse test dust slurried in a suitable fluid. A plot of dirt added versus increase in pressure drop at a constant flow is used to compare various filter media. Figure 3 is a typical dirt capacity/pressure drop curve for a 5 micron medium.

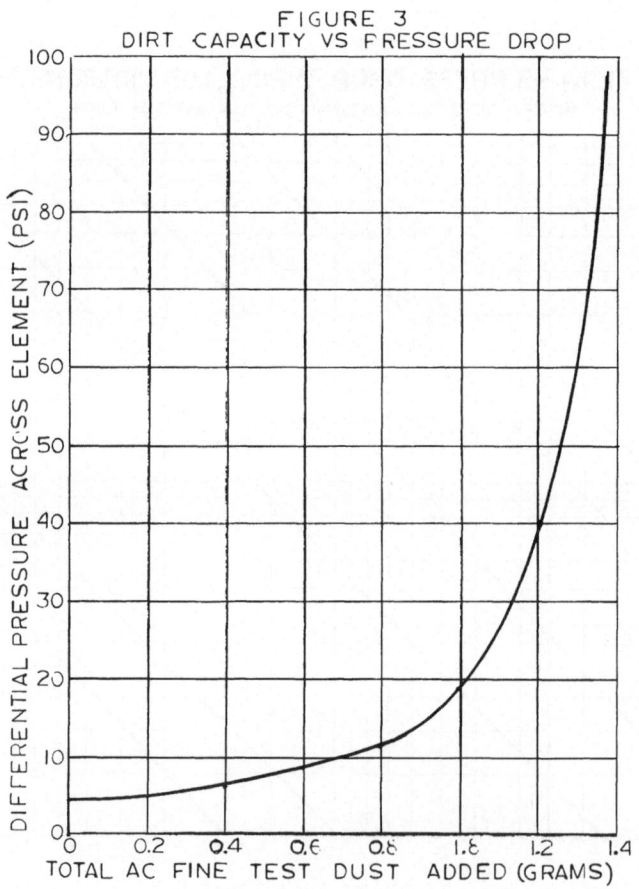

FIGURE 3
DIRT CAPACITY VS PRESSURE DROP

Electrical and thermal conductivity are always less than would be expected from the ratio of actual density to the density of solid material. This is because of the tortuous nature of the interconnecting path across the bond points of the powder and also the small size of bond points in relation to the particle size. Materials with porosities around 50% have conductivities about 1/3 that of solid material the same thickness. These properties are also useful in quality control as they can be correlated to strength (degree of sintering).

The strength, rigidity and ductility of porous stainless steel enables the fabrication of self-supporting structures for most applications and ease considerably the job of designing for use at extremely high pressures (to 10,000 psi). Absolute tensile strength values for materials in the 50% porosity range can reach 18,000 psi. Strength, everything else being equal, varies inversely with porosity. Ductility is such (elongations of approximately 5%) as to allow the forming of 1/2" diameter cylinders from 1/16" thick materials.

Porous stainless steel is not subject to spalling and catastrophic failure which frequently occurs with porous ceramics thereby allowing service under severe conditions of thermal cycling or mechanical shock.

Since no binders or filler metals are used the porous metal is as corrosion-resistant as the parent metal. However, two important facts must be remembered: a larger surface area is exposed to the corrosive agent and the small size of the bonds joining the particles. If these facts are properly considered when applying known corrosion rates to the design of porous metals there should be no nasty surprises popping up while the material is in service. For most materials in the 50% porosity range bond diameters are about .0003 to .0005 inches.

The corrosion and temperature resistance of porous stainless steel makes it 100% cleanable since the cleaning agents such as acids or concentrated caustic would not be harmful. As an alternate the material could be heated in a reducing atmosphere to over 2000°F (anneal) to remove organic materials. If only 95% of the original permeability is restored at each cleaning the material will still have 60% of its original permeability after 10 cleanings.

In many types of service a simple blowback operation is suitable thus eliminating the need for removal of the filter for cleaning. The strength of the material allows high blow-back pressures without danger of damage. The high corrosion resistance also prevents the porous metal from affecting the quality of the filtered liquid.

Freedom from media migration, which is so important in many critical applications is an inherent property of this material because of the strength and integrity of the sintered bond. Filter paper and cloth and other porous materials, shed fibers and particles which contaminate the filtrate they are supposed to protect.

Although we've been discussing porous stainless steel all the properties examined are typical of all porous metals. The choice of metal or alloy would depend upon cost, availability and ability to withstand the process conditions. Almost any metal or alloy can be manufactured in porous form but the most commonly used are the 300 and 400 series stainless steels, the Monels, the Inconels, the Hastelloys, nickel, copper, bronze, silver, gold and even platinum.

APPLICATIONS

The most obvious use for a material with a controlled pore size is as a filter medium. Porous stainless steel contributes its unique properties to the following industries: chemical, nuclear, polymer, beverage, brewing, food, pharmaceutical, cryogenic and synthetic fibers to name just a few. Filtration can be defined as the separation of solid contaminant (or in many cases a desirable product) from a liquid or gaseous stream.

In the majority of these applications, the primary function of porous stainless steel is to provide a completely reliable means of clarifying or "polishing" the process fluid. One example is the production of sparkling clear beverages making them more appetizing and easier to sell. There are also many applications where the solid product is valuable and must be collec - ed such as collection of platinum catalysts in fluidized catalyst processing of phthalic anhydride. In either case porous stainless steel provides an ideal filter medium. In addition to the high degree of reliability provided by its mechanical strength, the lower wall thickness that also results gives an improvement in pressure drop which in turn means more economical operation. The wide range of filtration ratings available allows the selection of a material which is an optimum in regard to life versus degree of filtration. Among specific areas of application are:

NUCLEAR: Porous stainless steel and porous monel media for the filtration of uranium hexafluoride and porous stainless steel for the filtration of liquid sodium and liquid sodium-potassium alloys used as heat exchange media.

CRYOGENIC: Use of porous metal media for the filtration of liquid gases such as oxygen, nitrogen, helium, and hydrogen. A specialized application involves the use of porous stainless steel for the removal of carbon dioxide crystals from liquid air prior to the production of liquid oxygen. This process eliminates the need for caustic scrubbing and provides a system that can readily be "cleaned" by simply raising the temperature of the filter and driving off the carbon dioxide particles by evaporation.

PHARMACEUTICAL: Here, the media find extensive application in the final polishing of solutions prior to packaging and in the separation of antibiotic crystals such as penicillin and streptomycin from their "mother liquor".

CHEMICAL: Porous metal media have been most successfully used in phthalic anhydride manufacture in fluidized beds to prevent the carry-over of valuable catalyst fines with the reacted gases. By an appropriate valving system, the operation can be maintained on a continuous basis, collecting fines on one section of the filter while discharging accumulated solids back into the bed from the other section. Such an installation can be maintained on-stream for as long as one year with no change in pressure drop across the filters, and in continuous operations at up to 1,000°F, 24 hours per day, 7 days a week.

POLYMERS & SYNTHETIC FIBERS: Porous metals are an ideal filter media for use with these products because they can easily be fabricated into reliable units capable of withstanding the high (3000 or 4000 psi) differential pressures encountered in filtration of high viscosity (250,000 centipoise) melts. Cleanability is another factor in this application.

MISCELLANEOUS: Almost any clarification operation can bene-
fit from the use of the media described in this paper. These
would include filtration of demineralized water and other fluids
used in the manufacture of television picture tubes and semiconductors; final polishing filtration of distilled products such as
whiskeys and beer; filtration of soft drinks, hydraulic fluids, and
any application where even the smallest quantity of contaminant
can be detrimental to the quality or sale of the final product or
the performance of a mechanical or hydraulic system.

The non-filtration applications of porous stainless steel
listed below utilize the properties of porosity and controlled pore
size to perform an almost unlimited range of tasks in all fields
of endeavor.

TRANSPIRATION COOLING: Turbine blades, jet engine parts
and missile nose cones made from porous metals can operate in
high temperature environments that would normally destroy un-
protected materials because of the internal cooling effect of the
injection of either liquid or gas. The porous metal can also be
used as a matrix for containing a given amount of low melting
material which evaporates thus protecting the structure for a
specific length of time.

SOUND ATTENUATION: Microphones, telephones and other
electro-acoustic devices use porous metals to improve frequency
response by smoothing out flow variations. Large mufflers for
air compressors and steam blow-down systems act in a similar
manner.

MATERIALS HANDLING: Hoppers full of powder or granular
material can be made free-flowing by incorporation of porous
metal pads which allow the controlled introduction of air which
fluidizes the particles increasing rate of emptying and avoiding
blockage of the outlet. Sheets of metal, plastic or paper can be
transported without contacting any surfaces by supporting them
on air film maintained at the surface of "air-bars" or even rollers
made of porous metal.

FLUID DISTRIBUTION: High speed continuous processing of
aircraft film is accomplished by passing the film over a series
of porous rollers containing the required processing chemicals.
Inks for printing may be applied in a similar manner. This
technique has also been used successfully for distribution of de-
icing fluid along aircraft wings.

GAS DISTRIBUTION: Porous metals with the proper pore size and surface treatment can be used to bubble gas through a liquid and still prevent flow back of liquid when the gas stream is stopped. Other applications are the sparging of nitrogen into mayonnaise and carbon dioxide into beer. Porous metal plates are also excellent support plates for fluidized beds in various processing operations.

The injection of air through porous metal inserts aids in controlling laminar flow on air foil surfaces improving lift at low forward velocities.

MISCELLANEOUS: Snubbers for protection of gages from pressure surges are a minor but important application of the controlled permeability properties of porous stainless steel. Flow restrictors can be similarly designed to meter gas flow to any number of analytical instruments. The rapid dissipation of heat by its large internal surface area makes porous metal an ideal flame arrester to eliminate the hazards of flashback in flammable gas lines and equipment. Low density porous metals are compressible enough to act as ultra-reliable containing rings to protect turbine casings from catastrophic damage in the event that a disc should fracture and the blades fly out.

Sublimator plates where an internal layer of ice sublimes into the vacuum of outer space provides a self-regulating static cooling system for NASA's Apollo program which only requires a supply of water for operation. In water desalinization by reverse osmosis porous metals provide a reliable support for the fragile membrane because perforated plates cannot be made with small enough holes to prevent the membrane from puncturing under the high operating pressures.

This list could go on forever for there is no limit in the application of porous metals to solve problems. New ideas are being constantly thought of and proven out.

To summarize, porous metals and porous stainless steel in particular offer another dimension to engineers and designers ...that of controlled porosity. The key word here is controlled because by proper choice of starting material and process conditions, a product can be made which will possess the optimum combination of porosity, permeability, strength and corrosion resistance and yet keep the cost within required limits.

ABOUT NITRIDES AND CARBONITRIDES AND NITRIDE-BASED CEMENTED HARD ALLOYS

R.Kieffer, P.Ettmayer and M.Freudhofmeier

Technical University Vienna, Austria

INTRODUCTION

Up till now nitrides of the transition metals have attained only cursory interest as compared with the hard and refractory carbides (1,2). Practical application of nitrides has been limited to the formation of hard and wear resistant surface layers in the nitriding of steels and other iron alloys. The relatively stable nitrides of the group IV of the periodic table (TiN, ZrN, HfN) have been used as refractory materials for the manufacture of crucibles with fair thermal shock resistance, but no larger technical success followed (3). Attempts to cement nitrides with binder-metals to form hard alloys have been up to now only partly successful (4-12). Niobium nitride has gained some reputation as superconducting material with high transition temperature and uranium carbonitride seems to be of interest as promising nuclear fuel because of its stability against graphite and greater chemical stability against humidity as compared with the uranium carbides UC and UC_2.

So far only the very stable nonmetallic ceramic-like nitrides Si_3N_4 and BN have won important technical application in the high-temperature field.

PROPERTIES OF NITRIDES

The transition metal nitrides are in many respects closely related to the corresponding carbides. They are of metallic nature, but pronouncedly less than the carbides, they are

usually isotypic with the corresponding carbides and
formation of solid solutions is common and governed by
Hume-Rothery's rule. By application of this rule the
miscibility of carbides and nitrides can be predicted
with fair reliability (1). Fig.1 gives a survey over the
miscibility of carbides and nitrides so far substantiated.
Hypothetical miscibilities are shown in brackets.

Similarities between nitrides and carbides exist with
respect to their hardness, the hardness values of nitrides
generally being lower. Nitrides have good electrical and
heat conductivity and relatively high melting points, provided the nitrides are not decomposed previously by evolution of nitrogen gas. Although thermodynamic stability
of the nitrides is in the same order of magnitude with
that of the carbides, nitrides are less stable than carbides due to the evolution of nitrogen at high temperatures and vacuum. Some of the more important properties
of metallic carbides and nitrides are summarized in Fig.2.

	TiC	ZrC*	HfC	VC	NbC	TaC
TiN	●	(●)	●	●	●	(●)
ZrN	●	(●)	●	○	●	(●)
HfN	●	●	●	○	●	●
VN	●	(○)	○	●	●	(●)
NbN	●	(●)	●	●	●	(●)
TaN	◑	(◑)	◑	(◑)	◑	◑

Fig.1: Miscibility of carbides and nitrides of the
       Transition metals

        ● Complete solid solution
        o Small solid solution or immiscible
      (●)Hypothetical solid solution
      (o)Hypothetical immiscibility
        ◑ Partial solid solution in the cubic phase

4 a	5 a	6 a	4 a	5 a	6 a
TiC ▫	VC ▫	$Cr_3C_2D5_{10}$	TiN ▫	VN ▫	Cr_2N ⬡
Fp 3160	2830	1895 z.	Fp 2950	2050	1500 z.
R 68	60		R 11,1	86	
H 3200	2950	2280	H 2450	+ 9 ?	
ZrC ▫	NbC ▫	Mo_2C ⬡	ZrN ▫	NbN ▫	Mo_2N ▫
Fp 3530	3500	2400 z.	Fp 2980	2300 z.	zers.
R 42	35	133	R 13,6	~ 200	
H 2560	2400	1950	H 1990	+ 8	
HfC ▫	TaC ▫	WC ⬡	HfN ▫	TaN ⬡ ▫	W_2N ▫
Fp 3890	3780	2600 z.	Fp 2700	3090	zers.
R 37	25	22	R < 26	135	
H 2700	1790	2080	H > 2000	3240	

Fig.2: Some physical properties of carbides and nitrides

▫	face centered cubic
⬡	hexagonal
Fp:	melting point °C
R:	electrical resistivity Ω-cm
H:	hardness (kg/cm^2 at 50-150 g load)

PREPARATION OF NITRIDES AND CARBONITRIDES

Nitrides used for our own investigations were prepared by reacting metal powder or metal sponge with nitrogen at temperatures between 1100 and 1300°C. This is a very convenient, but not necessarily inexpensive method for the preparation of pure nitrides. Reduction of oxides by carbon in the presence of nitrogen f.i. would yield cheaper carbonitrides but with relatively high oxygen content.

Carbonitrides have been prepared by mixing nitrides and carbides intimately and heating them in Ar-atmosphere together, until homogeneity is obtained. The oxygen content of these nitrides and carbonitrides was low, but oxygen pickup can be high during milling operations, even when organic solvents, such as n-hexane, as milling fluid, were used. Table 1 gives figures for the average oxygen content of titanium-nitride and -carbonitride during the various stages of production.

Table 1: Oxygen content of titanium-nitride and -carbo-
nitride

	TiN	$TiN_{0,8}C_{0,2}$
	wt %	
After preparation	0,3-0,5	0,3-0,4
Milled to 0,8 - 1,0 µ[+)	0,6-0,7	0,5
Sintered with 10 % wt.Ni-Mo(3:1)	>0,15	>0,15

[+) evaluated by Fisher Sub Sieve Sizer

Within the solid solutions of TiN with TiC the colour
changed from golden yellow for pure TiN to red-golden
for $TiC_{0,20}N_{0,80}$ to bronze coloured $TiC_{0,40}N_{0,60}$ until
at a composition $TiC_{0,5}N_{0,5}$ the compound is gray with a
violet tinge. This colour effects offer some interesting
by-effects for eventual nitride based hard alloys. To our
knowledge the watch making industry f.i. would be interested in golden coloured and scratch proof watch cases.

STABILITY OF CARBIDES AGAINST NITROGEN

Nitrides cannot be sintered in dynamic vacuum because of
the evolution of nitrogen and the resulting decomposition
of the nitrides. On the other side, carbides and carbonitrides tend to react with nitrogen (equation 1) under
formation of elemental carbon, the latter being known to
have an adverse effect on sintered hard alloys.

$$TiC_{1-x}N_x + \frac{y}{2} N_2 \longrightarrow TiC_{1-x-y}N_{x+y} + yC \qquad (Equ.1)$$

Thus during the sintering procedure of carbonitrides,
nitrogen pressure has to be adjusted within a certain
limit to the composition of the carbonitride and the
sintering temperature. Only few data have been published
concerning this special problem, and we have studied the
reaction of carbides of the IVth, Vth and VIth group of
the periodic table with nitrogen, carrying our experiments
to very high pressures of nitrogen (300-1000 atm) and to
temperatures up to 1800°C (13). The carbides of the IVth
group metals and VC_{1-x} reacted readily with nitrogen under
formation of carbonitride and graphite. The compositions
of the carbonitrides are dependent on nitrogen pressure
and temperature of reaction. As predicted by thermodyna-
mics, higher pressure of nitrogen led to carbonitrides
richer in nitrogen, raise in temperature tended under
identical conditions to decrease nitrogen content in the
carbonitrides. In Fig.3-6 the results of these investi-

gations are demonstrated. Niobium-carbide and molybdenum-carbide reacted only to a very small extent with nitrogen, whereas tantalum-carbide and tungsten-carbide proved to be stable against nitrogen even at very high pressures (13). Chromium-carbide Cr_3C_2 reacted with nitrogen of 300 at under formation of a ternary compound $Cr_3(C,N)_2$ as already previously reported by Kieffer, Ettmayer et al (14,15).

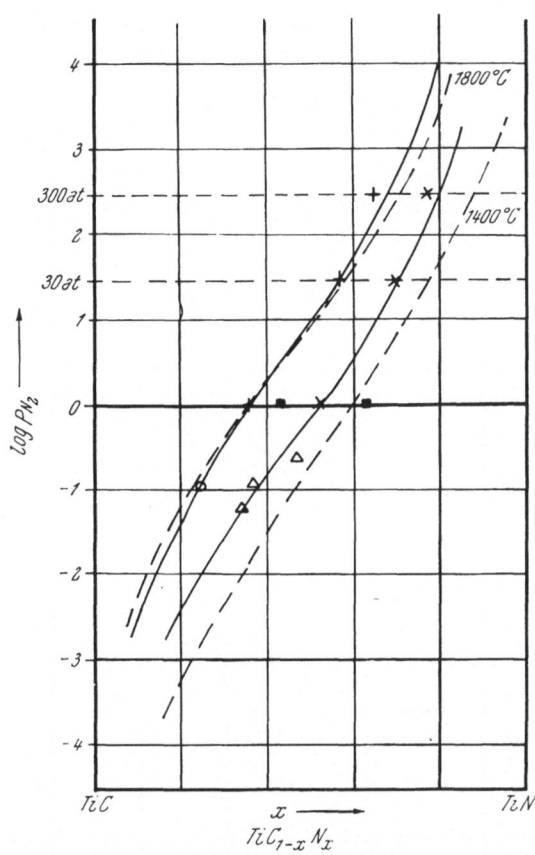

Fig.3: Composition of titanium-carbonitrides vs.pressure
of nitrogen

X	R.Kieffer et al (5) 1400°C
+	R.Kieffer et al (5) 1800°C
△	Zelikman and Gorowitz (21) 1500°C
o	Zelikman and Gorowitz (21) 1800°C
■	Portnoi and Levinskii (22) 1800°C and 1400°C
—	calculated from thermodynamic data = 8 kcal

This evaluation of the stability of carbides against
nitrogen of high pressure has been carried through by
using internally heated autoclaves.   The autoclave
for 30 at is shown in Fig.7, temperatures exceeding
2500°C have been attained therein. A second autoclave
for pressure up to 2000 at has been constructed by ourselves and is working basically on the same principles
(16). For safety purposes the walls of the pressure
vessel have been made of Nimonic (one of the best heat-
resistant alloys with high strength even at high temperatures). The upper and the lower part of the pressure
vessel are isolated by a PTFE-foil and are serving as
current leads (Fig.8).

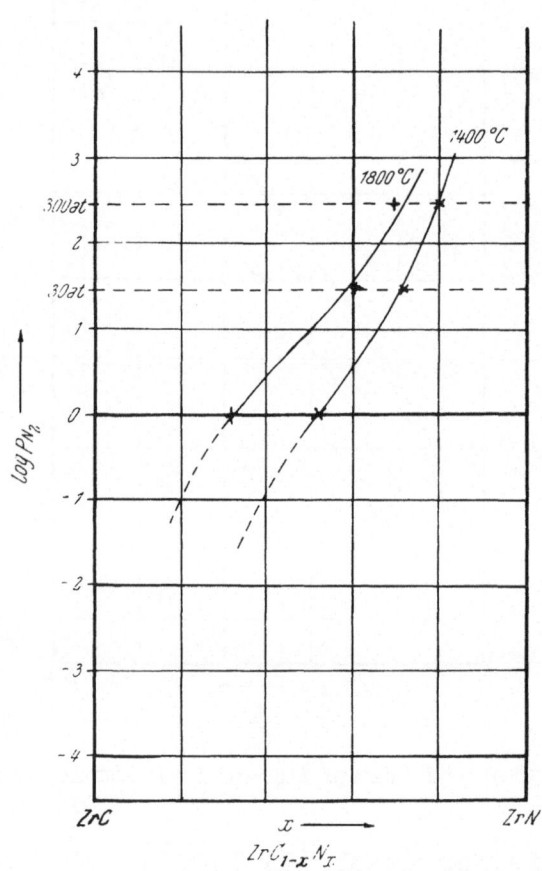

Fig.4: Composition of zirconium-carbonitride vs.pressure
                                              of nitrogen

        x   1400°C
        +   1800°C

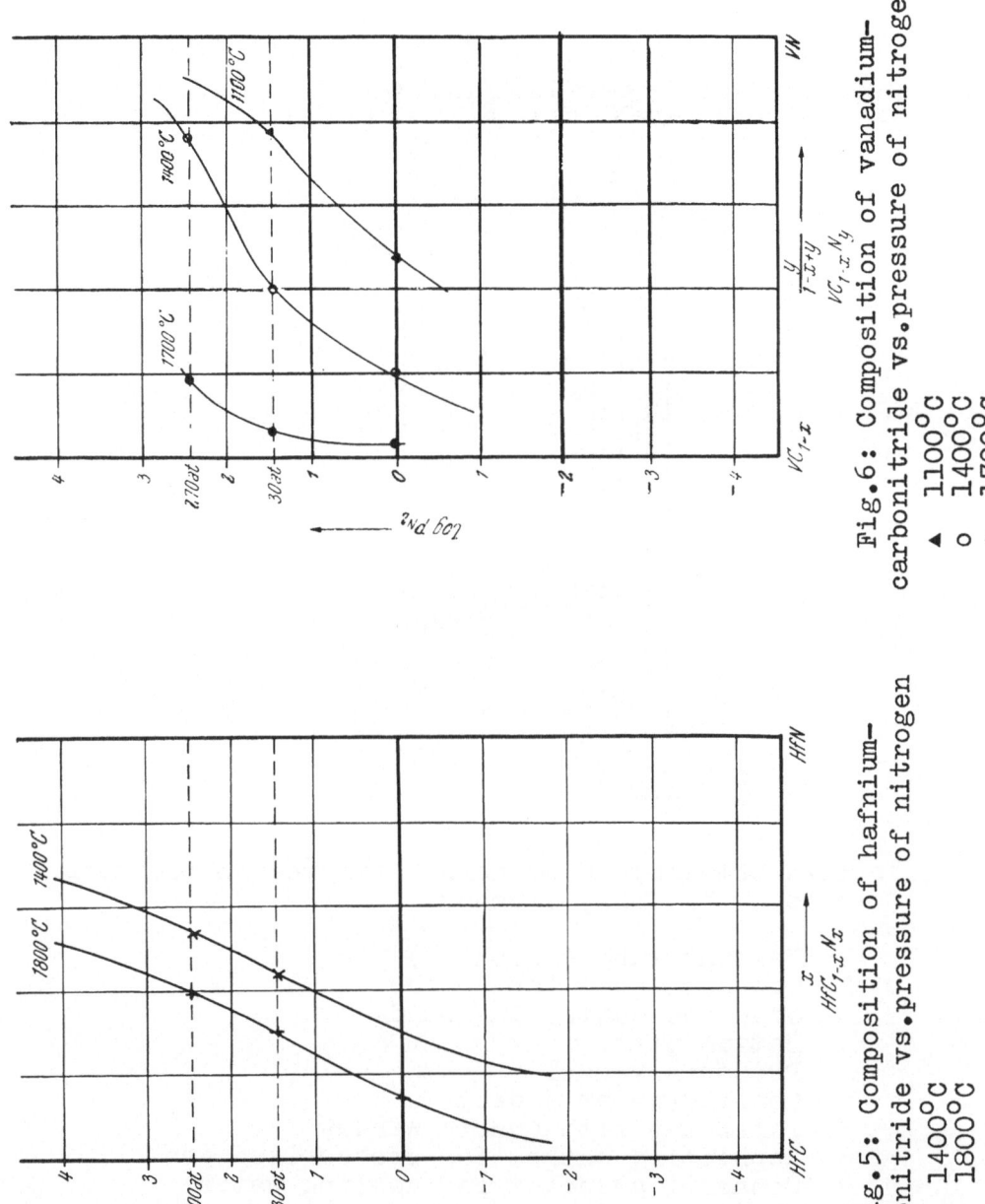

Fig.6: Composition of vanadium—
carbonitride vs. pressure of nitrogen

▲ 1100°C
○ 1400°C
● 1700°C

Fig.5: Composition of hafnium—
carbonitride vs. pressure of nitrogen

x 1400°C
+ 1800°C

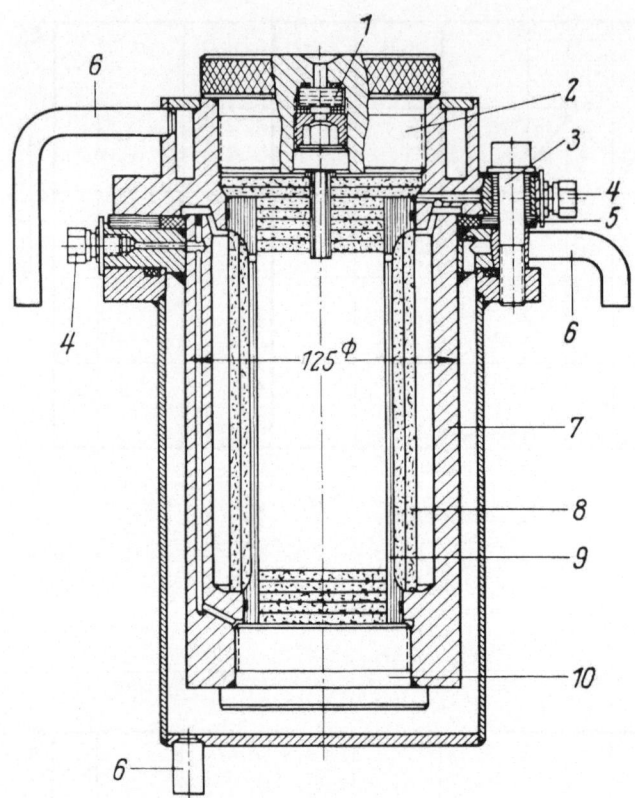

Fig.7: High temperature autoclave for medium pressure
       made by DEGUSSA, Germany

         1 Heating tube (graphite)
         2 Graphite-wool insulation
         3 Graphite contacting band
         4 Bottom plate
         5 Pressure vessel
         6 Cooling-water jacket
         7 Screw cap with quartz window
         8 Connecting socket for pressurized gas
         9 Connecting socket for cooling water
        10 Current lead

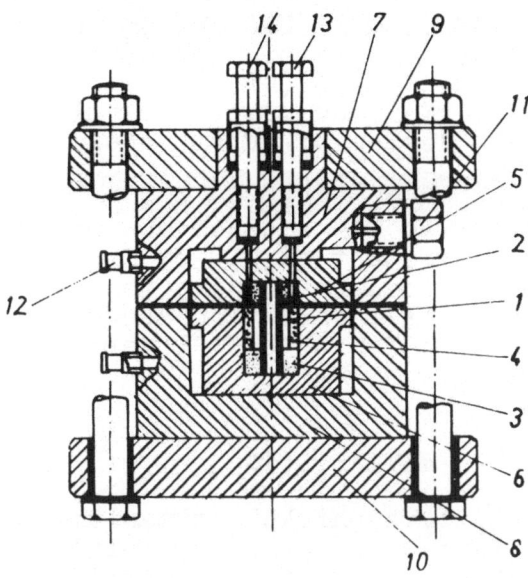

Fig.8: High temperature autoclave for high-pressure
                                    nitrogen

    1   Heating tube (graphite)
    2
    3) Contacting rings
    4   Graphite-wool insulation
    5   Pressure vessel
    6
    7 } Cooling-water jacket, designed for auxiliary
    8   support
    9
   10) Pressure plates
   11   Connecting socket for pressurized gas
   12   Connecting socket for cooling water
   13   Inlet
   14   Outlet) valves

          CEMENTED CARBONITRIDE-BASED ALLOYS

The system TiC-TiN is excellently suited for the study
of the properties of cemented carbonitride-based hard
alloys. Technical titanium-carbide is, chemically spoken,
in fact a nitrogen-poor titanium-carbonitride and has
been known to be a suitable base for tungsten free cemented alloys. Nickel-bonded titanium-carbide has been known
in Europe for more than 40 years (17) and has found a recent

revival by the Ford-Company in USA. We too have made some
experiments lately and have obtained cemented hard alloys
based on TiC-10/25 % Mo_2C-14 % (Ni,Mo) with excellent properties and cutting performance. Optimum values could be
obtained at a certain degree of substoicheiometry in the
solid solution TiC/Mo_2C (18).

       Since TiC forms with TiN a continous series of solid
solutions, gradual replacement of TiC by TiN should result
in a gradual change of properties. The objective of
our work was to gain some information about the amount of
TiC to be replaceable by TiN without lowering hardness
and strength too much.

WETTABILITY

To study the wettability by nickel, titanium-carbonitrides
of various compositions were tested by infiltration experiments. On top of preforms of carbonitride a small ball
of pure nickel was placed and the whole assembly was heated
in argon atmosphere of 400 torr to 1500°C. After cooling
down, the preforms were cut in two halves and the degree of
infiltration was examined visually. Pure titanium-nitride
was not wetted and nickel had not even penetrated the preform. By contrast to carbon-free TiN, the solid solution
$TiC_{0.2}N_{0.8}$ was completely infiltrated by liquid nickel or
a Ni-Mo-alloy. Iron and cobalt showed poor wetting properties. Wetting was achieved only at higher carbon contents
in the carbonitride, but some alloys of iron and cobalt have
proved to have better wetting properties. These results are
consistent with experiments by W.D.Kingery and F.A.Harden(19

MILLING AND SINTERING PROCEDURES

Titanium-carbonitride was wet-milled with nickel or a 75/25
nickel-molybdenum powder in cemented carbide lined ball
mills with cemented carbide balls to a average grain size
of 0,8-1 µ. As a milling fluid, acetone or n-hexane has been
used. Milling time between 5-6 days proved to be sufficient
After having removed carefully the milling fluid by evaporation, the carbonitride-mixture was mixed with 0,7-1,0 %
paraffin (dissolved in benzene) and pressed to preforms.
After presintering the preforms at 1000°C in stationary
hydrogen atmosphere the final sintering was performed in a
partial vacuum of 150-600 torr nitrogen at temperatures
between 1420 and 1550°C depending on the composition of the
carbonitride. Sintering times were held between 1 an 6 hours
Carbonitrides high in nitrogen need high sintering temperatures and long sintering times.

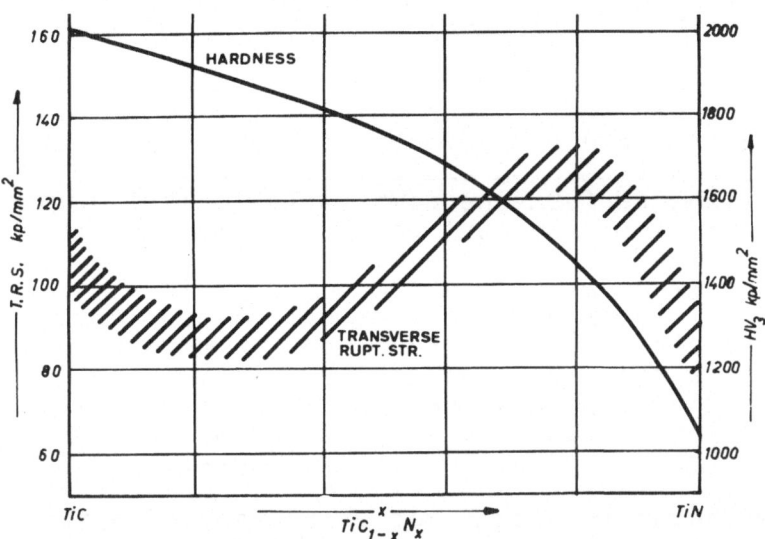

Fig.9: Hardness and transverse rupture strength of
cemented carbonitride plotted against composition
(10 % wt.binder)

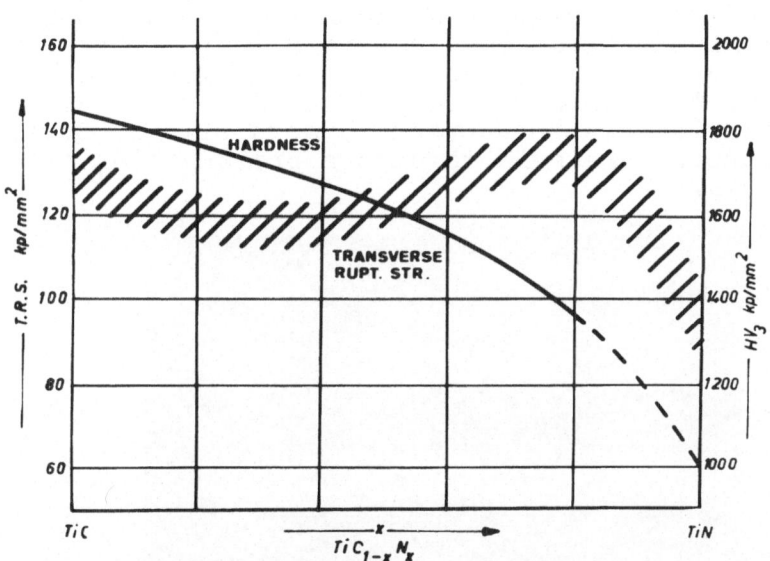

Fig.10: Hardness and transverse rupture strength of
cemented carbonitrides plotted against composition
(14 % wt.binder)

RESULTS

Table 2 and Fig.9 and 10 show hardness values and transverse rupture strenghts of cemented titanium-carbonitrides
with 10 and 14 % binder. Some preliminary tests showed the
cemented carbonitrides to exhibit satisfactory or even excellent cutting performance comparable to that of the cemented titanium-carbide, but combined with excellent wear
resistance and negligible cratering. Optimum properties
are to be obtained at medium TiC contents in the carbonitride, pure titanium nitride or carbon-poor carbonitride
just being a bit low in hardness. The metallographic section
(Fig.11) shows the fine grained Ti(C,N) particles embedded
in a nickel binder phase. The (gray coloured) titanium-
carbonitride particles have lighter seams, probably caused
by preferential solubility of TiC in nickel.

Not withstanding its lower hardness, titanium-nitride
has proved to be very efficient as a coating material on
cemented carbides. In joint research work with Schoof and
Mariacher (sponsored by Deutsche Edelstahlwerke Krefeld,
Western Germany) we have developed a process of chemical
vapor deposition of titanium-nitride and other nitrides and
carbonitrides on the surface of cemented carbides. This
process is based on work of Münster and Ruppert (20) who
have studied the deposition of wear resistant coatings on
steel. Titanium-nitride coatings seem to be even superior
to titanium-carbide with respect to resistance to cratering
and galling. Wear life can be prolonged 5 to 10 times
as compared with standard cemented hard alloys. Coatings of
titanium-carbide are said to have a prolongation effect of
about 3 times.

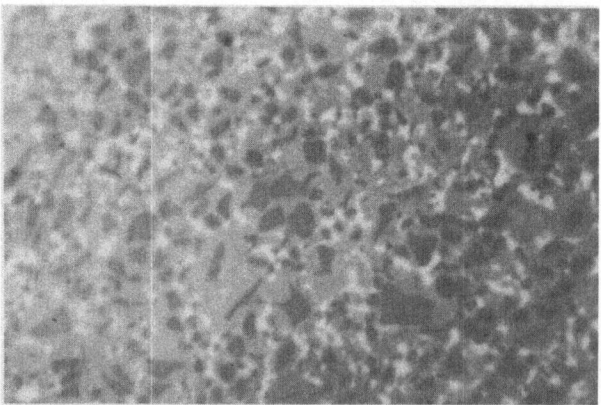

Fig.11: Metallographic section of a cemented titanium-
        carbonitride;   Magnification: 2000 x

Table 2: Transverse rupture strength and hardness of
          cemented titanium-carbonitrides

Composition (wt %)					
TiN	TiC	Ni–Mo	$T.R.S_2^{+)}$ kp/mm^2	$HV_{3.2}$ kp/mm^2	Colour
90	–	10	80	1050	yellow–golden
81	9	10	115	1230	yellow–golden
72	18	10	130	1490	red–golden
54	36	10	110	1700	bronze
36	54	10	85	1810	light–gray
18	72	10	85	1910	dark–gray
–	90	10	110	2020	dark–gray
69,0	17,0	14	135	1380	red–golden
51,5	34,5	14	125	1550	bronze
34,5	51,5	14	115	1630	light–gray
17,0	69,0	14	125	1720	dark–gray
–	86,0	14	130	1860	dark–gray
51,5	34,5	14 Fe	120	1500	bronze
51,5	34,5	14 Co	120	1500	bronze
51,5	34,5	14 Ni	125	1490	bronze

$^{+)}$ $\pm$ 10 kp/mm^2

SUMMARY

After a short discussion of the physical properties of
nitrides and carbonitrides of the transition metals some
comments on the stability of carbides against nitrogen of
high pressure are made. All carbides of the transition
metals react with nitrogen by formation of carbonitrides
with the exception of TaC and WC, which are stable against
nitrogen even of 300 at.

Sintering experiments have shown titanium–nitride and
carbonitrides to be suitable hard materials for cemented
hard alloys. 70 wt % nickel–30 wt % molybdenum alloy binder
phase gives optimum hardness and transverse rupture strength.
Cobalt and iron are suitable binder metals too, but are inferior to nickel–molybdenum alloys.

Strength and hardness values of cemented titanium–
carbonitride are comparable to those of TiC/Mo_2C–Ni alloys.
Cutting performance is excellent. Resistance to cratering
is believed to be superior to analogous cemented carbides.

LITERATURE

1) R.Kieffer, F.Benesovsky: Hartstoffe, Springer Wien 1963

2) H.J.Goldschmidt: Interstitial Alloys, Butterworths
London 1967

3) O.Meyer: Ber.dtsch.Keram.Ges. 11 (1930) 333

4) A.Schneider, R.Gehrke, M.Kretschmer, M.Wassermann:
Metall 23 (1969)

5) I.C.Kraitzer, I.E.Newnahm: Aust.J.Appl.Sci.Vol.7
(1956)215-224

6) V.K.Kazakov: Poroskovaya Metallurgiya 10 (1965)34,80-84

7) P.C.Yates, E.I. du Pont: U.S. 3,409.418 (1968)

8) P.C.Yates, E.I. du Pont: U.S. 3,409.419 (1968)

9) P.C.Yates, E.I. du Pont: U.S. 3,409.417 (1968)

10) C.Agte, K.Moers: U.S. 1,895.959 (1933)

11) J.Hinnüber: U.S. 2,077.239 (1937)

12) K.S.Seljesaeter: U.S. 2,108.618 (1938)

13) R.Kieffer, H.Nowotny, P.Ettmayer, M.Freudhofmeier:
Mh.Chemie 101 (1970)

14) R.Kieffer, P.Ettmayer, Th.Dubsky: Z.Metallk.58 (1967)560

15) P.Ettmayer, G.Vinek, H.Rassaerts: Mh.Chemie 97 (1966)1258

16) P.Ettmayer, H.Priemer, R.Kieffer: Metall 23 (1969) 307

17) P.Schwarzkopf, R.Kieffer, I.Hirschl: Ö.P. 160 172 (1931)

18) R.Kieffer, D.Fister: Paper presented at the BISRA/ICI
Conference at Scarborough 1970

19) W.D.Kingery, F.A.Harden: Ceramic Bulletin 39 (1955) 117

20) A.Münster, W.Ruppert: Z.Elektrochem.57 (1953) 564

21) A.N.Zelikman, N.N.Gorowitz: J.Prikl.Khim. 23(1950) 689

22) K.I.Portnoi, Yu.V.Levinskii: J.Fiz.Khim. 37 (1963) 2627

COMPOSITION AND PROPERTIES OF THE BINDER METAL IN COBALT

BONDED TUNGSTEN CARBIDE

O. Rüdiger, D. Hirschfeld, A. Hoffmann,
J. Kolaska, G. Ostermann, J. Willbrand

Fried.Krupp GmbH, Zentralinstitut für Forschung

und Entwicklung, Essen (BRD)

The properties of tungsten carbide-cobalt hard metals are determined by three parameters: a) properties of the carbide
phase, b) properties of the cobalt binder phase and c) interaction between carbide and binder. The present paper is intended as a contribution to the problem of the binder metal.
It is known that cobalt is not present in the hard metal in pure
form but as a solid solution with tungsten and carbon because
of its ability to solve considerable amounts of tungsten carbide at sintering temperature. Since the solubility of WC in Co
increases substantially with temperature, the composition of
the binder phase at room temperature can be expected to be determined by many features such as sintering temperature and time,
rate of cooling, particle size of the carbide used and,
consequently, the mean free path in the binder.

For investigating the composition of the binder phase in WC-Co
alloys the carbide was removed electrolytically so that the remaining binder could be tested analytically. The chemical analysis can be checked by determining the lattice constants of
powder specimens.

Moreover, the work is intended as a contribution to the question of the mechanical properties of cobalt-tungsten-carbon alloys. For this purpose, a number of alloys in the range of compositions of interest were prepared and mechanically tested.
The results of the measurements, however, must be regarded with
care when seen with respect to the conditions in sintered carbides, because the binder in hard metals is present in very thin
layers with considerable internal thermal stress. Nevertheless,
they give an idea of the excellent mechanical properties of the
binder.

INVESTIGATION OF THE AUXILIARY METAL OF WC-CO-ALLOYS

Electrochemical isolation of the binder phase was mentioned independently by Hinnüber and Rüdiger[1]as well as by Tumanov et
al.[2]at almost the same time. It is based on the different
electrochemical behaviour of carbide and binder. Tungsten and
cobalt of the isolate were determined by X-ray fluorescence
analysis. The carbon and oxygen contents were determined by
usual methods. X-ray analysis of the isolate showed that the
specimens neither contained impurities in the form of remnants
of WC nor of tungsten oxide hydrate. X-ray analysis further
showed that cubic as well as hexagonal Co were present.

Influence of the Carbon Content

The mechanical properties of WC-Co hard metal are known to depend, inter alia, on the carbon content. This question has been
examined by Gurland[3], for instance, and in recent years particularly by Suzuki and co-workers[4]. According to [5], [6] , the
tungsten content of the binder depends upon the total carbon
content. Therefore, some tests were carried out with WC-25Co-al-
loys sintered in same manner, but having different carbon percentages. The chemical analysis of the isolates shows that the
amount of tungsten in the isolate decreases as the carbon content of the specimen increases. The carbon-content of the isolate increases correspondingly from 0.12 % to 1.30 %. In Fig. 1,
the favourable influence exerted by tungsten dissolved in the
binder upon transverse rupture strength can clearly be seen in
the two-phase range WC-Co. In the three-phase regions WC-Co-η
and WC-Co-C, the strength decreases. These results are in good
agreement with the investigations of other workers[3] and [4].

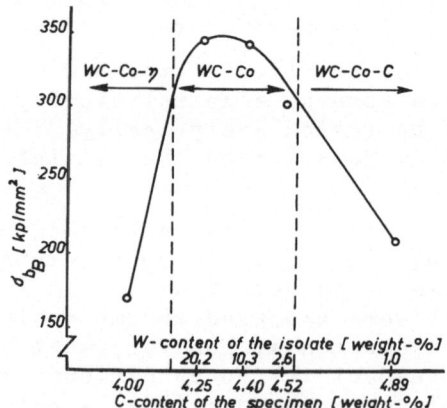

Fig. 1. Transverse rupture strength as function of
        C-content of WC-25Co sintered carbides.

Influence of the Cobalt-Content and WC-Grain Size

As already mentioned in the introduction the sintering temperature and, inter alia, the grain size of WC can be expected to
influence the composition of the binder. For this reason, alloys were prepared in which both the grain size of the tungsten
carbide used as well as the cobalt content were changed. WC-pow-
der was prepared either from ammonium paratungstate or from W-
powder, and the resulting WC-particle sizes determined by the
FSSS-method were 1.2 μm respectively 4.1 μm. After grinding the
surface of the sintered specimens containing 7, 11 and 15 % Co,
the binder was isolated. The amount of tungsten dissolved in the
binder is plotted in Fig. 2 against the Co-content, grain size
of WC, and heat treatment. It can be seen that if the coarse
tungsten carbide is used, the amount of tungsten in the binder
phase is always higher than in the case of the finer carbide.
With increasing cobalt-content, the amount of tungsten decreases
after sintering and subsequent furnace cooling (200 grd/h) while
it remains unchanged in the case of the additional heat treatment (50 h at 1250 deg. C) with subsequent quenching in water.
In the case of tungsten carbide of finer particle size, this additional heat treatment causes an increase in the amount of dissolved tungsten, while it decreases in the case of coarser WC
with the exception of the alloy containing 15 % Co.

The possibility that the different origins of the tungsten raw
materials influence the described results could not be excluded.
Therefore alloys containing 8 and 15 % Co were made, but using
the same tungsten powder for both a fine (2.45 μm) and a coarse
(5.2 μm) WC, made by different carburation temperatures. Sintering periods were 1, 3, 10, 30 and 50 h. Care was taken to
make sure that the carbon content was kept unchanged for comparable alloys and different sintering periods.

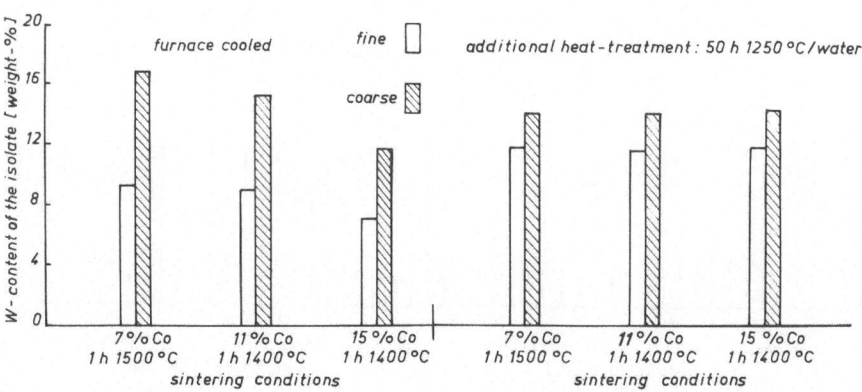

Fig. 2. W-content of the binder. Parameters: Co-content of the
alloys, WC grain size of the powder, heat treatment.

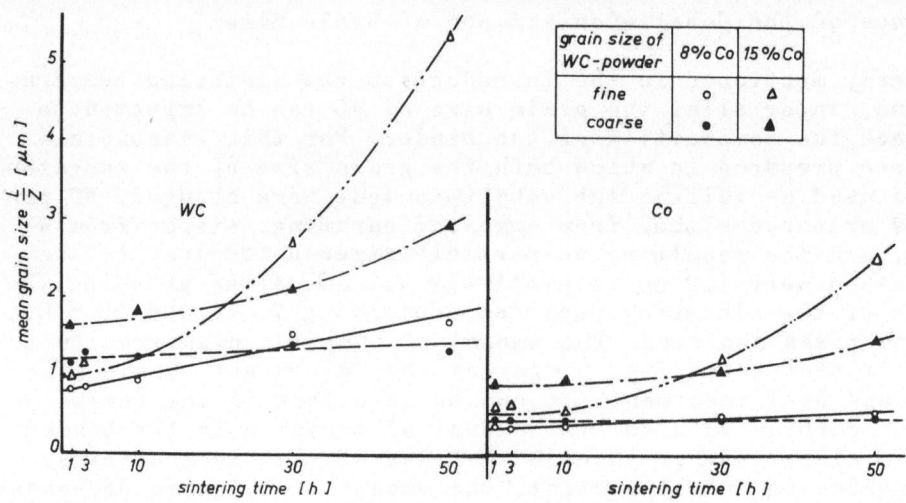

Fig. 3. WC and Co mean grain size as
        functions of sintering time.

Stereometrie analysis of the particle size of WC[7] shows that
the mean grain size increases with sintering time (Fig. 3). At
a low cobalt content, the average grain size of WC increases
linearly as a function of sintering time, while in the case of
high Co-contents a faster grain growth can be observed. Between
20 and 30 h the two curves representing the two WC-grain sizes
intersect each other, i.a. the finer WC-grains grows more rapidly than the originally coarser grains. The mean grain size of
cobalt shows always an analog change to that of the WC-crystals.

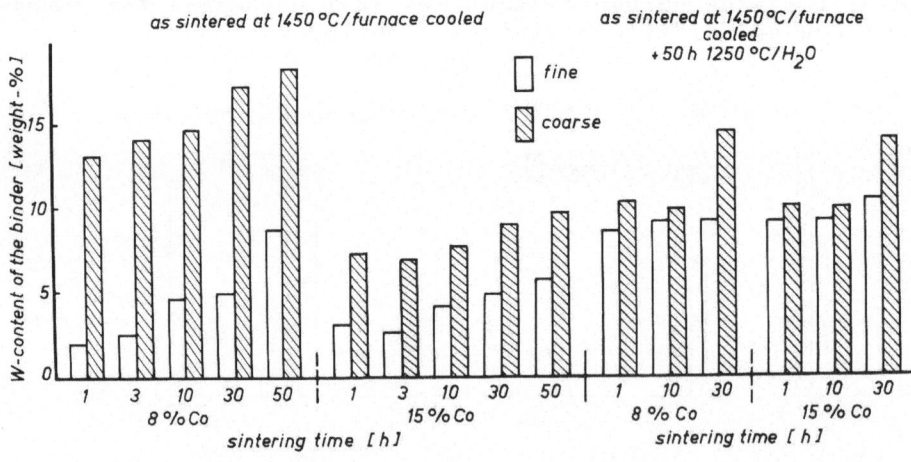

Fig. 4. W-content of the binder. Parameters: Co-content of the
        alloys, WC grain size of the powder, heat treatment.

The percentages of tungsten determined chemically in the binder (isolate) are shown in Fig. 4 as functions of grain size of tungsten carbide, cobalt content, sintering time and additional heat treatment. As described above, the specimens with initially coarse WC, when furnace-cooled from sintering temperature, again contain an appreciably higher content of W in the binder phase than the alloys with initially finer WC. In both cases, the amount of W increases with the sintering time. The result that the specimens having higher Co-contents generally exhibit a lower W-content in the binder also is in agreement with the results of the first series.

Even in the case of the specimens annealed for 50 h at 1250 deg. C and subsequently quenched in ice-water, there remains a slight difference in tungsten-content between the two initial grain sizes. The influence of different sintering times and cobalt-contents is almost eliminated by such a heat treatment.

Some isolates were tested by X-ray diffraction in order to determine the amounts of cubic and hexagonal cobalt and the lattice constants of the cubic Co-phase. Is was found that the percentage of cubic cobalt is higher in the 8 % Co specimens than in those containing 15 % Co. In the initially coarse-grained alloys, the amount of the cubic phase is larger than in the case of the fine-grained materials. The lattice parameters are in good agreement with calculated values based on the analysed composition of the isolates.

Using the methods available, neither free carbon nor carbon-deficient or intermetallic phases could be found in the WC-Co-alloys investigated. Assuming that such phases are in fact not present, the total amount of tungsten which should be dissolved in the binder phase, can be calculated from the total tungsten, cobalt and carbon content of the alloys and from the carbon percentages determined in the isolates. Comparison of the calculated with the determined values will show that there is good agreement between calculated and determined tungsten contents in the initially coarse-grained alloys regardless of sintering period and Co-content of the alloys. In contrast to this, the percentages of tungsten determined in the isolates of the fine-grained alloys are substantially below the calculated percentages. In 8 % Co alloys, this difference is on average 13 % absolute, in the 15 % alloys it is on average 6 % absolute. The following table shows some examples.

Type of alloy	Sintering time [h]	Co of alloy [%]	C of alloy [%]	C of isolate [%]	W in binder calc. [%]	W in binder exper. [%]
coarse	50	8	5.55	0.07	18.8	18.3
fine	50	8	5.52	0.06	22.4	8.8

This result makes it difficult to interpret the processes oc-
curing during sintering and cooling. One explanation might be
that the amount of tungsten lacking in the binder phase precipitates during cooling as an intermetallic or carbon-deficient
phase at the WC/Co interface and is solved during isolation.

The following results from the measurements can be stated:

1. Regarding only the mean grain size, without respect to the
   influence of the grain distribution, in the 8 % Co alloys a
   nearly linear and in the 15 % Co alloys a square growth law
   of the WC-crystals is found.
2. The amount of tungsten dissolved in the isolate is affected
   by the initial grain size. This applies both to the furnace-
   cooled and to the water-quenched specimens, although in the
   latter condition the difference in tungsten becomes smaller.
3. With increasing cobalt-content of the alloy, the amount of
   tungsten dissolved in the binder decreases in the slowly
   cooled specimens. Additional heat treatment for 50 h at
   1250 deg. C and subsequent quenching in water eliminates
   these differences.
4. With increasing sintering time, the W-content of the binder
   phase increases in all furnace-cooled specimens investigated, while the quenched specimens behave as described in 3,
   with exception of the coarse alloys sintered 30 h.

Discussion of the Results obtained in the Tests

Recently Exner and Fischmeister [8] have carried out investigations into the coarsening of grain structures and have found
that the mean grain size grows as $t^{1/2}$ or as $t^{1/3}$, depending
on whether continuous or discontinuous grain growth is considered. This is in disagreement with the present results ($D \sim t$
or $D \sim t^2$). Furthermore Exner and Fischmeister [9] observed
grain growth to be independent of cobalt content; we found –
in agreement with Gurland [3] – that the cobalt content exerts
a considerable influence.

The composition of the binder phase after cooling to room temperature is not only determined by the conditions in the specimen at sintering temperature, but also by the cooling rate from
sintering temperature.

Increasing supersaturation of the solution with decreasing temperature can only be eliminated by precipitation of excess
tungsten and carbon. The rate of precipitation may be determined by the diffusion paths in the binder phase or by the total
surface area of the WC grains adjacent to the cobalt phase,
because the existing WC-Co-interfaces can serve as precipitation nuclei. In both cases, less supersaturation should be expected in the case of the fine-grained alloys, i.e. a smaller

content of W in solution. If the mechanism suggested above is alone decisive, then the intersecting of the grain size/sintering time curves as found in Fig. 3 should again be found in the composition of the binder phase (Fig. 4). This is not the case, however. The dependence of the composition of the binder phase upon the content of cobalt does not fit into the above pattern either.

PROPERTIES OF COBALT–TUNGSTEN–CARBON ALLOYS

Pure cobalt alloys were produced in vacuum induction furnace by the carbon reduction process [10]. The binary alloys of cobalt and carbon contained 0.26 to 0.85 % C and those of cobalt and tungsten contained 0.21 to 13.1 % W. The ternary alloys contained 0.98 to 20.2 % W with 0.015 to 1.08 % C. The alloys were forged into 15 mm square bars and rods at temperatures between 1100 und 1180 deg. C. Three tensile test pieces conforming to B6 of DIN 50125 but with a gauge length of 41 mm were machined from the forged bars. These test pieces were annealed for 4 h at 850 deg. C in an argon atmosphere with furnace cooling. Tensile strength and elongation at room temperature were determined conventionally. The melting process, the purity, the micro-structure and the forgeability of the alloys discussed here will be described in detail elsewhere [11] .

Results of the Tensile Tests

In the cobalt–carbon–alloys, the carbon content was not observed to exert any influence on the mechanical properties. The values correspond largely to those of pure cobalt [12].

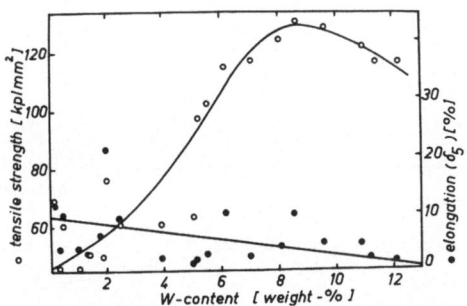

Fig. 5. Tensile strength and elongation as function of the tungsten content of Co–W–alloys with <0.01 % C (4 h 850°C argon/furnace).

The <u>cobalt-tungsten</u> alloys exhibit on average a clear dependence of the mechanical properties on the tungsten content after
annealing for 4 h at 850 deg. C in argon. In Fig. 5, the tensile strength and the elongation are plotted against the tungsten content. The tensile strength is increased by increasing
additions of tungsten up to about 130 kp/mm^2 at 9 % W, but beyond this drops distinctly to about 115 kp/mm^2 at 11.5 % W. The
elongation decreases from about 8 % to about 2 % as the tungsten
content increases from 1 % to 12 %. The initial increase in the
tensile strength with increasing tungsten content is attributable to the solid solution hardening of tungsten. At higher tungsten contents, the occurrence of the Co_3W phase after annealing
at 850 deg. C probably is of influence.^{3}

The addition of <u>carbon</u> to <u>cobalt-tungsten</u> alloys causes a particularly remarkable change in the elongation. The elongation
averages about 20 % at 1 % W and 10 % at 6 % W with carbon percentages from 0.015 % to 0.048 % (Cf. Fig. 5 for the binary alloys). As the carbon content increases to 0.050 till to 0.20 %,
elongation is about 25 % at 2 % W, about 15 % at 12 % W and
about 7 % at 20 % W. If the carbon content is further increased,
elongation as a function of tungsten content again decreases
(Figs. 6b and 6c for alloys with 0.21 to 0.40 and 0,41 to 0.60 %
C). For instance, elongation still is about 13 % and 9 % at 12 %
W for 0.21 to 0.40 and 0.41 to 0.60 % C, respectively. At carbon contents from 0.61 to 1.08 %, elongation is again down to
the values for the binary alloys, averaging 3.7 % for tungsten
contents between 9.7 and 15.5 %.

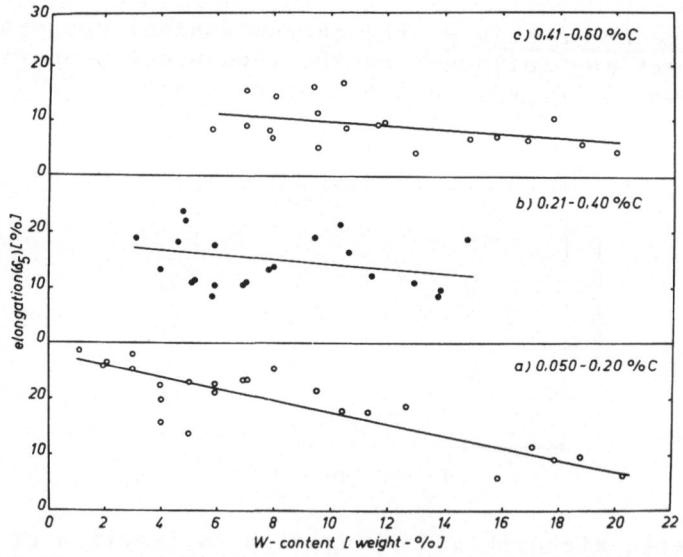

Fig. 6. Elongation as function of the tungsten content
    of Co-W-C-alloys (4 h 850°C argon/furnace).

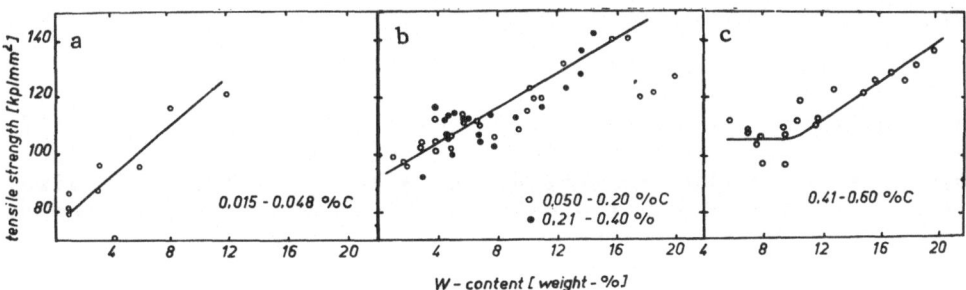

Fig. 7. Tensile strength as function of the tungsten content
of Co-W-C-alloys (4 h 850°C argon/furnace)

The tensile strength (Fig. 7a to 7c) slighty decreases for more
than 10 % W as the carbon content increases. In this case the
carbon values were combined for the range from 0.050 to 0.40 %.
The tensile strength is for 0.015 to 0.40 % C higher than in
the case of the binary cobalt-tungsten alloys containing up to
5 % W (Cf. Fig. 5) For carbon contents from 0.050 to 0.40 %,
Fig. 7b shows the mean-value curve only to about 17 % W, since
the values drop above this tungsten content. In the range from
0.61 to 1.08 % C, no dependence on tungsten content can be dis-
cerned. At 9.7 to 15.5 % W, the values vary between 86.1 and
137.3 kp/mm².

Discussion of the Results Obtained from the Model Alloys

The results of the experiments described above in combination
with the micro-structure [11] can be summarized as follows:

The tensile strength of pure cobalt is substantially improved
by the addition of W as an alloying element, while elongation
decreases (Fig. 5). By adding W and C as alloying elements,
elongation increases for small C-contents, but at higher carbon
values and a constant W-content it drops again.

The results obtained in the tensile testing of cobalt-tungsten-
carbon alloys (Cf. Figs. 6 and 7) cannot be explained generally
on the basis of the amounts of cubic and hexagonal crystals, of
the solubility of tungsten and carbon in the solid solutions,
as well as of the precipitated phases (WC, graphite, η-phase
and Co_3W). The decrease of the elongation with an increase in
carbon content may be attributable to the precipitation of WC
and graphite. The increase of the tensile strength beginning
from a tungsten content of about 10 %, as shown in Fig. 7c, is
possibly caused by the higher amount of cubic crystals and the
precipitation of WC and graphite. The occurrence of the η- or
the Co_3W-phase cannot be clearly related to the observed changes in the mechanical properties.

Summary

For investigating the composition of the binder phase in WC-Co-alloys, the carbide was removed electrolytically so that the remaining binder could be tested analytically and by X-ray diffraction. It was found that the tungsten content of the binder depends on the carbon content of the alloys, on the grain size of the carbide used, on the cobalt content and on the sintering and cooling conditions. The tungsten content of the binder increases as the carbon content of the alloys decreases, as the grain size of the carbide increases, as the sintering time increases, and as the Co-content of the alloy decreases. Additional heat treatment for 50 h at 1250 deg. C followed by water quenching almost eliminates the differences regardless of the initial grain size of the carbide, of the Co-content and of the sintering time. The W-contents calculated for the binder of binary WC-Co-alloys (free from other phases) from the total carbon contents of the alloys together with the C-values of the corresponding isolates, are in good agreement with the measured W-values for the coarse grained materials. However they differ for the fine grained alloys.

In the range of composition of interest here, the tensile strength and the elongation of some melted, forged and heat-treated model alloys of the Co-W-C-system were determined. It was found that tensile strength of more than 120 kp/mm^2 at elongations of about 10 % can be achieved.

An explanation for the results satisfactory in all respects is still outstanding. - The paper is to be published in detail in Techn. Mitt. Krupp (Forschungsber.) Vol. 28, 1970.

Literature

[1] J. Hinnüber, O. Rüdiger; Kobalt Nr. 19 (1963) 56/65 –
[2] V. J. Tumanov, Z. S. Truchanova, V. G. Ŝĉerbakow; Zavodskaya Laboratoriya (1963), Heft 3, 277/280 – [3] J. Gurland; J. of Metals (1954) 285/290 – [4] H. Suzuki, H. Kubota; Planseeber. Pulvermetallurg. 14 (1966) 96/109 – [5] A. Nishiyama, R. Ishida; Trans. Japan Inst. Metals 3 (1962) 185/190 – [6] H. Kubota, R. Ishida, A. Hara; Trans. Indian Inst. Metals 17 (1964) 132/138 – [7] E. Hillnhagen, J. Willbrand; Z. "Praktische Metallographie" 6, Heft 3 (1969) 135/144 – [8] H. E. Exner, H. Fischmeister; Z. Metallkde. 57 (1966) 187/193 –
[9] H. E. Exner, H. Fischmeister; Arch. Eisenhüttenwes. 37 (1966) 417/426 – [10] A. Hoffmann, W. A. Fischer; Arch. Eisenhüttenwes. 37 (1966) 221/226 – [11] O. Rüdiger, A. Hoffmann; Metall, in print – [12] O. Rüdiger, J. Burbach, A. Hoffmann; Techn. Mitt. Krupp · Forsch.-Ber. 24 (1966) 61/78

HIGH-STRENGTH TUNGSTEN CARBIDES

D. Moskowitz, M. J. Ford and M. Humenik, Jr.

Scientific Research Staff, Ford Motor Company

Dearborn, Michigan

INTRODUCTION

It is well known[1,2] that the mechanical properties of sintered tungsten carbide-cobalt alloys are critically dependent upon carbon content. Only when the carbon content corresponds closely to the theoretical value for WC, i.e. 6.13 weight %, are two-phase structures (WC and a Co-rich solid solution) observed. The optimum mechanical properties, as well as product performance, are obtained when only these two phases are present. It is found that deviation of the carbon composition results in the appearance of a third phase, with consequent inferior properties. In the case of a carbon excess, the third phase that appears is graphite, with a resultant lowering of strength and hardness. Deficiency of carbon below the stoichiometric value for WC, on the other hand, produces the double carbide $Co_3W_3C(\eta)$, with markedly inferior strength and impact resistance.

If similar procedures are employed in the preparation of iron-bonded tungsten carbide alloys, so that WC of theoretical carbon content is bonded with Fe, it is found that the analogous double carbide Fe_3W_3C still forms. This explains the lack of success in attempts to replace Co with Fe or Ni in WC-base materials. According to Schwarzkopf and Kieffer[3], "Iron or nickel-bonded tungsten carbide exhibited not more than about 40-60% of the transverse rupture strength of Co-bonded material, . . ."

Although sintered tungsten carbide-iron alloys are notable for their lack of commercial utilization, it has been reported[4] that additions of free carbon to Fe-WC alloys exceeding the stoichiometric amount required for the tungsten carbide can inhibit

225

the formation of large grains of the brittle "eta" phase. It has
also been shown by Agte[5] that cobalt could be successfully replaced
by 3:1 Fe-Ni alloys if "over-carburized" tungsten carbide is used
as the starting material. Comparable transverse rupture strengths
were reported by him for the Fe-Ni and Co-bonded WC alloys. However, studies by Suzuki, et al,[6,7] on the effect of carbon content
on the properties of Fe-WC and Fe/Ni-WC alloys have shown inferior
strengths for these systems compared to Co-WC.

In this paper, we will discuss the results of a more detailed
study of the influence of carbon content on the phase relationships
and properties of sintered Fe-WC alloys. In addition, studies on
Fe-Ni bonded WC alloys will be presented which show that strength
levels are attainable in this system which are considerably in excess of those reported for cobalt-bonded compositions.

EXPERIMENTAL PROCEDURE

Compositions were prepared by milling under benzene one
micron particle size tungsten carbide powder and hydrogen-reduced
electrolytic iron, with additions of electrolytic nickel and graphite powders as needed. Paraffin or polyethylene glycol lubricants were employed to facilitate pressing of the powder compacts.
The pressed specimens were dewaxed at 750°F under dry hydrogen, or
vacuum, and vacuum sintered for one hour at 2550-2600°F. During
sintering, the compacts were supported by crystalline WC grain on a
graphite substrate.

Transverse rupture strength was measured following the procedure outlined in ASTM No. B406-63T, with average values calculated using a 95% confidence factor[8]. Specimen density was
measured by immersion under tetrabromoethane. The procedure
followed for abrasion resistance is described in CCPA No. P-112.
Curie temperature was determined by measuring the force required
to separate a magnet from the specimen as a function of temperature[9].

IRON-TUNGSTEN CARBIDE MATERIALS

Composites prepared from Fe and WC with no excess carbon addition show the presence of large amounts of "eta" phase. From X-ray
analysis, the "eta" phase is identified as Fe_3W_3C[10], - a face-
centered cubic structure with a lattice parameter of 11.04 Å. In-
cremental additions of free carbon decrease the concentration of
"eta" and change its morphology from randomly-shaped particles to
polygonal agglomerates as large as 100 microns in size. Further
additions of graphite completely eliminate the formation of
Fe_3W_3C, until a point in carbon content is reached at which star-
shaped clusters of graphite make their appearance. Figures 1a

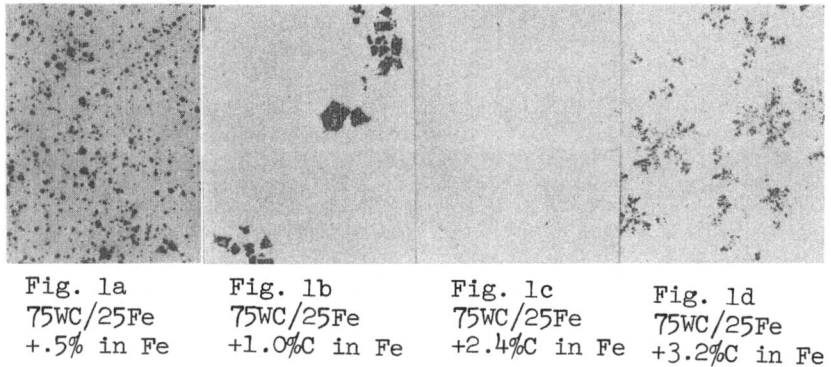

Fig. la   Fig. lb   Fig. lc   Fig. ld
75WC/25Fe   75WC/25Fe   75WC/25Fe   75WC/25Fe
+.5% in Fe   +1.0%C in Fe   +2.4%C in Fe   +3.2%C in Fe

through ld illustrate these structures for a composition containing 25 wt.% iron with 0.5, 1.0, 2.4, and 3.2 wt.% carbon respectively, relative to the iron. These carbon contents, which are
relative to the iron binder, represent excess carbon - over and
above the amount required to satisfy stoichiometric WC, and are
calculated from chemical analyses of the total carbon, tungsten,
and iron content of the sintered compositions.

Figure 2 shows the variation in hardness and density for
the 25 Fe-75 WC compositions as a function of excess carbon. It
can also be noted that the desired microstructure, free of "eta"
phase and graphite, is shown for a carbon range of about 1.4 to
3.0 wt.%. It is believed that the carbon associated with the
binder phase in this range is combined as Fe_3C, since Curie point
measurements (Figure 3) clearly show cementite to be present in
these materials. From these results, it is postulated that WC is
not stable in the presence of low-carbon austenite at elevated temperatures. Under these conditions, the more stable carbon-saturated
austenite would be obtained by solution of carbon resulting from
dissociation of WC to W_2C + C, subsequently producing Fe_3W_3C on
cooling. It can be noted in Figure 2 that the "eta" phase is completely inhibited at a carbon content in iron of about 1.4 wt.%.
This is in the range of maximum solubility of carbon in austenite,
particularly if the austenite contains some residual tungsten in
solid solution. The decrease in hardness observed from 0 to
1.4% C is attributed to decreasing amounts of the hard Fe_3W_3C phase,
while the increase in hardness from 1.4 to 3.0% C is due to increasing amounts of Fe_3C in the binder phase. At greater additions,
the sharp drops in hardness and density result from the formation
of graphite.

Iron-carbon bonded WC compositions, free of "eta" or graphite, were prepared covering the range from 6 to 30 wt.% Fe, with

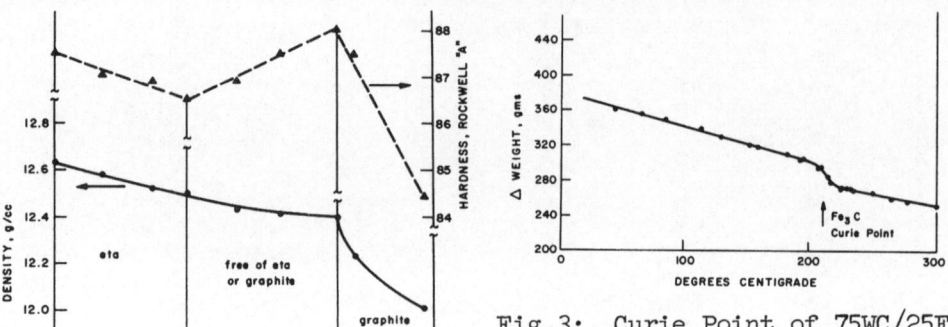

Fig.3: Curie Point of 75WC/25Fe
Composition Free of "Eta" or
Graphite

Fig. 2: Hardness and Density vs. Carbon Content Relative to Binder for
75WC/25Fe.

an average sintered carbide grain size of 1-2μ. Transverse rupture strength and hardness of these materials vs. binder content
are shown in Figure 4. The strength values obtained for the Fe-
WC system (Figure 4) are comparable to those reported for WC-Co
compositions of similar grain size[11]. It was also found that, for
Fe-WC materials free of "eta" or graphite, the strength was not
significantly influenced by carbon content.

IRON-NICKEL BONDED TUNGSTEN CARBIDE

(a) Hardness

If Fe-Ni alloy binders are used instead of Fe, the amount of
excess carbon required to prevent "eta" from forming is significantly lowered. For example, with an Fe20Ni binder alloy, only
0.5% carbon is required to completely inhibit "eta" formation, as
contrasted to 1.4% for iron. In addition, the upper limit for carbon contents at which graphite begins to appear is lowered from
3.0% for iron to about 1.0% for the Fe20Ni binder.

Alloying the iron binder with nickel has a pronounced hardening effect, which is particularly marked at high binder levels.
Figure 5 shows the hardening effect obtained for 25% binder compositions as a function of nickel contents ranging from 0 to 40
wt.% of the binder phase. The data is presented for specimens as
sintered (furnace cooled) and also after a refrigeration treatment
at -196°C, followed by tempering at 200°C. The influence of nickel
appears to be generally consistent with it's behavior in steels[12,13].
Considering the as sintered condition, the hardness increase with
up to 5% Ni is attributed to solid solution hardening of ferrite
and an increase in the amount and fineness of pearlite in the binder

phase. With increasing nickel contents up to about 10%, an austenitic binder phase begins to appear which transforms to martensite
during furnace cooling, thereby, maintaining a relatively high
hardness level. Further Ni addition **lowers the martensite** transformation temperature (M_s) to below room temperature, which results in
progressively larger amounts of retained austenite and consequently
decreasing hardness. The hardness remains constant in the nickel
range from 30 to 40 weight%, resulting from an austenitic binder that
is not subject to a martensite transformation at temperatures down to
-196°C. The additional hardening shown in Figure 5 for compositions
from 10 to 25% nickel that have been subjected to a low temperature
treatment at -196°C is due to additional transformation of the retained austenite in the furnace cooled specimens. However, the
decrease in hardening from 10 to 25% for the refrigerated specimens
probably results from incomplete transformation and hence increasing retained austenite.

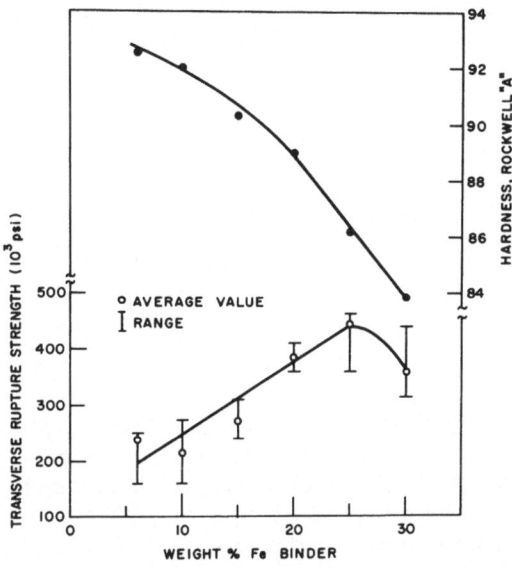

Fig. 4: Transverse Rupture Strength
and Hardness vs. Binder Content of
Iron-Bonded WC Free of "Eta"
or Graphite.

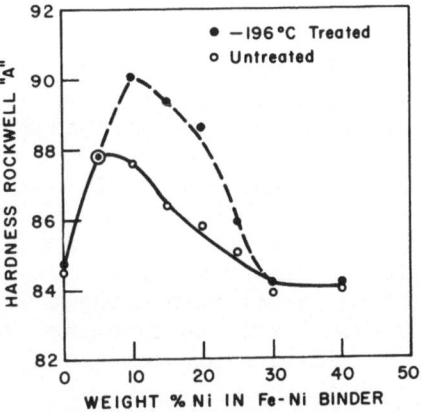

Fig.5: Hardness of 75WC/25
(Fe-Ni) vs %Ni in Binder for
As Sintered and -196°C
Treated Conditions

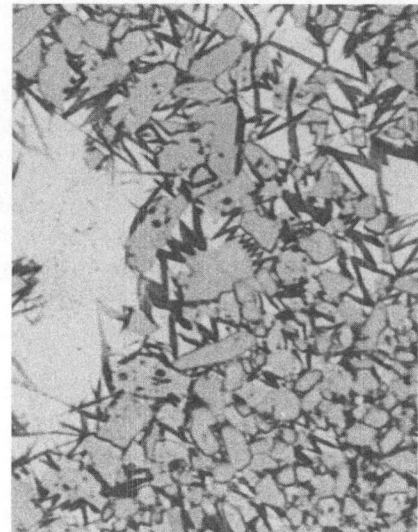

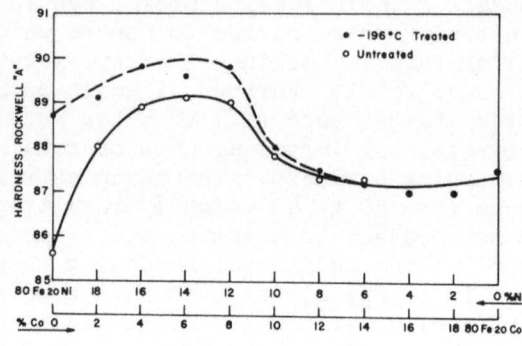

Fig.6: Martensite Needles in
75WC/25(Fe20Ni) Transformed at
-196°C and Tempered at 250°C.
Etch: 10% KoH:10% $K_3Fe(CN)_6$+4%
Nital.   x 4000

Evidence that a martensitic transformation of the binder is
indeed taking place is given in Figure 6, which is a photomicrograph of a coarse-grained 25% Fe20Ni-bonded composition that has
been transformed at -196°C and then tempered at approximately
250°C. The dark needle-shaped structures that show up after
etching with 4% Nital are typical of martensite plates. The fact
there is a measurable expansion after low-temperature treatment
is further evidence of this transformation. Finally, measurement
of M_S temperature using the method of saturation magnetization
gives results that compare reasonably with values calculated using
the relationship between M_S temperature and alloy content in
steel[14].

Hardness increases, similar to the effect of the low temperature treatment, can also be obtained in as sintered specimens by
the partial substitution of cobalt for nickel, as shown in
Figure 7. This results from the increase in M_S temperature with
cobalt additions, which enhances the martensite transformation
during furnace cooling.

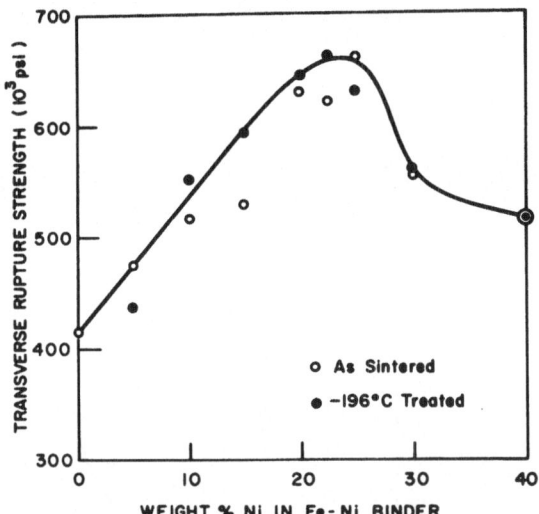

Fig.8: Transverse Rupture Strength vs. Ni Content of 75WC/25(Fe-Ni).

(b) Strength

In addition to its influence on hardness, alloying additions of nickel to Fe-WC also have a pronounced strengthening effect that is particularly effective at higher binder levels. Figure 8 shows the effect of a range of nickel contents in the binder, from 0 - 40 wt.%, on the strength of 25(Fe/Ni)-75WC compositions. The average sintered carbide grain size in these materials is 1 - 2μ, and the results are given for specimens both as sintered as well as in the -196°C treated condition for compositions that are subject to a martensitic transformation. It can be noted in Figure 8 that optimum average strength values in excess of 625,000 psi are obtained for compositions containing 20 - 25% nickel, with relatively small differences in strength for the as sintered and low temperature treated specimens. These values are about 50% greater than that obtained in the Co-WC system at a similar composition and carbide grain size[11].

The improvement in strength with up to 5% Ni addition is attributed to solid solution strengthening of ferrite, and possibly refinement of pearlite. With regard to the optimum strength at 20 - 25% Ni, we can only speculate at this time that it may result from the presence of an optimum amount of martensite and austenite, which is dependent on the nickel content and thermal treatment of the specimens. At nickel contents exceeding 25%, the decrease in strength is related to the formation of a lower strength austenitic binder that is not subject to a martensitic transformation even at -196°C.

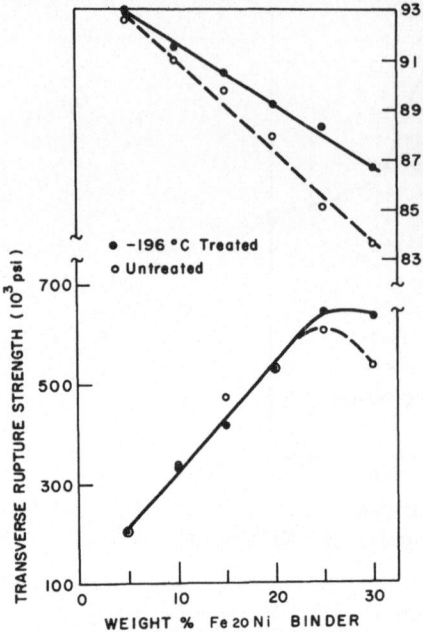

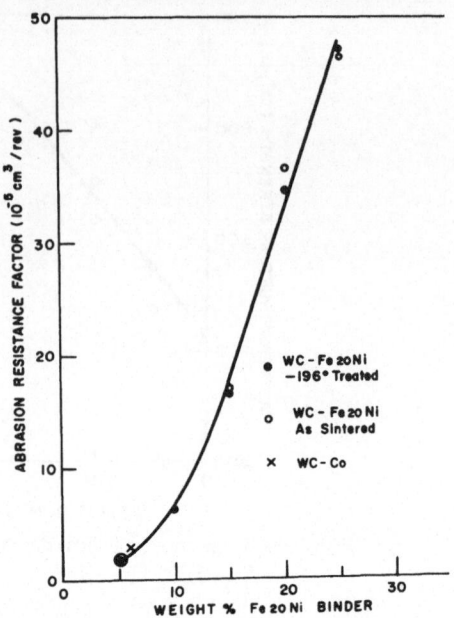

Fig.9: Transverse Rupture
Strength and Hardness
vs. Wt.% Fe20Ni Binder

Fig. 10: Abrasion Resistance
Factor vs. Weight% Fe20Ni Binder

Since optimum strengths were obtained with Fe/Ni binder compositions containing about 20 - 25% Ni, additional data was obtained on the influence of binder content on strength, using an
Fe20Ni binder alloy. The results are shown in Figure 9 for both
as sintered and low temperature (-196°C) treated specimens. In
addition, hardness data is also shown in Figure 9 for these compositions. The peak strength for this system is obtained at
about 25 wt.% binder, comparable to that observed in Co-WC with a
similar carbide grain size; however, the value of 625-650,000 psi
is considerably greater than that for Co-WC. It is also interesting to note in Figure 9 that the properties of the binder phase
appear to play a significant role on the strength of the composite
at binder contents exceeding about 25%, as evident by the higher
strength for the refrigerated specimens. Although some hardness
increases are obtained for the refrigerated specimens at all the
binder levels, this effect is also more pronounced at the higher
binder contents.

A brief study was also conducted to examine the influence of surface finish on the strength of Fe/Ni-WC. It is well known that the strength of brittle materials, in general, and cemented carbides in particular, is greatly affected by the surface condition of the test specimen. Gurland[15], working with 6Co-94WC samples, found an increase in average strength from 255 to 303 Kpsi when comparing 220 grit diamond-ground specimens with those polished with 1 micron diamond paste. We have noted similar behavior with 25(Fe20Ni)-75WC materials. One group of specimens, prepared by carefully lapping on a 15 micron diamond wheel using minimal pressure, showed an average transverse rupture strength of 780 Kpsi, with a range from 736 to 815 Kpsi. This compared with an average of 641 Kpsi and a range from 559 to 692 Kpsi for specimens that were prepared identically, except that their surfaces were ground more rapidly with greater pressure. All the above specimens had been refrigerated at -196°C and tempered at 200°C before the surface treatment.

(c)   Abrasion Resistance

Abrasion resistance was determined for Fe20Ni-WC materials as a function of Fe20Ni binder content in the range from 5 to 30 wt.%. The results shown in Figure 10 illustrate the typical decrease in abrasion resistance with increasing binder content. However, it was also unexpectedly found that no differences were obtained for as sintered (furnace cooled) and low temperature treated specimens at all binder levels, even though significant increases in hardness are obtained by the low temperature treatment, as shown in Figure 9. A possible explanation for this behavior may be that during the abrasion test a stress-induced martensite transformation takes place on the surface of the as sintered specimens. Although data have not yet been obtained for the same range of Co-WC compositions, the result for 6Co-94WC (Figure 10), with a comparable carbide grain size, shows a similar behavior to the Fe20Ni-WC system in this composition range. Further studies in this area with compositions containing a variable binder content and composition, along with a variable grain size, should provide a means for more precisely elucidating the role of the binder phase on the properties of composite materials.

CONCLUSIONS

It has been shown that the formation of the detrimental Fe_3W_3C phase in Fe-WC can be inhibited by proper control of the carbon content, resulting in mechanical properties for compositions in this system that are comparable to the Co-WC system. More significantly, it has been found that with alloying additions of

nickel to iron, bend strength values can be obtained in the
Fe/Ni-WC system that are substantially greater than comparable compositions in the Co-WC system. In addition, appropriate alloying
additions of nickel to iron provide a means of producing pearlitic,
austenitic, and martensitic-bonded WC, and further optimizing the
properties of compositions in this system aimed for specific applications.

ACKNOWLEDGEMENT

The authors wish to acknowledge the valuable assistance of
Messrs. J. L. Bell and E. J. Violante, who performed the chemical
analyses in this study.

REFERENCES

1. Gurland, J., Trans. AIME, 200, p. 285, (1954).
2. Brownlee, L. D., Edwards, R., Raine, T., Symp. on Powder
    Metallurgy, 1954, p.302, London, 1956, The Iron and Steel Inst.
3. Schwarzkopf, P. and Kieffer, R., Cemented Carbides, MacMillan,
    New York, p.188, (1960).
4. Ellis, J. L., British Patent No. 908,412.
5. Agte, C., Neue Hütte, 2, 537(1957).
6. Suzuki, H., Yamamoto, T., Kawakatsu, I., Jap. Soc. of Powder
    and Powder Met., Journ., 14 (7):308 (1967).
7. Ibid: 14, (2) 86 (1967).
8. Avery, H. S., Wear, 4, p.449 (1961).
9. Ault, G. M. and Deutsch, G. C., Trans. AIME, 200, p.1214 (1954).
10. Adelskold, V., Sundelin, A. and Westgren, A., Zeit. anorg.
    allgem. Chem., 212, p.401 (1933).
11. Gurland, J. and Bardzil, P., Trans. AIME 203, p. 311 (1955).
12. Eash, J. T. and Pilling, N. B., Trans. AIME, 150, p.294 (1942).
13. Metals Handbook, ASM, Cleveland, Ohio, p.473 (1948).
14. Grange, R. A. and Stewart, H. M., Trans. AIME, 167, p.467
    (1946).
15. Gurland, J., Powder Metallurgy, Leszynski, W. (ed.) Interscience,
    New York, p.661 (1961).

PHYSICAL PROPERTIES OF MONOCARBIDES OF ZIRCONIUM, NIOBIUM, AND THEIR ALLOYS IN HOMOGENEITY RANGE

G. S. Upadhyaya[*] and G. V. Samsonov[+]

[*]Department of Metallurgical Engineering, University of
Roorkee, Roorkee, U. P., India
[+]Institute of Materials Problems, Academy of Sciences
of Ukrainian SSR, Kiev, USSR

INTRODUCTION

At present although a number of investigations have been made on the physical properties
of monocarbides of transition metals of IV-V
groups in their homogeneity range, most of the
results do not agree with each other. In the
case of their binary alloys so far attempts have
been made to measure the properties of only
stoichiometric compositions and not many results
are available on the properties of the alloys in
their homogeneity range. With this in view, an
attempt is made in the present investigation to
measure the physical properties of monocarbides
of zirconium, niobium and their alloys in the
homogeneity range and also to explain the main
relationships of their electronic structure
with the changes in compositions.

REVIEW

Zirconium and Niobium Monocarbides

In Figs. 1 and 2 the results on variations
of some major electro-physical properties, such
as electrical resistivity, absolute thermo
e.m.f., Hall coefficient, magnetic susceptibility and high temperature electrical resis-

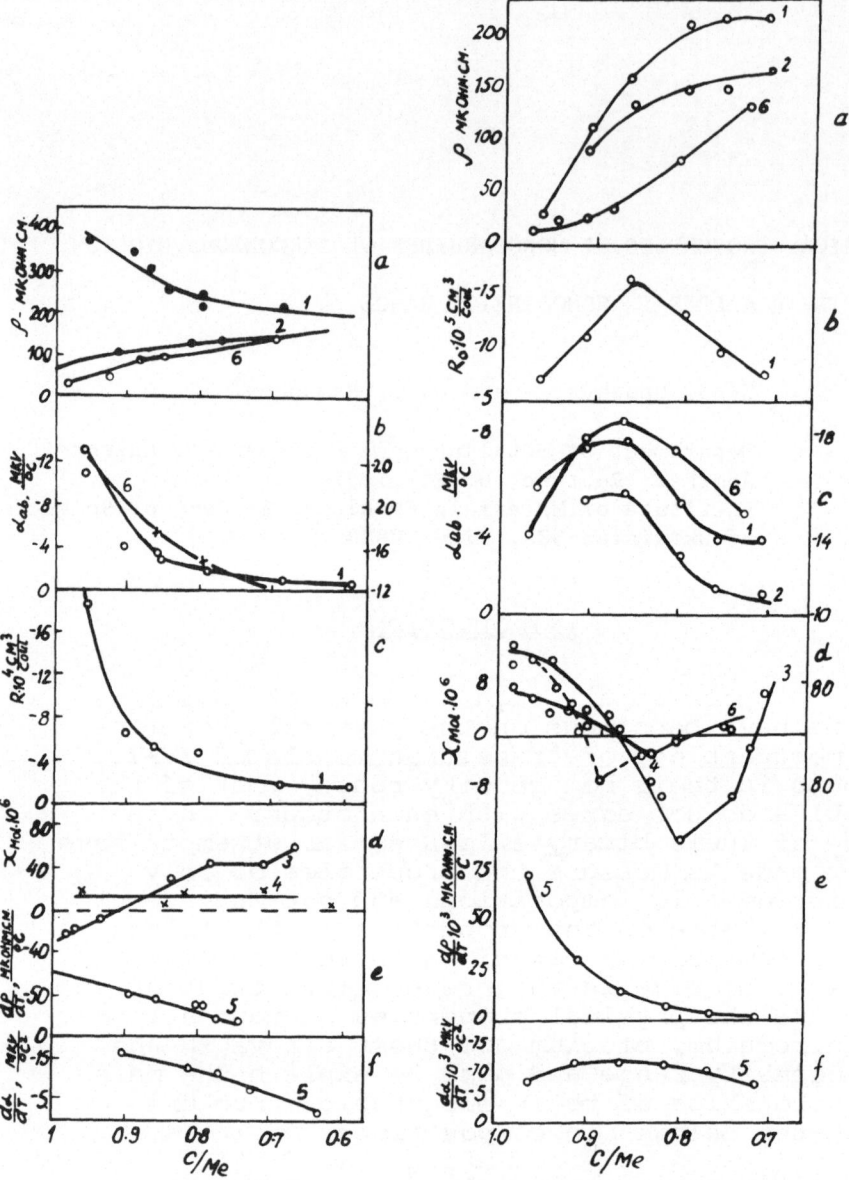

Fig. 1. Variations in electrical
resistivity (a), thermo e.m.f.
(b), Hall coefficient (c), magnetic susceptibility (d), $d\rho/dT$
(e), and $d\alpha/dT$ (f) of zirconium
monocarbide in homogeneity range.
1(3), 2(1), 3(6), 4(7), 5(11),
6(2).

Fig. 2. Variations in electrical
resistivity (a), Hall coefficient (b), thermo e.m.f. (c),
magnetic susceptibility (d),
$d\rho/dT$ (e), $d\alpha/dT$ (f) of niobium
monocarbide in homogeneity
range. 1(4), 2(1), 3(8), 4(6),
5(11), 6(2).

tivity and thermo e.m.f. changes in the homogeneity range of zirconium and niobium monocarbides have been collected from various
sources.

It is noticeable that the electrical resistivity variations of zirconium carbide as
reported by various authors are not similar in
nature. In the majority of the report (1-2)
a decrease in electrical resistivity with an
increase in carbon content is observed, although in another work (3) an opposite picture
exists. However, niobium monocarbide does not
show such an anomalous variation (1,3).

Zirconium carbide shows a decrease in its
absolute thermo e.m.f. values with a decrease
in carbon content (3), while niobium monocarbide is accompanied with a maximum in such plot
(1-2,4).

Variation of Hall coefficient of zirconium
carbide in homogeneity range shows a decrease
in its value with a decrease in carbon content
(3). The plotted values of Hall coefficient
corresponding to the stoichiometric composition
is approximate to the value obtained by Tsuchida
et al. (5)   In case of niobium monocarbide, a
maximum in Hall coefficient value is noticed at
the composition $C/_{Me}=0.85$ (4), which has been
explained by authors in terms of the formation
of two sub-zones of the electronic state in
transition metal.

The magnetic susceptibility of zirconium
carbide decreases with increase in carbon content
and attains a diamagnetic value near the stoichiometric composition (6). This, however, is not
confirmed by Samsonov and Kuchma (7). For
niobium monocarbide, it was observed that the
value of magnetic susceptibility decreases with
decreasing carbon content, reaching a minimum
at an intermediate composition and with further
decrease in carbon the magnetic susceptibility
again increases (6-8).

The temperature dependence of electrical
resistivity of transition metal refractory carbides in their homogeneity range at high tem-

peratures has been studied by various authors
(3,9-15). With increase in temperature electrical resistivity is found to increase, which
reflects the metallic type of conduction in these
carbides. Similar behaviour is also seen in the
case of the temperature dependence of thermo
e.m.f. of the said carbides. Golikova et al.
(13), however, have shown that at higher temperatures in the range of 200-2000°C titanium and
zirconium carbides resemble semiconductors.
Neshpor and others (11) established that the
slope of curves of electrical resistivity variation with temperature for the monocarbides in
homogeneity range decreases with decreases with
decrease in carbon content. The slope of curves
of temperature dependence of thermo e.m.f. for
the monocarbides of IV group transition metals
continuously decreases with decrease in C/$_{Me}$
ratio, but in the case of monocarbides of V group
transition metals there appears a maximum within
the homogeneity range (11).

ZrC-NbC Alloy Carbides

     Rudy and Benesovsky (16) studied the elec-
trical resistivity changes in ZrC-NbC alloys and
established a smooth maximum at equimolecular
composition. However, results of Avgustinik et
al. (17-18) do not tally with the previous
result.

     Barantseva et al. (20) have observed
maximum in the Hall coefficient variation for
the alloy with 60 mole % ZrC.

     On magnetic properties of ZrC-NbC alloys the
single work of Samsonov et al (21) is so far
available, but due to the presence of impurities
in their specimens, the results are difficult to
evaluate.

     Kovalskii and Petrova (22) measured the room
temperature microhardness of alloys of ZrC-NbC
and other similar binary carbide-systems. However, unlike other binary systems ZrC-NbC did not
reveal the maximum on the microhardness vs. composition plot.

EXPERIMENTAL METHODS

Preparation of Carbides

The powders of zirconium and niobium monocarbides in their homogeneity range were prepared by synthesis of metal powder with carbon soot in the vacuum furnace at the temperature of 1600 and 1800°C respectively. The compounds thus obtained were thoroughly mixed in calculated amounts to obtain the alloy carbides containing 25, 50 and 75 mole% NbC respectively and were subsequently hot pressed at temperatures 2300-2500°C under pressure of the order of $\sim$ 150 kgm/cm^2. The specimens of 8 mm dia and 15-20 mm height thus obtained were given homogenizing treatment under vacuum at 2000°C. Chemical compositions of the specimens are given in Table-1. X-ray analysis established that all the alloys were single phase with F.C.C. structure of NaCl type.

Properties Measurement

Microhardness measurements were done on the PMT-3 unit with 50 gm load. Compensation method was applied to determine the electrical resistivities. Hall coefficients were measured on the unit with pole field of 18333 oersteds (23). Magnetic susceptibility measurements were done using Guy's principle (24). High temperature electrical resistivity and thermo e.m.f. measurements were performed on the apparatus described elsewhere (25). Emissivity measurements at monochromatic conditions within the temperature range 1200-2100°K were done on compact specimens using absolutely black body conditions as described by Fomenko et al. (26). Wetting angle by liquid copper was measured on the apparatus described elsewhere (27).

RESULTS

Zirconium and Niobium Monocarbides

As shown in Fig. 3a the lattice parameter of ZrC_{1-x} and NbC_{1-x} increases with increase in $C/_{Me}$ ratio.

Table 1:   Chemical composition (weight %) of ZrC, NbC and their alloys in homogeneity range.

Zr	Nb	C Total	C Free	C Combined	Formula
89.1	-	10.1	0	10.1	$ZrC_{0.85}$
88.1	-	11.2	0.5	10.75	$ZrC_{0.9}$
89.2	-	11.0	0.15	10.15	$ZrC_{0.93}$
88.7		11.5	0.4	11.14	$ZrC_{0.96}$
-	90.6	8.55	0.15	8.41	$NbC_{0.73}$
-	89.3	9.50	0.23	9.29	$NbC_{0.81}$
-	89.5	10.21	0.19	10.03	$NbC_{0.88}$
-	87.6	10.84	0.52	10.37	$NbC_{0.91}$
63.3	24.50	11.0	0.5	10.55	$Nb_{0.27}Zr_{0.73}C_{0.93}$
41.5	46.0	10.6	0.3	10.33	$Nb_{0.52}Zr_{0.48}C_{0.91}$
22.2	63.0	10.45	0	10.45	$Nb_{0.74}Zr_{0.26}C_{0.93}$
65.2	22.3	10.70	0	10.70	$Nb_{0.25}Zr_{0.75}C_{0.94}$
43.4	42.7	10.75	0	10.75	$Nb_{0.49}Zr_{0.51}C_{0.96}$
21.5	63.1	10.5	0	10.50	$NbC_{0.75}Zr_{0.25}C_{0.96}$
77.2	9.8	11.9	1.1	10.92	$Nb_{0.11}Zr_{0.89}C_{0.98}$
45.7	41.2	11.3	0	11.30	$Nb_{0.47}Zr_{0.53}C$
25.3	62.0	11.7	0	11.70	$Nb_{0.7}Zr_{0.3}C$

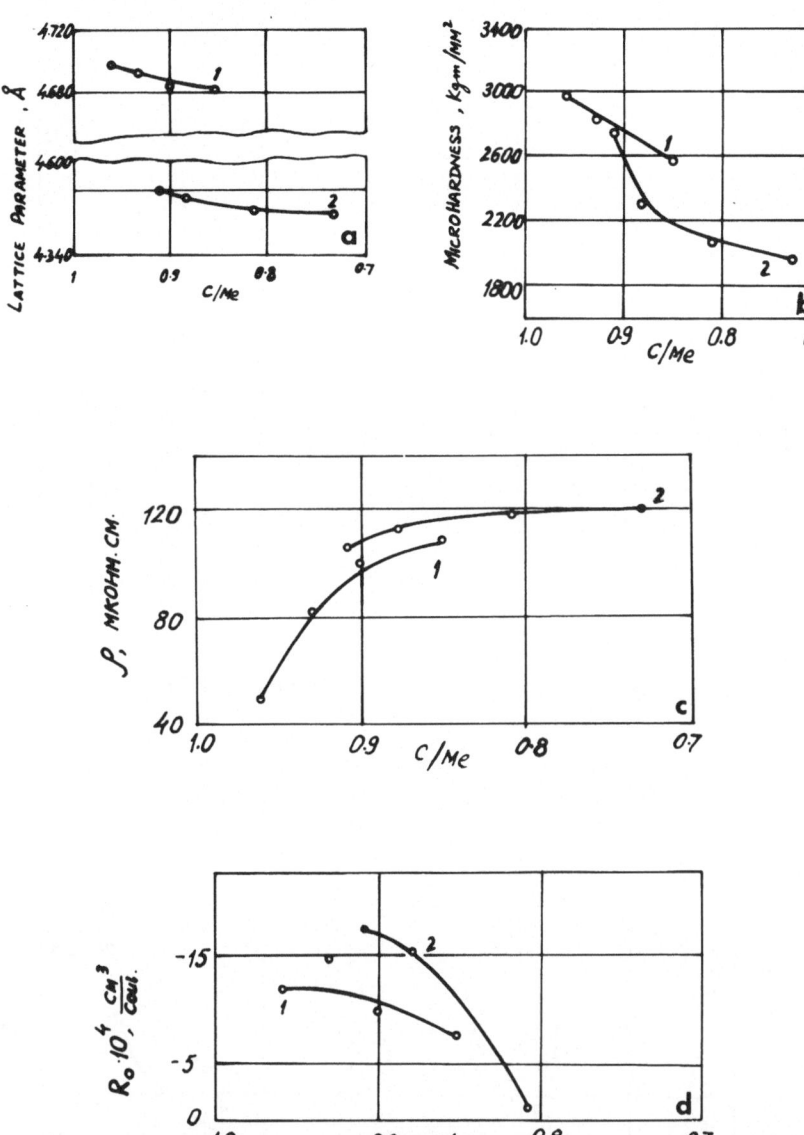

Fig. 3. Variations in lattice parameter (a), microhardness (b),
electrical resistivity (c), Hall coefficient (d), carrier
current concentration (e), carrier mobility (f), absolute thermo
e.m.f. (g), magnetic susceptibility (h), $d\rho/dT$ (i), $d\alpha/dT$ (j),
emissivity (k), and liquid copper wetting angle (1) of zirconium
(1) and niobium (2) monocarbides in their homogeneity range.

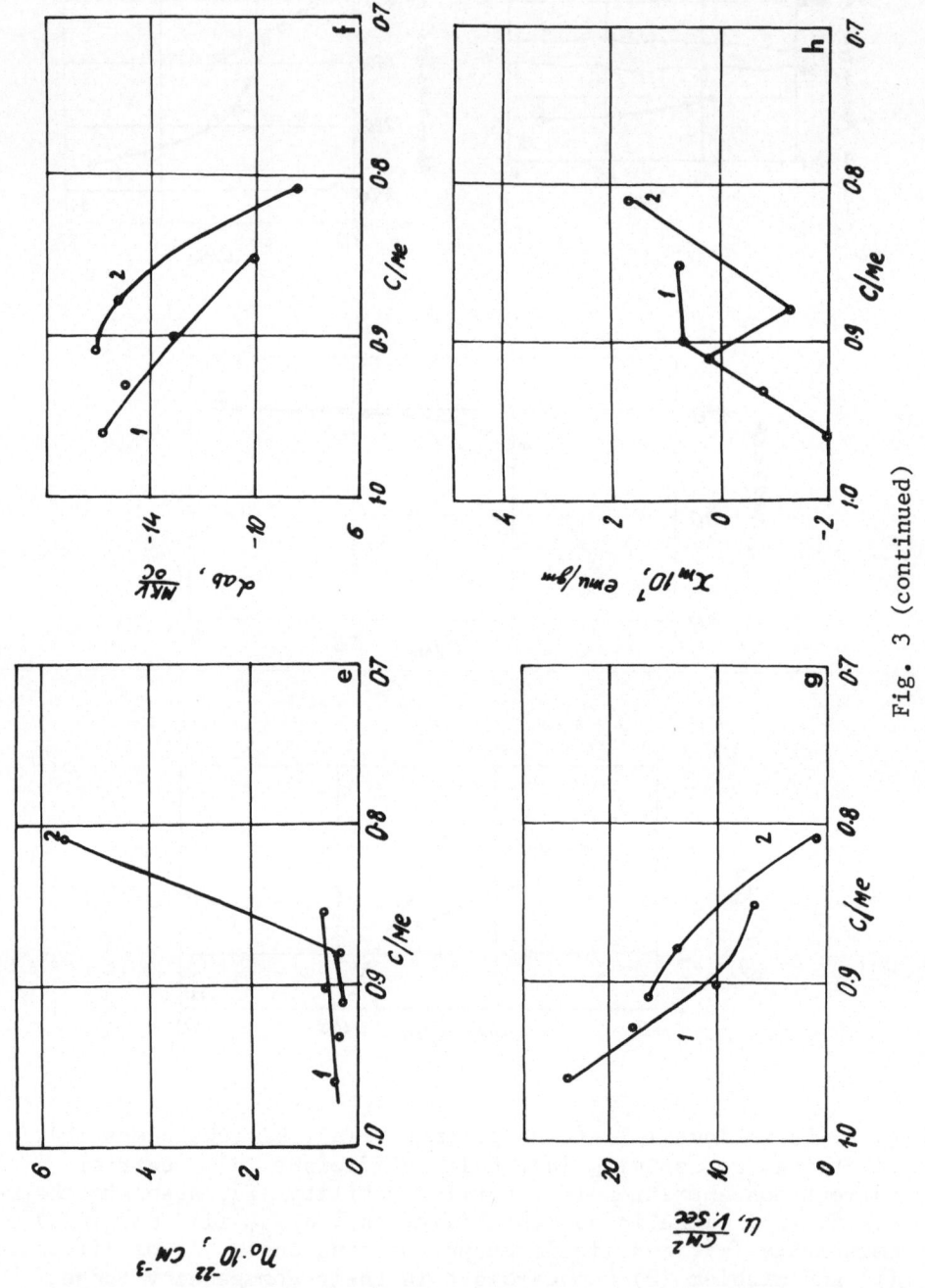

Fig. 3 (continued)

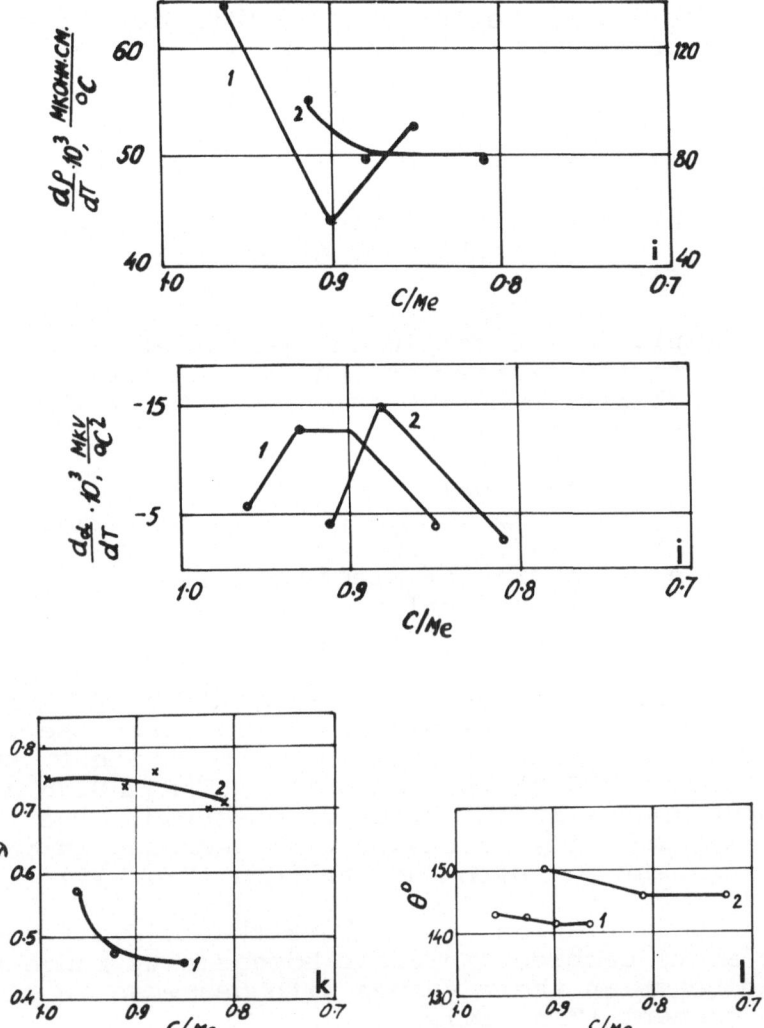

Fig. 3 (continued)

Microhardness of both the carbides in homogeneity range increases with increasing C/Me
(Fig. 3b). However, in case of NbC_{1-x} a clear
deviation from linear plot is observed at composition $NbC_{0.81}$.

Electrical resistivity in both cases falls
with increasing C/Me. However, at high carbon
compositions increase in electrical resistivity
is comparatively higher (Fig. 3c).

Hall coefficient values indicate that for
both carbides in their homogeneity range electron type conduction occurs. Like other properties their variation is also not a linear one
(Fig. 3d).

Applying the single carrier model in these
carbides, the carrier current concentration n and
their mobility u can be calculated using the following formulae:

$$n = 1/eR_x \qquad \qquad .. (1)$$
$$u = R_x/\rho \qquad \qquad .. (2)$$

where e = charge of electron,
        ρ = electrical resistivity,
        R_x - Hall coefficient.

The plot of carrier concentration of these
carbides (Fig. 3e) shows its linear increase in
case of ZrC with decrease in carbon content, but
in case of NbC at the composition $C/Me \angle 0.88$ a
sudden change in the slope is observed. The
carrier mobility decreases with decrease in carbon content for both the carbides (Fig. 3f).

The variation in absolute thermo e.m.f. of
the above carbides in their homogeneity range shows
a decrease in their values with decrease in carbon content (Fig. 3g).

Magnetic susceptibility values of ZrC at
high carbon compositions show diamagnetic properties. But in case of NbC with decrease in carbon
content magnetic susceptibility first falls and
reaches a maximum diamagnetic value in the homogeneity range, but with further decrease in car-

bon content the susceptibility again rises and carbide phase changes into paramagnetic one (Fig. 3h).

The variation of electrical resistivity and thermo e.m.f. at high temperatures (50-1000°C) shows a linear or nearly linear plot (Fig. 4). The change in $d\rho/dT$ with composition for the above carbides in their homogeneity ranges is not a regular one, i.e. with increasing carbon content initially $d\rho/dT$ decreases, but later on increases (Fig. 3i). The variation of $d\alpha/dT$ with composition shows a maximum at $C/Me = 0.8-0.9$ (Fig. 3j).

As shown in Fig. 3k emissivity at 1600°K of ZrC and NbC increases with increasing C/Me ratio, while monocarbide zirconium has higher values of emissivity than for NbC.

The wetting angles of ZrC and NbC by liquid copper at 1150°C show lower values for ZrC as compared with NbC. But in both cases with increase in C/Me ratio the wetting angle increases (Fig. 3l).

ZrC-NbC Alloy Carbides

The lattice parameter variation in ZrC-NbC system (Fig. 5a) does not show a linear relationship of Vegard's law. With increase in ZrC content in the alloys microhardness linearly increases (Fig. 5b). The electrical resistivity vs. composition plot shows the presence of maximum near the equimolecular composition (Fig. 5c). Similarly the variations of Hall coefficient, carrier current concentration and its mobility reflects the presence of extreme values near the 50 mole% NbC composition (Fig. 5d, e, f). However, the extreme value in the thermo e.m.f. plot is observed not at equimolecular but at 50 mole% ZrC composition (Fig. 5g). This does not tally with the data of Avgustinik et al. (18), who found a regular decrease in the thermo e.m.f. values from ZrC to NbC.

The paramagnetic properties of the alloys decrease with increase in ZrC and for higher carbon contents in the homogeneity range, similar to TiC-NbC system (28), paramagnetic susceptibility lowers (Fig. 5h).

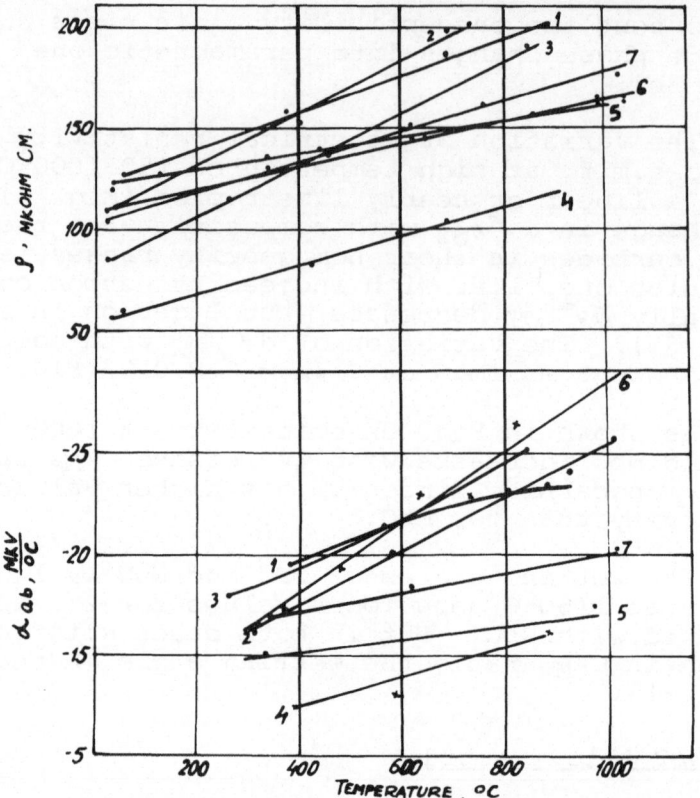

Fig. 4. High temperature electrical resistivity
and thermo e.m.f. variations in monocarbides of
zirconium and niobium in their homogeneity
range. 1, $ZrC_{0.85}$; 2, $ZrC_{0.9}$; 3, $ZrC_{0.93}$; 4,
$ZrC_{0.96}$; 5, $NbC_{0.81}$; 6, $NbC_{0.88}$; 7, $NbC_{0.91}$.

Like simple carbide phases the alloy carbides show a linear or nearly linear increase
of electrical resistivity and thermo e.m.f. at
high temperatures 50-1000°C (Fig. 6). It is
interesting to note that with increase in carbon
content $d\rho/dT$ increases (Fig. 5i), while
$d\alpha/dT$ shows an anomalous behaviour (Fig. 5j).
However, in both cases, the change is not a
linear one and near equimolecular composition
there exists a deviation.

The wetting angle variations of liquid copper
on the above alloy carbide system show that with increase in carbon content, like simple carbides,

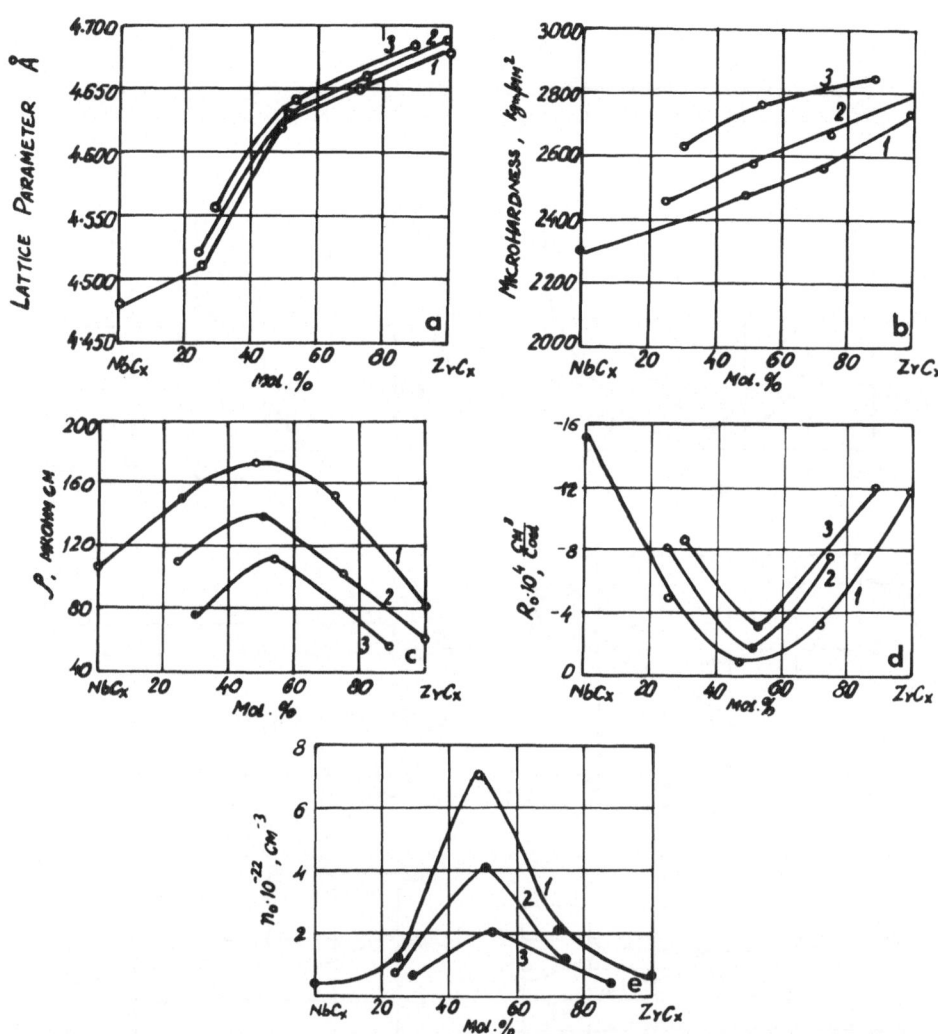

Fig. 5. Variations in lattice parameter (a), microhardness (b), electrical resistivity (c), Hall coefficient (d), carrier concentration (e), carrier mobility (f), absolute thermo e.m.f. (g), magnetic susceptibility (h), $d\rho/d_T$ (i), $d\alpha/d_T$ (j), and liquid copper wetting angle (k) of ZrC-NbC alloys in their homogeneity range. 1, $Nb_yZr_{1-y}C_{0.9}$; 2, $Nb_yZr_{1-y}C_{0.95}$; 3, $Nb_yZr_{1-y}C$.

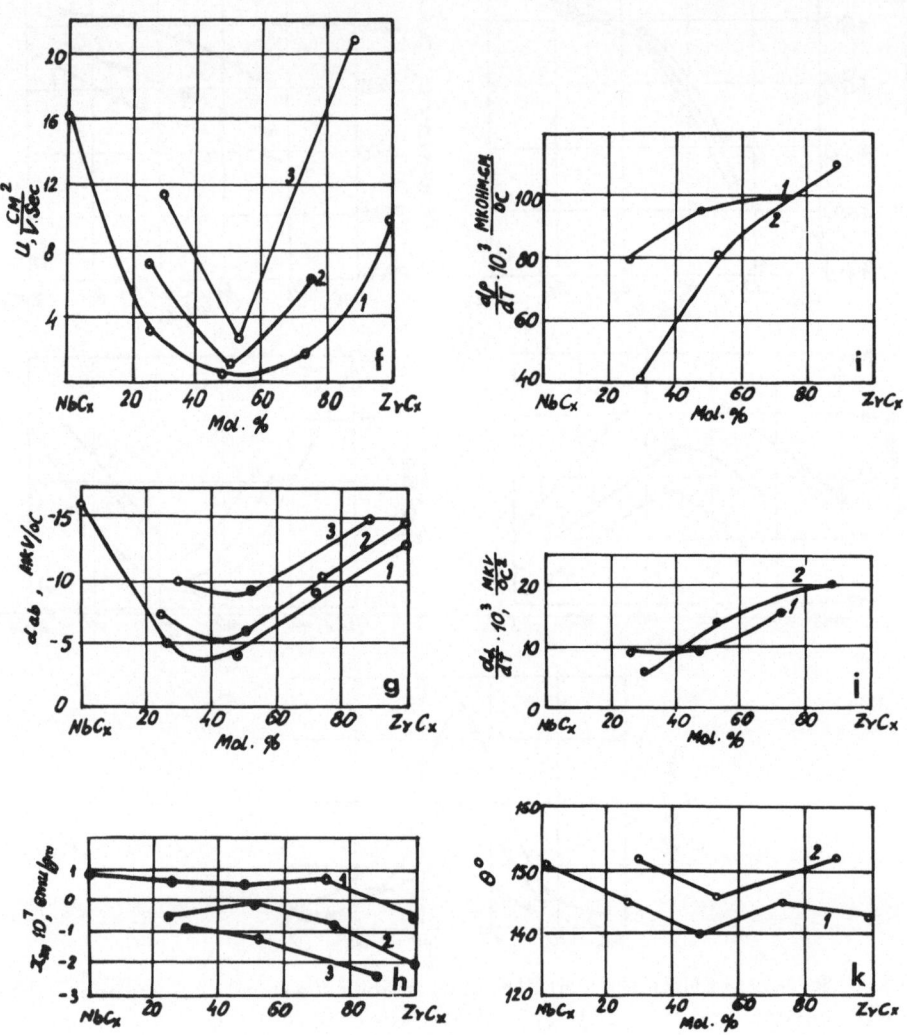

Fig. 5 (continued)

the wetting angle increases and also at 50
mole% ZrC composition it has a minimum value
(Fig. 5k).

DISCUSSION

The present discussion is based on the consideration of the condensed state model of materials. During the formation of condensed
state from the isolated atoms, valence electrons
may be divided into localized and nonlocalized
parts (29-31). The proposal has been experi-

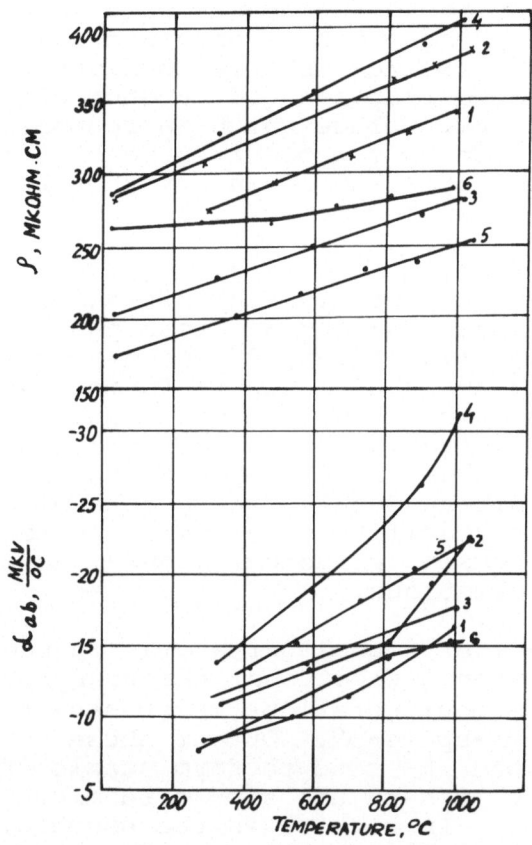

Fig. 6. High temperature electrical
resistivity and thermo e.m.f. variations in ZrC-NbC alloys in their
homogeneity range. 1, $Nb_{0.27}Zr_{0.73}C_{0.93}$;
2, $Nb_{0.52}Zr_{0.48}C_{0.91}$; 3, $Nb_{0.74}Zr_{0.26}$
$C_{0.93}$; 4, $Nb_{0.11}Zr_{0.89}C_{0.98}$; 5, $Nb_{0.47}$
$Zr_{0.53}C$; 6, $Nb_{0.7}Zr_{0.3}C$.

mentally verified by X-ray spectroscopy (32-35) and also by Hall constant data (36) of the transition metals.

In the further development of this approach it was proposed (37-40) that the localized fraction of the valence electrons form a fairly broad spectrum (statistical collection) of configurations in which the maximum statistical weight is possessed by the energetically most stable electronic configuration corresponding to the minimum store of free energy. Such stable electronic configurations for d-metals are d^0, d^5 and d^{10} and according to the degree of descending energy stability, these can be arranged in the order d^5-d^{10}-d^0. Between stable configurations and nonlocalized fractions of valence electrons an exchange takes place which is responsible for the bonding. In this process, some of the localized electrons appear to be in a rather free state and the remainder become involved in the actual exchange.

Assuming that for transition metal atoms in the free state having d-electrons $n_d \nless 5$, only two stable configurations, d^0 and d^5, of localized electrons are predominantly formed, using X-ray spectroscopic and Hall coefficient data the statistical weight of atoms having stable d^5-configurations (SWASC) in the metallic crystal of the transition metals can be determined. The energetic stability of d-configurations increases with increasing principal quantum number of the valence electrons.

In case of nonmetal atoms, characterized by the sp-valence electrons, the energetic stability of $s^x p^y$ - configurations increases in the order s^2p-sp-sp^2-sp^3-s^2p^6. Out of these configurations s^2p, sp and sp^2 tend to form stable sp^3 configurations, while s^2p^4 and s^2p^5 predominantly tend to form s^2p^6. In both cases the nonmetal atoms act as acceptors for electrons. The electronic configuration s^2p^3 provides the opportunity of equally forming either sp^3 or s^2p^6 stable configurations.

A detailed review of electronic configurational approach to understand crystalline solid state has been made by Samsonov (41).

Transition Metal Monocarbides (ZrC, NbC)

The previously described model of solids adequately fits up in understanding the nature of carbide phases. Since all the transition metals are donors of valence electrons, their carbide formation can be understood in terms of the transfer of nonlocalized valence electrons so as to stabilize the sp^3-configurations of carbon atoms in different degrees. Thus the reaction $sp^3 \rightleftharpoons sp^2 + p$ is pushed to the left hand side. The donor capacity of the transition metals decreases with increase in their valence electron localization, i.e. from III group metals to VI group metals and correspondingly the stabilization of sp^3-configuration of carbon atoms decreases. Arbuzov and Khaenko (42) have justified the case of transition metal donation of valence electrons and the consequent appearance of some negative charge on titanium atoms. The recent X-ray and electron spectroscopy studies of Zhurakovskii (43-44), Ramquist (45-46) and Holliday (47) satisfactorily show the approximation of CKα- band in carbides of IV-V group transition metals to that with sp^3-state in diamond.

Most of the early transition metal carbides have a wide range of homogeneity, in which the change in carbon concentration does not bring a change in their crystal structure. The filling of the transition metal lattice with interstitial carbon atoms, consequently, gives rise to the appearance or destruction of stable electronic configuration of both metals and nonmetal atoms.

The formation of transition metal carbides encounters a competition between two main processes - localization of sp^3-configuration of carbon atoms directed to strengthen the Me-C bond and localization of d^5-configurations of metal atoms to form stable Me-Me bonds. For the IV group transition metal carbides the stabilization of sp^3-configuration is maximum, which makes the Me-C bond stronger in comparison with Me-Me bond. This explanation justifies the availability of only monocarbides in case of titanium, zirconium and hafnium, whereas the V group transition metals in addition to monocarbides form

semicarbides also. Moreover, the homogeneity
range in case of IV group metal monocarbides
(TiC, ZrC, HfC) is wider as compared with the
homogeneity range of V group transition metal
monocarbides (VC, NbC, TaC).

The higher SWASC value of sp^3-configurations
of carbon atoms in case of ZrC, due to higher
donor capacity of the metal atoms, contributes in
its higher microhardness value as compared with
that of niobium monocarbide.

A decrease in carbon content in the carbides
is, therefore, accompanied by the increase in
Me-Me bonds and corresponding decrease in Me-C
bonds. The ratio between these two types of
bonds changes in the homogeneity range of the
respective carbides. As the stability of sp^3-
configuration of carbon atoms in diamond is very
much greater than that of d^5-configurations of
transition metals, the intensity in change of Me-C
bond, determined by the formation of sp^3-configur-
ation of carbon atoms, for IV-V group transition
metal monocarbides is not appreciable, whereas
the Me-Me bond in V group transition metal carbides
is expected to significantly increase with deficiency of carbon in its sublattice. This uneven
change in both types of bonds is thus mainly
responsible for the appearance of non-linear
changes in the physical properties of V group
transition metal monocarbides in the homogeneity range. This is confirmed by our present
results on the physical properties of niobium
monocarbide in the homogeneity range.

In case of zirconium carbide, where the
SWASC d^5-localized electrons is not fairly high
(52%), higher statistical weight of sp^3-confi-
gurations of carbon atoms results in decrease in
the electrical resistivity due to a lower degree
of carrier electrons scattering. Thus the de-
localization of valence electrons, participating
in forming Me-C bonds, increases the conduction
electron concentration.

The magnetic susceptibility changes in the
present investigation also confirm our electronic
model. The formation of covalent bonds brings
to the pairing of electrons of opposite spins and

consequently a decrease in the paramagnetic
susceptibility. Thus magnetic studies can very
well supply us information about the relative
changes in the bond strengths of the carbides
in their homogeneity range. The presence of
diamagnetism in NbC at C/Me=0.8 can be explained
on the grounds of the higher total energies of
Me-C and Me-Me bonds, which exist at that
particular composition. It is interesting to
note that in titanium carbide diamagnetism is
observed near stoichiometric composition while
in case of ZrC and HfC it appears at carbon
deficient compositions. On further passing to
NbC this point moves to the composition still
poorer in carbon.

    In addition, it can be noticed that there
is no physical property so far observed in NbC
which varies linearly in the homogeneity range.
The variation in superconducting transition
temperature (48), density (49), coefficient of
thermal expansion (50) and the heat of formation
of niobium monocarbide in homogeneity range gives
ample evidence in its support.
    The temperature dependence of electrical
resistivity ($d\rho/dT$) characterizes the intensity
of resistivity increase owing to the scattering
of carrier currents on thermal lattice vibrations due to increased temperatures. The
Matheason's Rule is not applicable to carbon
deficient compositions of carbides, as the
increase in number of carbon vacancies not only
changes the electrical resistivity, but also the
slope in resistivity vs. temperature plots. The
change in $d\rho/dT$ is thus conditioned with the
formation of stable electronic configurations
and in case of their higher statistical values,
$d\rho/dT$ must be low. But as is clear from Fig.
3i in case of ZrC this is not followed, as the
curve corresponding to ZrC has a higher position
in comparison with that of NbC. However, the
possibility of 4d → 4f transition of the valence
electrons in zirconium in the crystalline state
may not be ignored, which may give rise to such
anomalous behaviour. A general theoretical
interpretation of the temperature dependence of
electrical resistivity and thermo e.m.f. becomes difficult to be elucidated at this stage,
as sufficient data on the change in carrier con-

centration of carbides with temperature are still not available.

The anomalous behaviour in the change in emissivity and wetting angles of ZrC in its homogeneity range as compared with TiC and NbC can also be explained in a similar manner (51), i.e. 4d → 4f transition of zirconium metal.

Transition Metal Alloy Carbides (ZrC-NbC)

The lattice parameter variations in ZrC-NbC system clearly indicate a deviation from Vegard's law and therefore the possibility of the presence of two phase mixtures for alloy compositions poorer in ZrC. As the statistical weight of sp^3-configurations of carbon atoms in TiC is greater than in ZrC, its stabilizing effect on the niobium carbide lattice during alloying should be higher. This has been confirmed in our previous work (28) on TiC-NbC alloy system, where a deviation in Vegard's law was not so noticeable.

The existence of the extreme values in the measured properties of ZrC-NbC alloys at about equimolecular composition is not possible to explain only on the basis of Me-C bonds. If the total statistical weight of Sp^3-configurations of carbon atoms, while passing from ZrC to NbC, decreased regularly, then the physical properties of the above system must also have revealed a regular change. The influence of Me^{I}-Me^{II} bonds, particularly when there is wide difference in SWASC d^5 values of transition metals, e.g. zirconium and niobium, which play a considerable role in comparison with Me^{I}-Me^{I} and Me^{II}-Me^{II} bonds, must also be incorporated.

However, a detailed comparison of the present results on ZrC-NbC alloys, with those already existing on TiC-NbC alloys (28) indicates some shifting in the composition at which the maximum properties are observed towards the alloy poorer in ZrC. Evidently this shows a higher capacity of niobium atoms to destroy the d^5-configurations of titanium atoms having a lower SWASC value, than of zirconium atoms.

On the basis of earlier discussions, the $d\mathcal{S}/dT$ of the alloy carbides rich in ZrC must have lower values as compared with alloys rich in NbC. But this is not likewise noticed in our results, although it is confirmed by our previous work on TiC-NbC alloys (28). Such behaviour might well follow up with the possible assumption of 4d $\rightarrow$ 4f transition of valence electrons of zirconium atoms in the crystalline state.

The significance of Me^{I}-Me^{II} type bond in the alloy carbides ZrC-NbC has been thermo-dynamically established by Hoch (52) recently.

REFERENCES

1. V.S. Neshpor, S.S. Ordanyan, A.E. Avgusti-nik, M.B. Khusidman, Zh. Prikl. Khim, 37, 2375, 1964.

2. G.V. Samsonov, A.D. Parasyuk, Taplophyz. Visok. Temperatur, 4, 207, 1966.

3. A.E. Avgustinik, O.A. Golikov, G.M. Klimashin, V.S. Neshpor, S.S. Ordan'yan, V.A. Snetkova, IZV.AN SSSR, Neorg. Mater., 2, 1499, 1966.

4. A.E. Avgustinik, O.A. Golikova, V.S. Neshpor, S.S. Ordan'yan, IZV.AN SSSR, Neorg. Mater., 3, 286, 1967.

5. T. Tsuchida, Y. Nakamura, M. Mekama, J. Sakurai, T. Takaki, J. Phy. Soc. Japan, 16, 2453, 1961.

6. H. Bittner, H. Goretzki, Mh. Chem., 93, 1000, 1962.

7. G.V. Samsonov, A.Ya. Kuchma, IZV. AN SSSR, Neorg. Mater., 4, 1361, 1968.

8. P.P. Matveenko, L.B. Dubrovskaya, P.V. Ghel'd, M.G. Tretnikova, IZV. AN SSSR, Neorg. Mater., 1, 1061, 1965.

9.   P. Costa, R.R. Conte, In "Compounds of
     Interest in Nuclear Reactor Technology",
     Ed. J.T. Waber, P. Chiotti, A.I.M.E.,
     New York, 1964.

10. S.N. L'vov, V.F. Nemchenko, T. Ya. Kosola-
     pova, G.V. Samsonov, DAN SSSR, 157, 408, 1964.

11. V.S. Neshpor, S.V. Airapetyants, S.S.
     Ordan'yan, A.E. Avgustinik, IZV. AN SSSR,
     Neorg. Mater., 2, 855, 1966.

12. O.A. Golikova, A.E. Avgustinik, T.M.
     Klimashin, L.V. Kozlovskii, Fiz. Tver.
     Tela, 7, 2860, 1965.

13. O.A. Golikova, A.O. Zhafarov, A.E. Avgus-
     tinik, G.M. Klimashin, Fiz. Tver. Tela,
     10, 168, 1968.

14. O.A. Golikova, F.L. Feigeliman, A.E.
     Avgustinik, G.N. Kolimashin, Fiz. i Tekh.
     Poluprovodnikov, 1, 293, 1967.

15. O.A. Golikova, N.N. Matveeva, A.E. Avgustinik,
     S.S. Ordan'yan, Teplofiz. Visok. Temperatur,
     5, 1001, 1967.

16. E. Rudy, F. Beresovsky, Plansee ber, Pulver-
     met, 8, 72, 1960.

17. O.A. Golikova, A.O. Dzhafarova, A.E.
     Avgustinik, L.V. Kudryashova, S.S. Ordan'yan,
     Fiz. Tverd. Tela, 9, 1557, 1967.

18. A.E. Avgustinik, O.A. Golikova, L.V. Kudrya-
     shova, S.S. Ordan'yan, V.A. Snetkova, IZV.
     AN SSSR, Neorg. Mater., 3, 1823, 1967.

19. A.E. Avgustinik, O.A. Golikova, L.V. Kudrya-
     shova, S.S. Ordan'yan, V.A. Snetkova, IZV.
     AN SSSR, Neorg. Mater., 4, 904, 1968.

20. E.G. Barantseva, V.N. Paderno, Yu. B.
     Paderno, Poroshkovaya Metallurgia, No. 2,
     70, 1967.

21. G.V. Samsonov, V.S. Neshpor, N.S. Strel'ni-
     kova, DAN Ukr. SSSR, 8, 838, 1958.

22.   A.E. Kovalskii, L.A. Petrova, In "Microtver-
      dost", Izd. AN SSSR, Moscow, 1951, p. 170.

23.   S.N. L'vov, V.F. Nemchanko, V.E. Marchenko,
      Pribori i Tekhnika Experimenta, No. 2, 159,
      1961.

24.   V. F. Chichernikov, Magnitnie Izmereniya,
      Izd. MGU, Moscow, 1963, p. 91.

25.   S.V. L'vov, P.E. Mal'ko, V.F. Nemchenko,
      Poroshkovaya Metallurgia, No. 9, 89, 1966.

26.   V.S. Fomenko, G.V. Samsonov, Yu. B. Paderno,
      Ogneupori, No. 1, 40, 1962.

27.   V.N. Eremenko, Yu. V. Naidich, Zmochuvannya
      Z Ridkimi Metallami Poverkhn Tugoplavkikh
      Spoluk, Izd. AN Ukr. SSSR, Kiev, 1958.

28.   G.V. Samsonov, G.S. Upadhyaya, In "High
      Temerature Materials" VI Plansec Seminar,
      Metallwerk Plansec. Tirol. Ed. F. Beresovsky,
      Springer-Verlag, Vienna, New York, 1969.

29.   R. Weiss, J. De Marco, Rev. Mod. Phys., $\underline{30}$,
      59, 1958.

30.   J. Wood, Phy. Rev., $\underline{117}$, 714, 1960.

31.   N. Mott, K. Stevens, Phil. Mag., $\underline{3}$, 1304, 1957.

32.   M.I. Korsunskii, Ya. E. Genkin, Dokl. Akad.
      Neuk. SSSR, $\underline{142}$, 1276, 1962.

33.   M.I. Korsunskii, Ya. E. Genkin, IZV. Akad.
      Neuk. SSSR, Ser. Fiz., $\underline{28}$, 832, 1964.

34.   M.I. Korsunskii, Ya. E. Genkin, IZV. Akad.
      Neuk, Ser. Neorg. Mater., $\underline{1}$, 1701, 1965.

35.   A.Z. Menshikov, S.A. Nemnonov, Fiz. Metall.
      Metalloved., $\underline{19}$, 57, 1965.

36.   G.V. Samsonov, Yu. B. Paderno, B.M. Rud, IZV.
      Vuzov, Fiz. No. 9, 129, 1967.

37.   G.V. Samsonov, Ukr. Khim. Zh., $\underline{31}$, 1233, 1965.

38. G.V. Samsonov, Poroshk. Metall., 12, 49, 1966.

39. I.R. Kozlova, V.N. Gurin, A.P. Obukhov, Poroshk. Metall., 12, 69, 1966.

40. I.F. Prydko, Poroshk. Metall., 12, 61, 1966.

41. G.V. Samsonov, In "High Temperature Materials", VI Plansec Seminar, Metallwerk. Plansec. Tirol, Ed. F. Beresovsky, Springer-Verlag, Vienna, New York, 1969.

42. M. P. Arbuzov, B.V. Khaenko, Porosh. Metall. No. 4, 74, 1966.

43. E.A. Zhurakovskii, DAN SSSR, 180, 1088, 1968.

44. E.A. Zhurakovskii, DAN SSSR, 184, 1317, 1969.

45. L. Ramquist, B. Ekstig, E. Kallne, E. Noreland, R. Manne, J. Phy. Chem. Solids, 30, 1849, 1969.

46. L. Ramquist, K. Hamrin, G. Johnson, U. Gelius, C. Nordling, J. Phy. Chem. Solids, 30, 1835, 1969.

47. J.E. Holliday, Advances in X-ray Analysis, Vol. 9, Plenum Press, New York, 1966.

48. A.L. Giorgi, F.G. Szklarz, E.K. Storms, A.L. Bowman, B.T. Matthias, Phy. Rev., 125, 837, 1962.

49. S.S. Ordan'yan, A.E. Avgustinik, L.V. Kudrya-shova, Poroshkovaya Metallurgia, No. 8, 26, 1968.

50. C.P. Kempter, E.K. Storms, J. Less Common Metals, 13, 443, 1967.

51. G.S. Upadhyaya, "Physical Properties of Monocarbides of IV-V Group Transition Metals and their Alloys in Homogeneity Range", Ph.D. Thesis, Kiev Institute of Technology, USSR, 1969.

52. M. Hoch, In "Anisotropy in Single Crystal Refractory Compounds", Vo. I, Plenum Press, New York, 1968, p. 163.

TZM MOLYBDENUM ALLOY VIA P/M

L. P. Clare   A. J. Vazquez

Chemical & Metallurgical Div.   Ladish Co. - U.S.A.

Sylvania Electric Products Inc. - U.S.A.

INTRODUCTION

For a number of years TZM molybdenum alloy has been made by arc-casting. Because of the coarse grain structure of the cast billet, it has to be extruded before it can be directly worked by forging, rolling, etc. As a result, unless back extrusion techniques are resorted to, the maximum arc-cast billet diameter is limited to approximately 7-1/2", mainly because of extrusion press limitations.

Several years ago we became interested in fabricating TZM via powder metallurgy. A successful process was developed and physical and mechanical testing of the wrought P/M rod indicated it to be comparable to the properties of arc-cast TZM.

Our first products consisted mainly of some small diameter rod, wire, sheet and small forging billets. The material was worked at various temperatures from 1260°C (2300°F) to 1540°C (2800°F).

A requirement arose in a chemical industry for hot pressing dies that would require forgings 16" to 17"D x 650 pounds. Problems were encountered when large forgings were attempted. Billets were fabricated for the application by isostatic pressing and sintering, and then they were hammer forged in the 1430°C (2600°F) to 1540°C (2800°F) range. The forging temperatures were chosen because of previous success on smaller billets. Results were catastrophic. The failures appeared to be similar to the hot-shortness that results in high-carbon steel and other alloys when they are worked at too high a temperature.

FORGING RESPONSE STUDY

General

In order to determine what was causing the catastrophic failures, a decision was made to do a forging response study. A. J. Vazquez of Ladish Co. undertook this study for us. A 6"D x 24-3/8" billet of MT-104 (P/M-TZM) and a recrystallized arc-cast TZM billet 6"D x 7-3/4" were supplied for the program.

MATERIAL QUALIFICATION

Non-Destructive Testing and Material Preparation

Both billets were dye penetrant and ultrasonically inspected and found to be free of defects. Four 1/2" thick slabs were cut from the P/M billet one at the top, bottom, one-third and two-thirds length positions. One similar slab was cut from the arc-cast billet. These slabs were used subsequently for destructive testing. The balance of the billets were cut into wedges for test forging. The wedges are described in greater detail further on in this article.

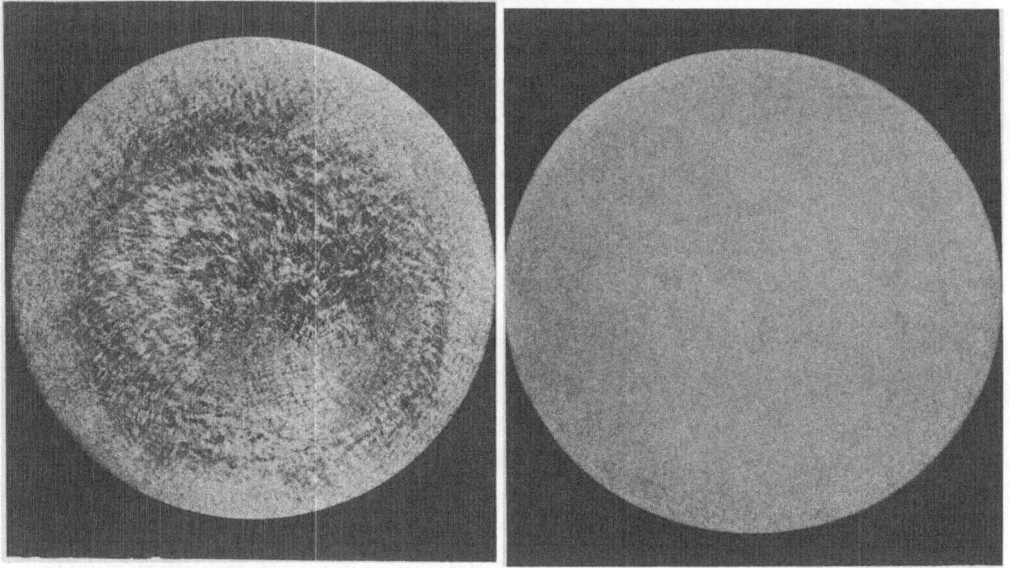

Arc-Cast TZM   P/M TZM

Figure 1.   Macrostructure of the Slabs Sectioned from the Billets.   (Both pictures were taken originally at three-quarters of full size.)

Macrostructure and Destructive Testing

Figure 1 shows the macrostructure of the slabs sectioned from the billets. The arc-cast billet shows a duplex structure, and "P/M Material" represents the very fine uniform structure throughout the sections of the P/M billet. Table I shows a comparison of ASTM grain size across and through the P/M billet and on the top of the arc-cast billet. Table II has the results of a diametric hardness survey on the cut slabs showing the uniformity of both the P/M and arc-cast material. Table III shows the wedge identity from each section and a comparison of their material densities before and after forging and the forging temperature for each wedge. Chemical analyses on both the P/M and arc-cast material found them to conform nominally to the requirements of ASTM Specification B385.

TABLE I

STRUCTURAL GRAIN SIZE OF TZM ALLOY BILLETS
(ASTM Values)

Slab No.	Surface Range	Surface Predominant	Mid-Radius Range	Mid-Radius Predominant	Center Range	Center Predominant
P/M-1	4 - 8	6	4 - 8	6	4 - 8	6
P/M-2	5 - 8	7	5 - 8	7	5 - 8	7
P/M-3	4 - 8	6	5 - 8	7	5 - 8	7
P/M-4	5 - 8	7	5 - 8	7	5 - 8	7
Arc Cast	2 - 7	5	2 - 7	5	2 - 7	5

TABLE II

HARDNESS SURVEY ON TZM ALLOY BILLETS

Slab No.	Hardness (Rockwell "B") Edge		Center	Edge	
P/M-1	83	84	85	84	83
P/M-2	83	84	88	85	83
P/M-3	83	84	87	85	83
P/M-4	83	84	85	84	84
Arc Cast	89	89	89	89	90

TABLE III

COMPARISON OF WEDGE DENSITY BEFORE AND AFTER
FORGING AND THE FORGING TEMPERATURE USED

Wedge Identity	Percent of Theoretical Density		Forging Temperature	
	Prior to Forging	After Forging	°C	°F
P/M-1-1	93.86	97.48	1095	2000
P/M-1-2	93.90	98.26	1205	2200
P/M-2-1	93.78	97.91	1315	2400
P/M-2-2	93.82	97.03	1420	2600
P/M-3-1	93.87	97.13	1540	2800
P/M-3-2	93.98	97.41	1650	3000
Arc Cast-4-1	96.42	98.26	1205	2200

Wedges for Test Forging and Forging Temperatures

From each third section of the P/M billet and from the arc-cast billet two wedges were electro-chemically machined. Figure 2 shows the dimensions of the wedges. The six P/M wedges were hammer forged; one each at 95°C (200°F) intervals starting at 1095°C (2000°F) and running to 1650°C (3000°F). One of the arc-cast wedges was forged at 1205°C (2200°F) to be used as a control.

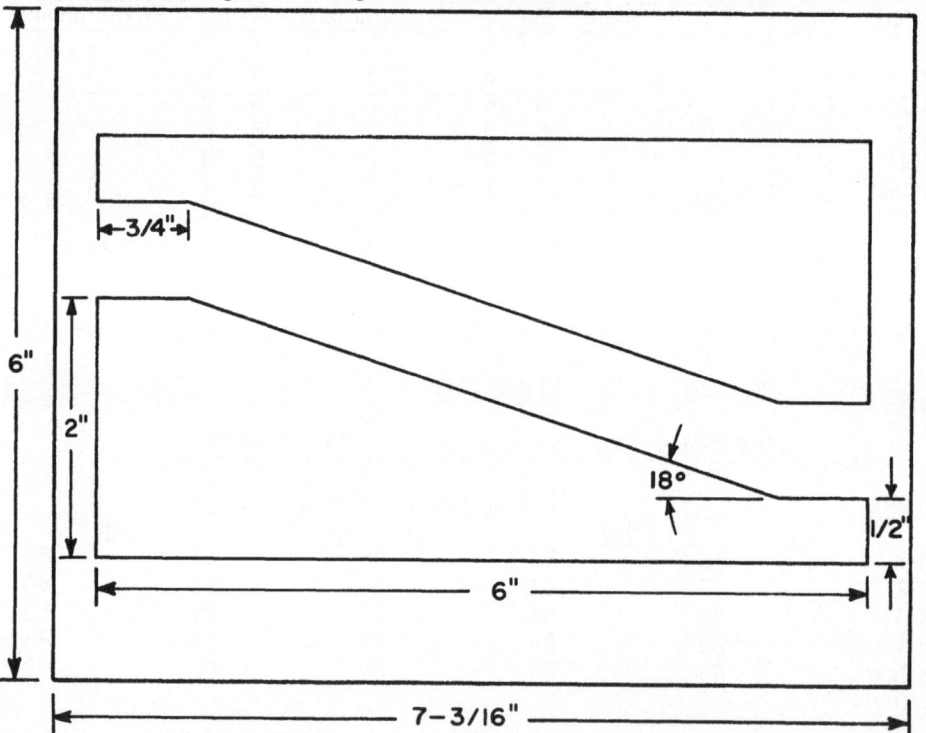

Figure 2. Sectioning Diagram for Forgeability Test Wedges

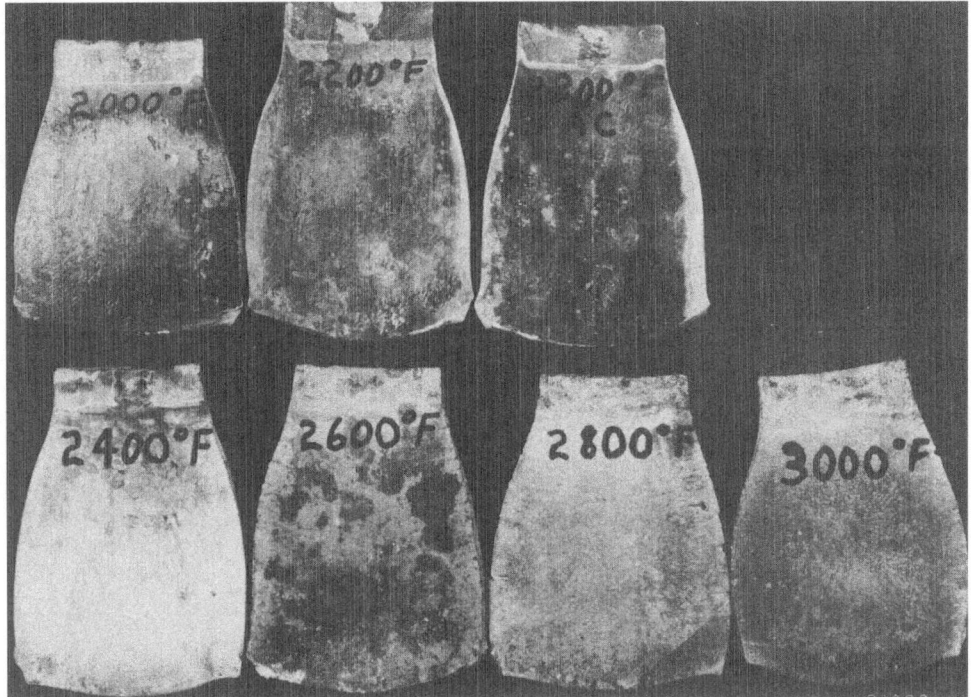

Figure 3.   Plan View of As-Forged Wedges of TZM. Forging Temperatures are Indicated.

Figure 4.   Edge View of As-Forged Wedges of TZM. Forging Temperatures are Indicated.   All Pieces are P/M Material Except the one Marked AC, which is Arc-Cast.

Forging Results

Forging results can be seen in Figures 3 and 4. These figures indicate the P/M material is readily forgeable between 1095°C (2000°F) and 1315°C (2400°F). However, the test wedge forged at 1315°C (2400°F) showed that minute cracks initiated at the edges of the "pancake". Wedges forged in the range from 1430°C (2600°F) to 1650°C (3000°F) showed extensive edge cracking with the condition seemingly less severe at 1650°C (3000°F). Therefore, the possibility exists that the material could be forged successfully at a temperature higher than 1650°C (3000°F) in the hot working range of the alloy. No study was made in plus 1650°C (3000°F) area, nor below 1095°C (2000°F).

EVALUATION OF RESULTS

Sample Selection and Preparation

The purpose of forging the wedge shaped pieces was to get a number of reductions for a given forging temperature. In order to establish the true reduction pattern of the forged wedges, the remaining arc-cast wedge was scored with the

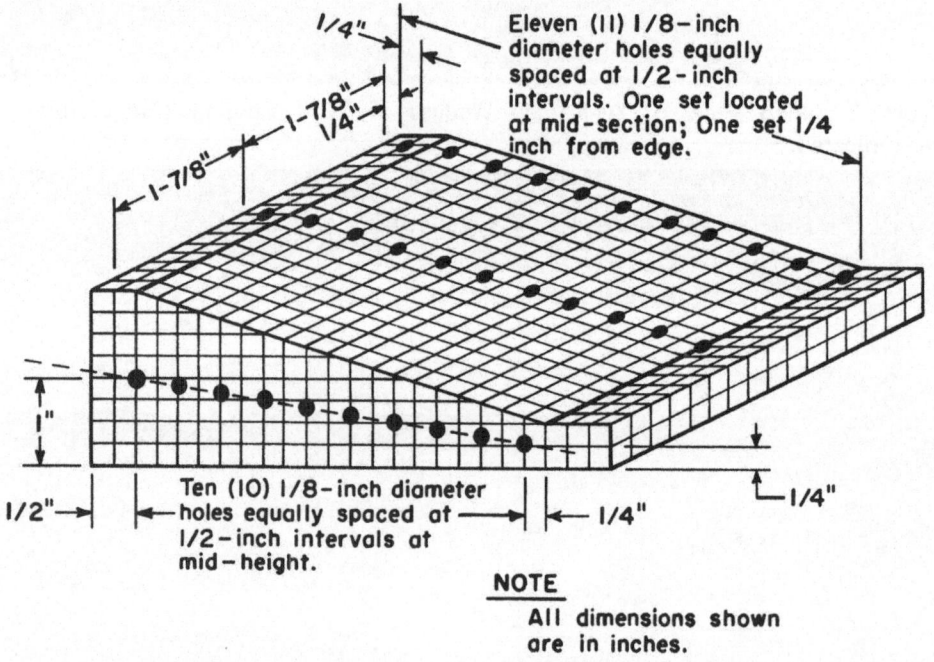

Eleven (11) 1/8–inch diameter holes equally spaced at 1/2–inch intervals. One set located at mid–section; One set 1/4 inch from edge.

Ten (10) 1/8–inch diameter holes equally spaced at 1/2–inch intervals at mid–height.

NOTE

All dimensions shown are in inches.

Figure 5. Sketch Showing The Surface Scoring at 1/4" Intervals and Holes Drilled for Insertion of Rods Into the Arc-Cast TZM Wedge Prior to Forging at 1205°C (2200°F) to Determine Flow Pattern.

grid pattern and holes drilled and plugged with 1/8"D rod as shown in Figure 5. This wedge was then forged at 1205°C (2200°F). The flow pattern of the resulting forging was analyzed based on the scored pattern and the rod plugs, and the reductions impacted to the wedge at the points coinciding with the vertical rod plugs were calculated and are presented in the isograph shown in Figure 6.

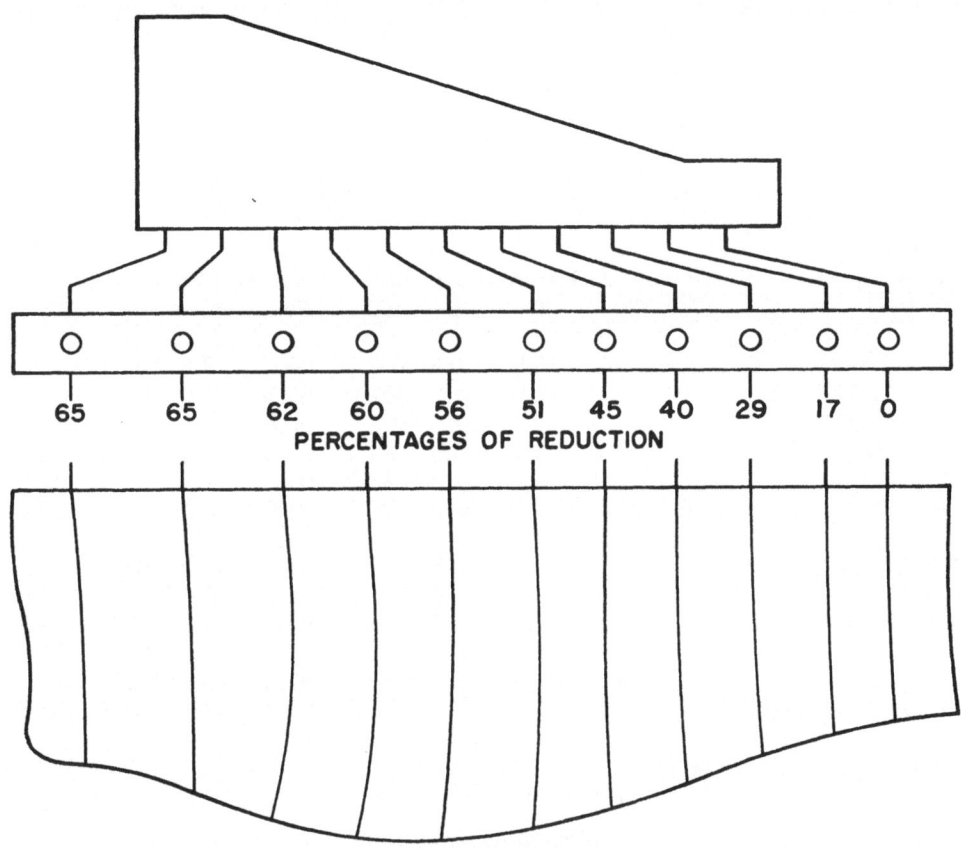

Figure 6.   Isograph Depicting Constant Reduction Lines

Subsequently each of the forged seven test wedges was sectioned longitudinally through the whole reduction pattern to yield six test bars. A test bar from each of the seven wedges was annealed at each of the six forging temperatures (seven test bars per temperature) from 1095°C (2000°F) to 1650°C (3000°F).

After annealing, each test bar was plotted with the constant reduction pattern as shown previously in Figure 6. Ten microspecimens, one from each constant reduction point, were removed from each test bar for the metallographic investigation of the recrystallization temperature from 0 to 65% reduction.

Hardness Survey

Hardness surveys were performed on each microspecimen. They show that for all forging temperatures studied, a drop in hardness, indicative of the beginning of recrystallization, had occurred in the specimens from the test bars annealed between 1310°C (2400°F) and 1425°C (2600°F). Table IV shows the portion of the hardness survey for the 1205°C (2200°F) forging temperature for the P/M and arc-cast material.

TABLE IV

HARDNESS SURVEY FOR THE
1205°C (2200°F) FORGING TEMPERATURE

Annealing Temperature (°F)	Rockwell "B" Hardness for Various Reduction Percentages									
	0	17	30	40	45	50	55	60	62	65
P/M-2000	93.0	95.3	99.3	95.5	100.3	102.3	102.7	103.5	103.1	103.5
P/M-2200	91.8	94.0	91.5	97.5	101.8	102.0	101.5	97.5	103.3	102.5
P/M-2400	90.2	98.8	100.3	99.3	102.0	103.0	100.8	102.0	101.8	100.3
P/M-2600	89.8	97.8	95.4	93.5	94.5	91.8	93.3	92.8	90.8	92.8
P/M-2800	91.8	92.5	92.0	90.1	91.3	92.0	91.8	91.1	92.0	92.0
P/M-3000	88.3	91.8	91.3	90.8	89.2	90.7	91.0	90.0	90.0	90.8
*A/C-2000	88.0	93.5	98.0	100.5	100.5	102.5	103.5	104.0	103.5	104.3
A/C-2200	86.0	93.2	97.8	101.8	96.0	102.3	102.5	100.8	103.0	102.5
A/C-2400	85.8	91.4	97.3	97.8	98.3	98.3	95.5	98.0	95.7	96.5
A/C-2600	80.0	90.6	81.0	89.0	88.5	90.5	89.7	90.0	89.7	90.4
A/C-2800	85.2	87.3	85.0	88.3	83.0	89.5	90.3	87.8	89.8	89.8
A/C-3000	84.5	85.7	82.8	88.8	88.0	88.3	89.4	89.5	88.5	88.8

*A/C - Arc Cast Material

Metallographic Evaluation

Using the "hardness drop" as a criterion for recrystallization, specimens from the test bars annealed between 1205°C (2200°F) and 1430°C (2600°F) for all the forging temperatures - 1095°C (2000°F) through 1650°C (3000°F) - were selected for the metallographic studies. This review revealed that the following conditions were prevalent for all forging temperatures for the various degrees of deformation:

1.   No recrystallization had occurred at an annealing temperature of 1205°C (2200°F).
2.   Partial recrystallization had begun at an annealing temperature of 1315°C (2400°F).
3.   Complete recrystallization was attained at the 1430°C (2600°F) annealing temperature.

Forged At 1315°C (2400°F)
Annealed at 1205°C (2200°F)
No Recrystallization

Forged at 1315°C (2400°F)
Annealed at 1315°C (2400°F)
Partial Recrystallization

Forged at 1315°C (2400°F)
Annealed at 1430°C (2600°F)
Full Recrystallization

All microspecimens were magnified originally at 100X; the etchant was 1.25g NaOH and 5.75g $K_3Fe(CN)_6$ in 20 ml of water.

Figure 7. Photomicrographs of P/M TZM Depicting Structure from the Test Specimen Reduced 30% at 1315°C (2400°F) and Annealed at Three Different Temperatures.

A typical metallographic survey supporting the above statements is shown in Figure 7.

The metallographic analysis indicated that, regardless of the forging temperature, full recrystallization of the P/M material can be accomplished at an annealing temperature of 1430°C (2600°F). As might be expected, the degree of recrystallization attained is dependent upon the forging temperature and the amount of deformation imparted. In other words, when the material is forged at 1095°C (2000°F) full recrystallization can be accomplished at an annealing temperature of 1430°C (2600°F) by imparting reductions of 25% or more. As the forging temperature increases through the range to 1430°C (2600°F), the amount of reduction necessary to get full recrystallization at the 1430°C (2600°F) annealing temperature increases proportionately until it reaches 50% at the 1430°C (2600°F) forging temperature. At forging temperatures above 1430°C (2600°F), the process reverses wherein only reductions above 35% are required to fully recrystallize material forged at a temperature of 1650°C (3000°F).

METALLOGRAPHIC STUDY OF THE HOT-SHORTNESS CONDITION

Indicated earlier, the P/M alloy exhibited a hot-short condition particularly above the 1315°C (2400°F) forging temperature. A metallographic analysis was conducted to establish the cause of this condition. Microspecimens reduced zero percent at forging temperatures from 1315°C (2400°F) to 1650°C (3000°F) and annealed at 1650°C (3000°F) were reviewed. Results indicated precipitate(s) had formed at the grain boundaries. When the precipitate(s) were detected, the zero-reduction microspecimens forged at 1095°C (2000°F) and annealed at temperatures from 1315°C (2400°F) to 1650°C (3000°F) were also reviewed. Again the grain boundary precipitation was evident, exhibiting the most severe condition at the 1650°C (3000°F) annealing temperature.

At this point, it could not be established whether the high forging or high annealing temperature was responsible for the phenomenon. As-forged microspecimens representing forging temperatures from 1315°C (2400°F) to 1650°C (3000°F) were prepared. These revealed that at a forging temperature of 1430°C (2600°F) and above the precipitation is very pronounced. The photomicrographs representative of these evaluations are shown in Figure 8. The crack-like appearance along some of the grain boundaries is a result of the precipitate being pulled out during metallographic preparation.

We concluded that the formation of the precipitate(s) causes the hot-short condition. There is strong evidence to indicate that the phenomenon is strain-induced, since subsequent annealing returns the precipitate to solution until an annealing temperature of 1650°C (3000°F) is approached when the precipitate reappears.

This condition encountered in the P/M TZM alloy is similar to that reported by Vorontsova and Morgunova[1] in their work on alloying molybdenum with small quantities of zirconium and titanium in the arc-cast condition. They

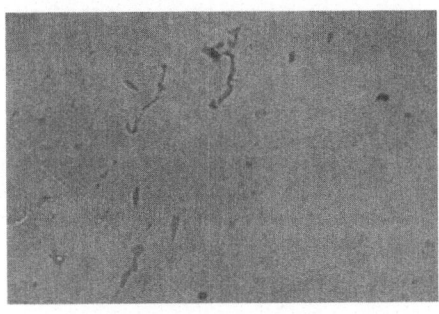

As-Forged at 1315°C (2400°F)

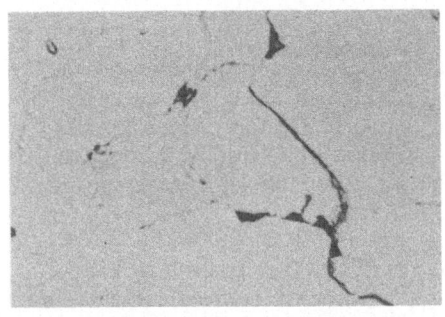

As-Forged at 1430°C (2600°F)

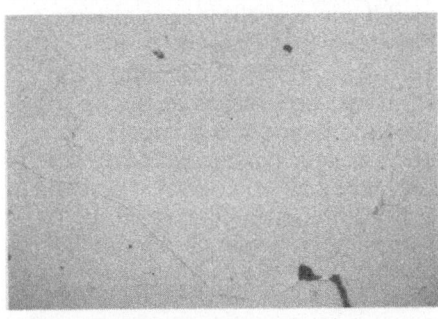

As-Forged at 1540°C (2800°F)

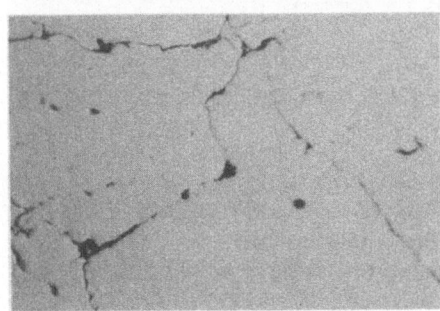

As-Forged at 1650°C (3000°F)

Figure 8. Photomicrographs Depicting the Results of the Metallographic Study to Determine the Cause of the Hot-Shortness Condition Encountered in the P/M TZM. (All Photographs originally at 1000X Magnification; All Specimens Unetched.)

reported an abrupt change in formability above a temperature of 1300°C (2372°F), with a slight improvement noted above 1600°C (2192°F). Those results correlate with the findings of this study, wherein the P/M material exhibited good forgeability up to 1315°C (2400°F) and poor forgeability above that forging temperature.

Vorontsova and Morgunova[1] attributed the poor forgeability to the formation of a fusible molybdenum oxide, which decreases the bonding between grains. Their subsequent work on the formability of the alloy indicated that it is greatly improved if it is partially worked, annealed, and then reworked at higher temperatures.

Comparable behavior of this alloy at higher temperatures (2000°C or 3630°F) was reported by Bonchak[2] of Westinghouse in his study on arc-cast ingots. He also encountered poor forgeability, which was overcome by partially forging (establishing 25% as the maximum reduction) and annealing, and subsequently reworking the material at higher temperatures. Bonchak attributed the poor forgeability to the formation of carbides, which later entered into solution upon annealing.

Both of the references studied in this report indicate that there is a reaction at the grain boundaries, whether it be the formation of oxides, carbides or something else (workers at Allegheny Ludlum[3] did some qualitative microprobe analyses on the precipitates in the grain boundaries of both arc-cast and P/M TZM, and found the same amount of titanium but more zirconium in the precipitates than in the matrix). The reaction causes the hot-shortness of the alloy. It is surmised that this phenomenon is strain-induced and occurs when the forging temperature exceeds 1315°C (2400°F), causing nucleation of the precipitate(s) at the grain boundaries. This strain-induced condition can be corrected by partially forging at a lower temperature and annealing, followed by finish forging at a higher temperature, thus increasing the alloy's forgeability.

CONCLUSIONS

The following conclusions were drawn as a result of the investigative work described in this report:

1. The P/M TZM exhibits good forgeability in the forging temperature range from 1095°C (2000°F) to 1315°C (2400°F).
2. Poor forgeability (hot-shortness characteristics) was encountered at forging temperatures above 1315°C (2400°F).
3. P/M TZM can be fully recrystallized by annealing for one hour at 1430°C (2600°F). Partial recrystallization can occur at 1315°C (2400°F).
4. The poor forgeability (hot-shortness) is attributed to unidentified precipitate(s) which form at the grain boundaries at all forging temperatures studied. These precipitate(s) become severe and detrimental as the forging temperature exceeds 1315°C (2400°F).

PRESENT WORK AND APPLICATIONS

Many successful forgings have been made subsequent to this investigation by not exceeding a forging furnace temperature of 1260°C (2300°F). Subsequently the manufacture of P/M TZM forging billets has been upscaled from the 650 pound level to 2200 pound billets, and even now we have been asked to consider fabricating a 3200 pound billet. We feel we are limited only by the availability of isostatic presses large enough in diameter and the necessary sintering equipment with which to fabricate almost any size billet. Contrast the size of these P/M billets with the twenty to thirty pound P/M parts being attained only today in the ferrous P/M industry.

We see applications of P/M TZM ranging from large dies for hot-pressing metal powders and chemicals, to piercer points for making stainless steel tubing, to inserts and dies for aluminum die casting and for development work on ferrous die casting.

REFERENCES

(1)   T. V. Vorontsova and N. N. Morgunova, "Forging Molybdenum and Alloys On Its Base", Russian Periodical, U.S.A.F. Foreign Technology Division, Translation FDT-TT-62-1789, February 14, 1963.

(2)   J. Bonchak, "Molybdenum Forging Process Development", Contract AF33(600)-41419, Westinghouse Materials Manufacturing Department Project 7-756 for the U. S. Air Force.

(3)   Allegheny Ludlum, "Extruding and Drawing Molybdenum to Complex Thin H-Sections", Contract AF33(657)-11203, 1963 to 1965.

THE FABRICATION OF TWO PHASE ALLOYS IN THE

RUTHENIUM-GOLD-PALLADIUM SYSTEM

R.SAVAGE[a] AND V.A. TRACEY[b]

(a) BRITISH NON-FERROUS METALS RESEARCH ASSOCIATION, LONDON

(b) INTERNATIONAL NICKEL LIMITED, BIRMINGHAM, ENGLAND

ABSTRACT

Pure ruthenium and the majority of ruthenium rich alloys have negligible ductility at room temperature. Such a deficiency makes the pure metal and these alloys less attractive for a number of applications one of which is electrical contacts. Cold ductility is required for this application so that wire can be cold upset to make snap headed rivets or strip can be cold rolled on to base metals to produce ruthenium clad strip. It has been shown that considerable cold ductility is obtained with a ruthenium rich alloy containing 14% gold and 11% palladium which has a duplex structure consisting of ruthenium grains dispersed in a ductile gold-palladium matrix. This paper describes investigations made to determine methods of preparing the alloy and to evaluate its workability.

The alloy was produced by infiltration of a ruthenium skeleton with a palladium-gold alloy or by liquid-phase sintering a compact of elemental powders. These treatments were carried out at temperatures in excess of the solidus of the palladium-gold alloy-1470°C and preferably between 1520 and 1550°C. Several atmospheres were tried for the infiltration and sintering operations and the highest densities and best ductilities were achieved using vacuum. However, because considerable loss of metal occurred above 1350°C when heating in vacuum, the final stage of heating from 1350°C to the infiltration or sintering temperature was therefore made in an argon atmosphere.

The densities of the infiltrated billets were in excess
of 90% of theoretical and those for liquid phase sintered
billets in excess of 85% of theoretical.

Infiltrated billets were fabricated into 0.13mm (0.005 in.)
thick strip by hot forging, hot rolling and cold rolling.
Liquid phase sintered billets of circular section were processed
to wire of 1.5mm (0.06 in.) diameter by hot die forging, hot
rod rolling and hot and cold drawing.   The hot processing of
both wire and strip was done by heating to 1100°C in hydrogen
prior to each pass.   In the final stages of processing into
both wire and strip, the reductions in area per pass were
below 5%.

Processing caused elongation of the ruthenium grains and
it was found necessary to periodically heat above the solidus
temperature of the palladium-gold matrix to spheroidise the
ruthenium grains before further deformation.

The tensile strength of the ruthenium palladium-gold alloy
was similar to pure ruthenium.   The superior ductility of the
alloy was demonstrated by the fact that strip was cold bent
to an angle of 40 degrees and wire was upset into snap headed
rivet shapes typical of those used for electrical contacts.

1.INTRODUCTION

Ruthenium is attractive for several applications because
of its wear, erosion and corrosion resistance[1].   One
application, that of electrical contacts, requires material
which has all three properties but, because electrical contacts
are cold formed by roll cladding and wire upsetting, the material
must have considerable cold ductility so that it can be formed
into contacts by the accepted techniques of production.

Pure ruthenium lacks room temperature ductility[2] but
recent researches have shown that cold ductility can be obtained
in a ruthenium rich alloy containing 14% gold and 11% palladium.
The cold ductility is achieved by having ruthenium grains
dispersed in a ductile gold-palladium matrix[3],[4].

This paper describes the powder metallurgical procedures
employed to produce the duplex alloy billets and discusses
the techniques used to fabricate the billets into wire and strip.

2. EXPERIMENTAL PROCEDURE

Billets were prepared either by infiltrating a pressed
ruthenium skeleton with molten palladium-gold alloy, or by

TABLE 1

RAW MATERIALS

Metal	Form	Mean Particle Size, * μm	Source
Ruthenium	Powder	4.0	International Nickel Limited
Gold	Grain Powder(tan)	20.0 3.6	International Nickel Limited Engelhard Limited
Palladium	Sponge Powder	10.0 4.0	International Nickel Limited International Nickel Limited

* Determined by Fisher Sub Sieve Sizer

liquid phase sintering a compacted mixture of powders of the three constituent elements.   Details of the raw materials used are given in Table 1.

One grade of ruthenium powder was used throughout, but gold and palladium were employed in two forms.   Gold grain and palladium sponge were used to prepare the palladium-gold alloy in strip form for use in the infiltration technique.   Fine powders were used to give the best initial distribution of the constituents necessary for the liquid phase sintering technique.

(a) Preparation of Material by the Infiltration Route

Infiltrated billets were produced by placing appropriate weights of a 56% gold - 44% palladium alloy strip on unsintered ruthenium compacts of rectangular cross-section, and then raising the temperature of the compacts above the solidus of the gold-palladium alloy - 1470°C, so that the latter alloy penetrated the ruthenium skeletons to form the ruthenium - 14% gold - 11% palladium billets.

The ruthenium skeletons were prepared by isostatically compacting ruthenium powder in rectangular latex bags. A skeleton density of 32%, achieved by using a compaction pressure of 46.3 hbar (67 k.s.i.), was found to be suitable for obtaining ease of penetration of the gold-palladium alloy into the compact whilst providing sufficient rigidity to the compact to avoid collapse during infiltration.

After infiltration, the ruthenium-gold-palladium billets were processed to strip by hot forging, hot rolling and cold rolling.   During the hot working processes the materials were reheated to 1100°C prior to each deformation step.   Throughout both hot and cold processing the materials also had to be periodically liquid phase annealed by heating them above the solidus of the gold-palladium alloy to promote rapid solution and reprecipitation of the ruthenium to restore the ruthenium grains to the rounded form produced on infiltration.

(b) Preparation of Material by the Sintering Route

The circular section billets were prepared from mixtures of powders of the three constituent elements by isostatically pressing at 39.2 hbar (57 k.s.i.) and sintering by raising the temperature of the compacts above 1470°C, the solidus of the final 56% gold-44% palladium matrix composition.

The billets were processed to wire by hot die forging, hot rod rolling and hot and cold wire drawing.   As in the production of strip, the materials were heated to 1100°C prior to each hot processing step and periodic liquid phase anneals were required.

The final few wire drawing passes were done at room temperature to provide a good surface finish.

(c) Determination of Structure and Properties

The shape and size of the ruthenium grains, and the distribution of the gold-palladium matrix were determined by metallographic examination of sections taken from material at various stages of processing.

A Cambridge Microscan Analyser was employed to determine the extent of palladium and gold solution in the ruthenium grains and the extent of solution of ruthenium in the gold-palladium matrix.

The mechanical properties of the ruthenium-gold-palladium alloy were determined by tensile testing samples on the Hounsfield tensometer.   The hardnesses of the ruthenium grains and palladium-gold matrix were compared by doing microhardness measurements on sections prepared for microexamination.

The degree of cold deformation obtainable at room
temperature was assessed by bend testing samples of strip, and
preparation of snap headed rivets from wire using the usual
wire upsetting procedures employed to produce electric contacts.

3. EXPERIMENTAL RESULTS AND DISCUSSION

(a) Billet Preparation

After the rapid infiltration of the ruthenium skeleton,
the remainder of the infiltration time was taken up in
reorientating the two phases through growth of the ruthenium
grains as in liquid phase sintering.   As the infiltration
cycle is similar to the liquid phase sintering cycle, the
majority of the factors influencing billet preparation are
common to both processes.

In preparing the billets by both processes it was
necessary to heat above the solidus temperature of the 56%gold-
44% palladium alloy - 1470°C.   The typical structure developed
after such treatment is shown in Figure 1.   Failure to heat
into the liquid phase region resulted in very porous billets
of poor ductility.

The atmosphere used for both procedures was found to be
critical. When heating in vacuum considerable metal loss
occurred by evaporation on heating above 1350°C.   The use of
argon and nitrogen throughout instead of vacuum resulted in
lower densities.   Infiltration and liquid phase sintering in
hydrogen produced billets containing gross porosity and having
poor hot and cold ductilities.   The structure also appeared to
be more angular after heating in hydrogen than when using
argon (Figure 2).   The preferred atmosphere was, therefore,
vacuum changing to argon at a temperature when loss of metal
became appreciable.   The most suitable temperature to change
from vacuum to argon was found to be 1350°C.

Although the solidus temperature of the gold-palladium alloy
is 1470°C, the minimum satisfactory temperature for infiltration
was found to be 1500°C and the minimum temperature for liquid phase
sintering was 1520°C.   The maximum temperature for billet
preparation in argon had to be restricted to 1600°C owing to the
loss of matrix elements by volatilisation at higher temperatures.

The optimum temperature range for infiltration was judged
from the densities achieved to be between 1520 and 1550°C.
Infiltration in this temperature range resulted in densities
in excess of 90% of theoretical (Table 2).   When sintering
compacts of mixtures of elemental powders, the optimum

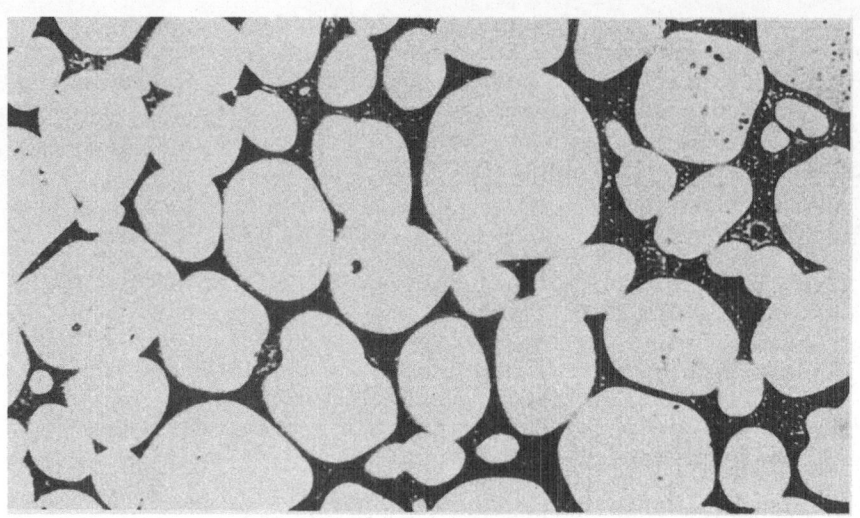

Figure 1 x 300.   19/96/36.
Infiltrated by heating in vacuum to 1350°C and then in argon
to 1550°C and holding for 3h.
Etched sodium hypochlorite in aqua regia.

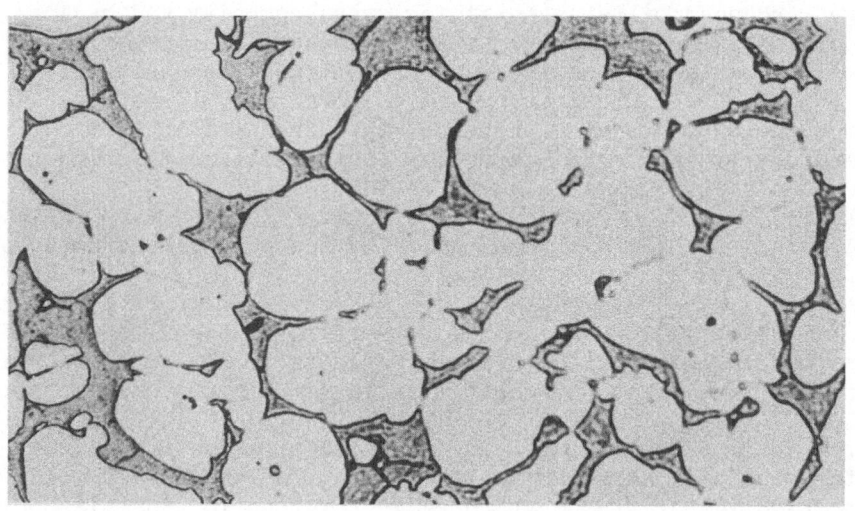

Figure 2 x 300.   20/83/34.
Infiltrated for 3h at 1550°C in hydrogen. Etched sodium
hypochlorite in aqua regia.

TABLE 2

EFFECT OF INFILTRATION TEMPERATURE ON DENSITY

Temperature*, °C	Percentage of Theoretical Density
1470	75
1520	93
1550	91
1570	88
1600	85

* held in all cases for 3h.

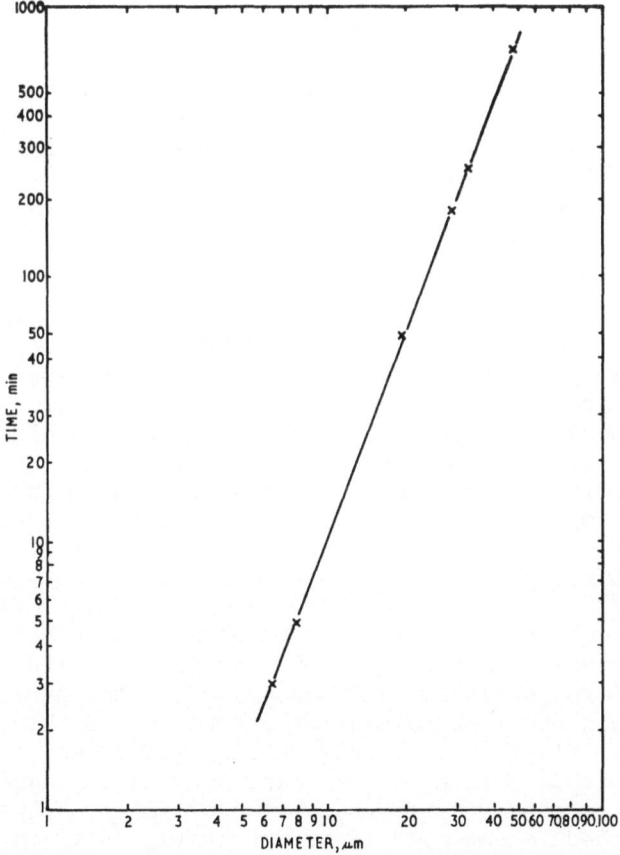

Figure 3 Growth of ruthenium phase during
infiltration at 1520° C

temperature range was 30°C higher, but the final density was
only slightly in excess of 85% of theoretical.   The holding
time after infiltration had a pronounced effect on the ruthenium
grain size.   After 5 minutes at temperature the grain diameter
had increased by a factor of two.   Growth continued with time
at a rate obeying a cubic law (Figure 3), indicating that the
growth of the ruthenium-rich phase was by solution and
precipitation and was controlled by diffusion through the molten
palladium-gold matrix in agreement with findings of Price et al(5)
and Wagner(6).

The ruthenium grains became more spherical with increasing
time at temperature up to 1600°C.   At 1600°C considerable
loss of metal occurred by volatilisation, which appeared to alter
the solution and reprecipitation characteristics and result in
the formation of irregular ruthenium grains.   The ruthenium
grains, in all cases, appeared to be single crystals as no
grain structures were revealed on deep etching.

Microprobe analysis revealed that the extent of solution
of one phase in the other was limited.   The ruthenium phase
had palladium and gold contents of 5% and 0.7% respectively.
The solubility of ruthenium in the matrix phase was about 1%.
Electron microscopical evidence did not reveal any precipitates
indicating that the ruthenium in the matrix phase and
palladium and gold in the ruthenium grains were in solution.

(b) Billet Processing

The ruthenium grains became progressively more elongated
as the degree of deformation increased (Figure 4).   Failure
occurred finally by cracks forming in and propagating through
the gold-palladium phase (Figure 5).   The degree of
deformation possible appeared to be limited by the elongation
of the grains and it was found necessary to periodically
restore the ruthenium grains to the spherical shape by
raising the temperature above the solidus of the gold-palladium
phase under conditions similar to those employed for liquid
phase sintering.

When producing strip, the hot forging and the hot rolling
reductions between reheats at 1100°C had to be limited to
10-15% and 3% respectively.   Processing by both operations
required liquid phase anneals after reductions in thickness
of about 30%.   In the final cold rolling stage reductions
had to be limited to about 2% per pass and liquid phase
anneals were necessary after reductions of 30%.   A typical
processing schedule employed to produce strip is given in
Table 3.

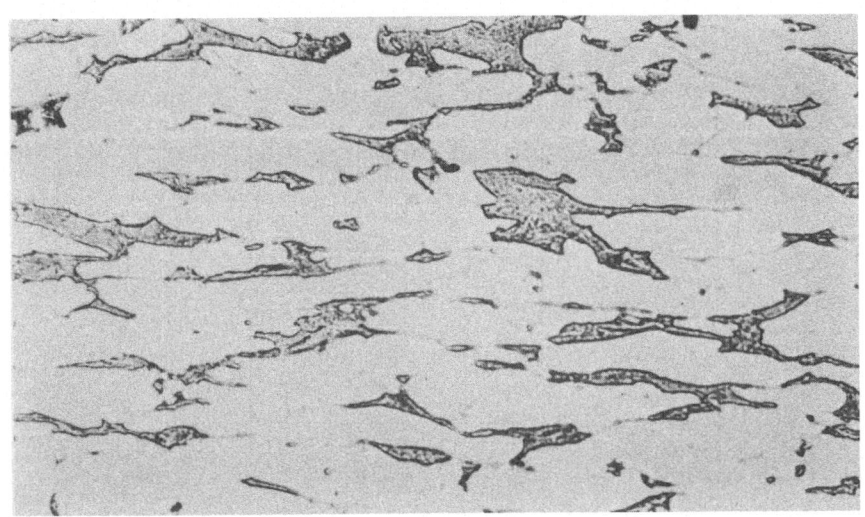

Figure 4 x 300.   21/58/12.
Hot rolled; 80% reduction
Etched sodium hypochlorite in aqua regia.

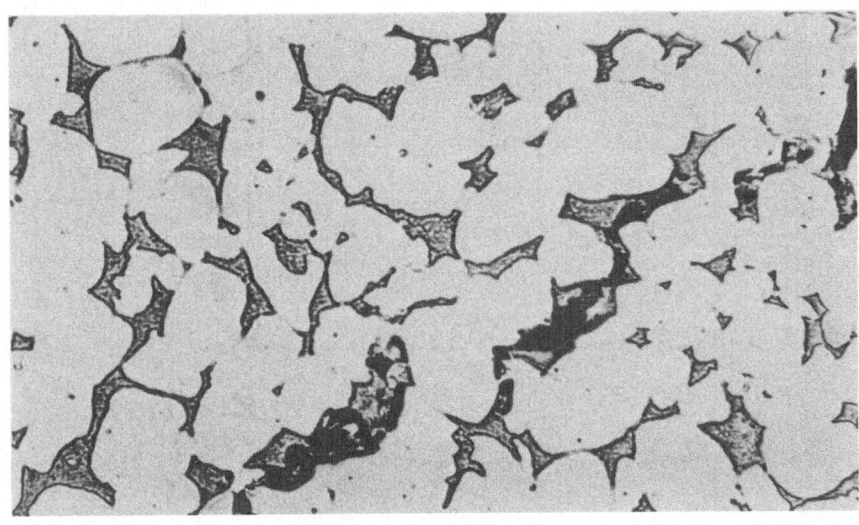

Figure 5 x 300.   20/26/9.
Fracture path through matrix.
Etched sodium hypochlorite in aqua regia.

TABLE 3

TYPICAL SCHEDULE FOR PRODUCTION OF SHEET

Process	Thickness, mm(in)	Reduction for Each Operation %	Number of Passes Between Each Liquid Phase Anneal	Average Reduction in Thickness per Pass, %	Number of Liquid Phase Anneals for Each Process	Average Reduction in Thickness Between Liquid Phase Anneals
Infiltration	9.0 (0.35)	-	-	-	1*	-
Hot Forging	5.0 (0.2)	40/45	2***	15	2**	30
Hot Rolling	1.3 (0.05)	75	8***	3	4**	30
Cold Rolling	0.13(0.005)	90	13	2	6**	30

* Infiltration by heating in vacuum to 1350°C and then in argon to 1520°C and holding for 3h.

** Liquid phase annealed by heating in vacuum to 1350°C and then in argon to 1550°C and holding for 2h.

*** Heated in hydrogen at 1100°C prior to each pass.

TABLE 4

TYPICAL SCHEDULE FOR PRODUCTION OF WIRE

Process	Diameter mm (in)	Reduction in Area for Each Operation %	Number of Passes Between Each Liquid Phase Anneal	Average Reduction in Area per Pass, %	Number of Liquid Phase Anneals for Each Process	Average Reduction in Thickness Between Liquid Phase Anneals, %
Liquid Phase Sinter	7.0(0.28)	-	-	-	1*	-
Hot Die Forge	5.0(0.2)	45/50	3***	5-10	3**	20
Hot Rod Rolling	3.3(0.13)	55/60	12***	2	4**	20
Hot Wire Drawing	1.6(0.063)	75/80	10***	2	8**	20
Cold Wire Drawing	1.52(0.06)	10	5	2	-	-

* Sintered by heating to 1350°C in vacuum and then to 1550°C in argon and holding for 3h.

** Liquid phase annealed by heating to 1350°C in vacuum and then to 1550°C in argon and holding for 2h.

*** Heated in hydrogen at 1100°C prior to each pass.

TABLE 5

MECHANICAL PROPERTIES OF RUTHENIUM AND RUTHENIUM-
14% GOLD - 11% PALLADIUM ALLOY

Property		Ruthenium* Annealed	Ru-Pd-Au* Annealed	Ru-Pd-Au* Cold Worked 40%
U.T.S., hbar (k.s.i.)		49[+] (71)	48 (70)	84
Y.S., hbar (k.s.i.)		37[+] (54)	25 (36)	-
Angle of bend to cracking, degrees		3	40	30
Hardness	Ru m HV50	450/600	~400	~550
	Matrix m HV20	-	~190	~250

+ Ref.7.
* Determined on 0.5mm (0.020 in.) thick strip
  and 1.52mm (0.060 in.) diameter wire.

Figure 6 x 10.   B12134

Snap head rivets processed from 1.5mm (0.060 in.) dia. wire

Wire bar could not be produced by hot swaging the billets at 1100°C and consolidation was achieved by hot die forging at 1100°C with reduction in area per pass of only 5-10%. The wire bar was further processed by rod rolling at 1100°C with reductions in area of 2% per pass. Liquid phase anneals were found to be necessary after reductions in area of 20%. Most of the wire drawing had to be done at 1100°C and cold drawing was confined to the last few passes to improve the surface finish. The reductions in area per pass were limited to about 2%, but again liquid phase anneals were necessary after reductions in area of 20%. Colloidal graphite was used for lubrication throughout drawing. The typical processing procedure employed for wire production is given in Table 4.

Sheet proved easier to produce than wire because fewer steps at each stage were required and it was possible to conduct much of the processing at room temperature. Sheet, 0.13mm (0.005 in.) thick, was successfully prepared, but greater difficulty was encountered in producing wire to 1.5 mm (0.060 in.) diameter. The problems were greatly increased when processing wire to less than 2mm (0.080 in.) diameter.

The tensile strength of the ruthenium - 14% gold- 11% palladium alloy in wire and strip was similar to that of pure wrought ruthenium, Table 5. The higher cold ductility developed in the ruthenium-gold-palladium was borne out by the greater bend ductility, Table 5, and by cold heading tests on wire in which an area spread as high as 180% was obtained, indicating that snap headed rivets, as used for electrical contacts, can be prepared (Figure 6).

CONCLUSIONS

The investigation showed that the ruthenium-palladium-gold alloy can be fabricated into strip or wire by using hot and cold working treatments, provided that liquid phase anneals are given periodically throughout processing. Production of wire was found to be more difficult than sheet because of the many more steps required for fabrication and the need for more of these to be done at elevated temperatures. The ductility would appear to be satisfactory for the preparation of rolled strip contacts and perhaps sufficient for some rivet type contacts.

ACKNOWLEDGEMENTS

The authors wish to thank International Nickel Limited for permission to publish this paper.

REFERENCES

1.   Betteridge W. and Rhys D.W.
      Metal Ind. 26 Aug. and 2 and 9 Sept. 1960.

2.   Rhys D.W., Jnl. Less Common Metal, 1959. Vol.1. 269-291.

3.   Holtz F.C. and Laemont T.G.
      Armour Research Dept. No. ARB 885-6.

4.   Holtz F.C. and Vines R.F., U.K. Patent Specification
      1,025,400.

5.   Price G.H.S., Smithells S.V. and Williams S.V.,
      J.I.M. 1938. Vol. 62. 239-254.

6.   Wagner C.Z. Elektrochem. 65 1961. p.581.

7.   Ruthenium. Inco Ltd. Publication 2161a.

OXIDES AND NUCLEAR MATERIALS

REDUCED MICROGELS - A NEW FORM OF POWDERED METALS AND METAL OXIDES

R. H. Lindquist

Chevron Research Company

Richmond, California 94802

A new process for producing fine powders of metals and oxides in intimate mixtures has been developed in which the particle size of the metals and oxides can be easily controlled over a wide range of particle sizes. The key to the technique is the simultaneous and homogeneous gelation of mixed metal halides in alcohol solution. The gelling agent is an epoxy compound, such as propylene oxide, which abstracts halide ions such as chlorine to form chlorohydrins. A small amount of water present produces hydroxyl groups as the chloride ions are removed. Thus, a homogeneous metal hydroxide gel results with the cations uniformly distributed throughout the gel. When dried, a powdery material is formed consisting of metal oxy hydroxide particles. The chloride ions are volatilized as chloro- hydrins. Depending on the initial metal halides used, materials can be gelled which will finally result in metal alloy powders, mixed metal-metal oxide powders, and complex metal oxides such as ferrimagnetic garnets.

INTRODUCTION

A number of approaches have been used to mixing fine metal and metal oxide particles for powder metallurgy applications.[1,2] Probably the earliest technique was comminuting metal and oxide particles in a ball mill, sometimes with solvent. Difficulties in

1. R. L. Sands and C. R. Shakespeare, Powder Metallurgy, Newnes, London (1966), p. 20.

2. A. R. Kaufmann, U.S. Pat. 3,440,042, Apr 22, 1969.

maintaining a uniform dispersion arise from gravity settling the
heavier particles and the lack of interaction of the oxide particles
with the metal particles. When a final alloy is desired between
two or more metal particles, interparticle metal diffusion at high
temperatures is necessary to form such an alloy with little assurance that a uniform alloy would result.

Improved physical methods for forming small particles of uniform composition include atomization of molten metal resulting in
small alloy particles that can be internally oxidized and selective
reduction of mixed metal oxides.[3]

Chemical techniques aimed at forming mixed metal-metal oxide
powders mainly involve variations on the theme of coprecipitation.
Salts of the desired metals and metal oxides are dissolved in water
or their own melts, and then various techniques are used to minimize
segregation of the precipitates on drying. A summary of some con-
temporary processes and the reported minimum powder sizes is given
in Table I. The incentives to produce dispersion-strengthened
nickel base alloys for jet engine applications have resulted in
most examples being 2-4% thoria or alumina in nickel. Several of
the techniques involve grinding of the mixed oxide powders prior
to reduction to increase the homogeneity.

The technique described here develops a uniform gel of metal
cations in a homogeneous mixture prior to any comminution. As a
result, low temperature heat treatments may be used for particle
size control with assurances that the final result will be microscopically uniform.

THE MICROGEL PROCESS

The chemical reactions involved in the Microgel process are
the following for nickel-thoria:

$$Ni^{+2} + Th^{+4} + 6Cl^{-} + 6H_2O + 6H_2C\underset{O}{\overset{H}{-}}C\text{-}CH_3 \longrightarrow$$

$$(Ni,Th)(OH)_6 + 6H_2C\overset{H}{-}\underset{\substack{Cl \\ OH}}{C}\text{-}CH_3$$

in alcohol solution.

3.   J. G. Rasmussen and J. J. Grant, Powder Met., 8, 92 (1965).

Table I

Chemical Methods for Producing Metal-Metal Oxide Powders

Technique	Metal	Size, μ	Metal Oxide	Size,* Å	Vol %	Ref.
Coprecipitation of nitrates with ammonia	Copper		Alumina	200-1000*	2	4
	Nickel	0.05-0.250	Thoria	100-500	2	10
Salt fusion of metal oxides and nitrates	Nickel		Thoria		2	5
			Alumina		2	
			Magnesia		2	
Ignition of metal oxide doped with stable metal oxide nitrate	Nickel		Alumina	50-1500*	2	5
	Copper		Magnesia	50-2000*	0.4-4.0	
Spray dry aqueous metal oxalates, acetates, and nitrates	50% Cobalt 20% Chromium 11% Nickel 8% Molybdenum	0.6-2.0	Thoria	110-280	2	6
Flash decomposition in fluidized bed of metal nitrates	Iron		Alumina		6	7
Flash drying Selective reduction	Cobalt 20% Chromium 10% Nickel 15% Tungsten		Thoria or Lanthana	1200-4100 2800-7700	4 4	9
Pressure hydrometallurgy reduction of aqueous suspension	Nickel	0.2-0.5	Thoria	120-630	0.5-5	11
Microgel formed by gelation of metal halides	Nickel	0.25-0.7	Thoria	25-75	2	8

*Indicates metal oxide size in compacted, annealed material.

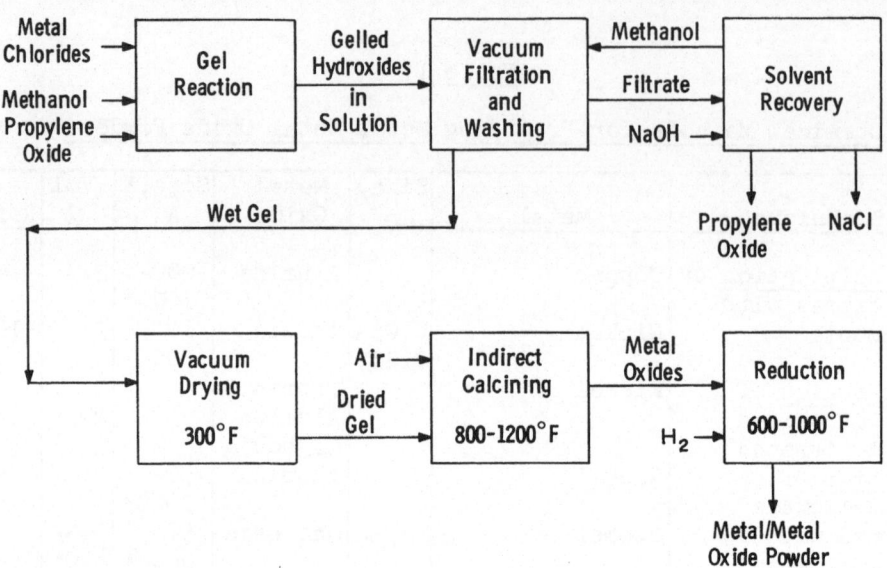

FIGURE 1.   BLOCK FLOW DIAGRAM OF REDUCED MICROGEL PROCESS
FOR TRANSITION METALS AND INERT METAL OXIDES.

Upon drying, the alcohol solvent and propylene chlorohydrins
vaporize, leaving a glassy, dried metal hydroxide gel of homogeneous
composition.

Figure 1 is a block flow diagram of the Microgel process showing the gel reaction loop which generates homogeneous wet gel to
the drying, calcining, and reduction sections.

4.   D. H. Desy, Bu. Mines RI 7228 (1969).

5.   M. F. Grimwade and K. Jackson, Powder Met., 6, 13 (1962).

6.   R. F. Cheney and W. Scheithauer, Jr., NASA CR 54599 (1968).

7.   B. Bovarnick and H. W. Flood, U.S. Pat. 3,305,349, Feb 21, 1967.

8.   R. H. Lindquist, U.S. Pat. 3,458,306, July 29, 1969.

9.   B. H. Triffleman, NASA CR 54516 (1967).

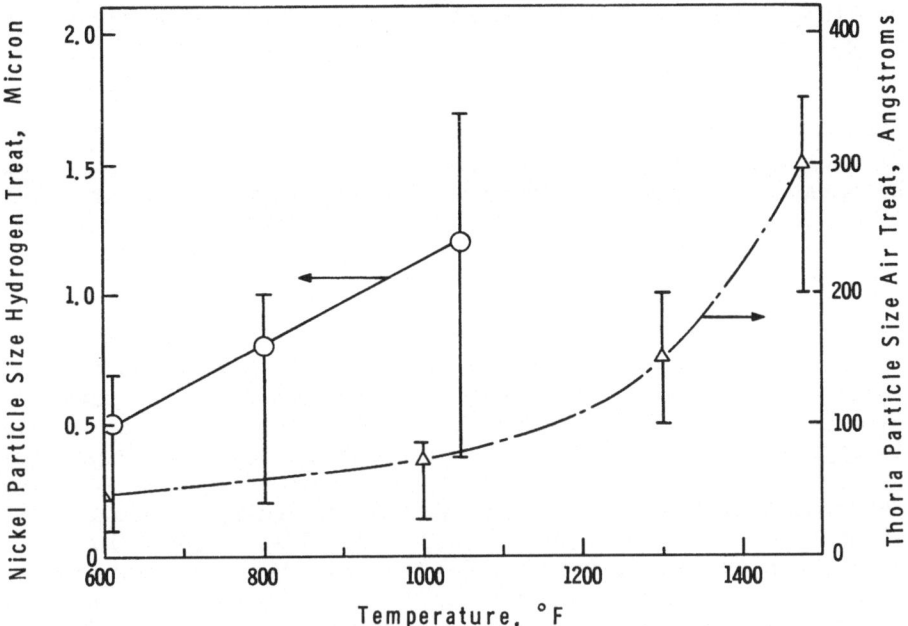

FIGURE 2.   PARTICLE SIZE DISTRIBUTION AND MEAN PARTICLE SIZE
FOR THORIA AND NICKEL PARTICLES PREPARED BY MICROGEL
TECHNIQUE AFTER FOUR HOURS IN ATMOSPHERE INDICATED.

Particle size is controlled for the metal oxides by the time
and temperature of calcining. An illustration is shown for thoria
powder in Figure 2. Of the metal oxides prepared, yttria and thoria
show the smallest particle size for given heat treatments followed
by alumina and silica. The particle size of the metal constituent
is determined by time and temperature in a hydrogen reducing atmosphere. Experiments with fluidized beds versus static beds show no
differences in the particle size or oxide distribution. The effect
of temperature on nickel particle size during hydrogen reduction is
also shown in Figure 2. The size distribution is weighted as shown
by the position of the median size point on the range line at

10. G. B. Alexander and W. H. Pasterfield, U.S. Pat. 3,019,103,
     Jan 30, 1962.

11. R. W. Fraser, B. Meddings, D. J. I. Evans, and V. N. Mackiw,
     Proc. Int. Powder Met. Conf., 2nd, N.Y., 2, 87 (1965).

several temperatures. Little particle growth is seen in the thoria
and yttria systems during the reduction step. Apparently, particle
growth is inhibited by the presence of hydrogen rather than air.

The distribution of metal and metal oxide was followed for the
nickel-thoria system from powder to sintered metal. Figure 3 shows
in Photograph A the powder size distribution. Notice how the small
thoria grains, 100 Å in size, cluster around the nickel particles
1 micron in diameter. This powder was cold pressed at 60 ksi without lubricant.

The billet was hydrogen sintered at 2000°C and then hot rolled
at 1800°F. A plastic replica removal process stripped thoria particles from an electropolished region of the metal sheet, giving
the replica shown in Photograph B of Figure 3. The regions without
thoria approximate the size of nickel particles in Photograph A.

A segment of the metal sheet was thinned using Bollmann's
technique[12] to a thickness less than 200 Å. Photograph C in
Figure 3 shows the appearance of the thinned specimen by transmission electron microscopy. Particularly interesting is the degree
of strain lines and the dispersion of thoria particles. Such a
degree of dislocation lines indicates high temperature stability
for nickel-thoria superalloys.

DISPERSION-STRENGTHENED POWDERS

Tensile tests were obtained for a sample sheet fabricated of
nickel-thoria Microgel powder similar to that shown in Figure 3.
Dr. M. Quatenetz of NASA Lewis Research Center hot pressed the
Microgel powder and then hot rolled it twice. After seven cold
reductions and anneals, test specimens were made and tested at
2000°F. The ultimate tensile strength observed was 9700 psi for a
specimen containing thoria particles 200-500 Å in size. A specimen
made from powder with thoria particles 60-100 Å in size gave a tensile strength of 9100 psi at 2000°F. Both samples contained 2 vol %
thoria. It is probable that neither sample had the optimum thoria
content for the given nickel particle size (0.5-2.0 microns).

The smaller size thoria particles were in such abundance that
clustering occurred, with resulting lower tensile strength. The
optimization involves reducing the nickel particle size or decreasing the amount of thoria.

12. W. Bollmann, Phys. Rev. 103, 1588 (1956).

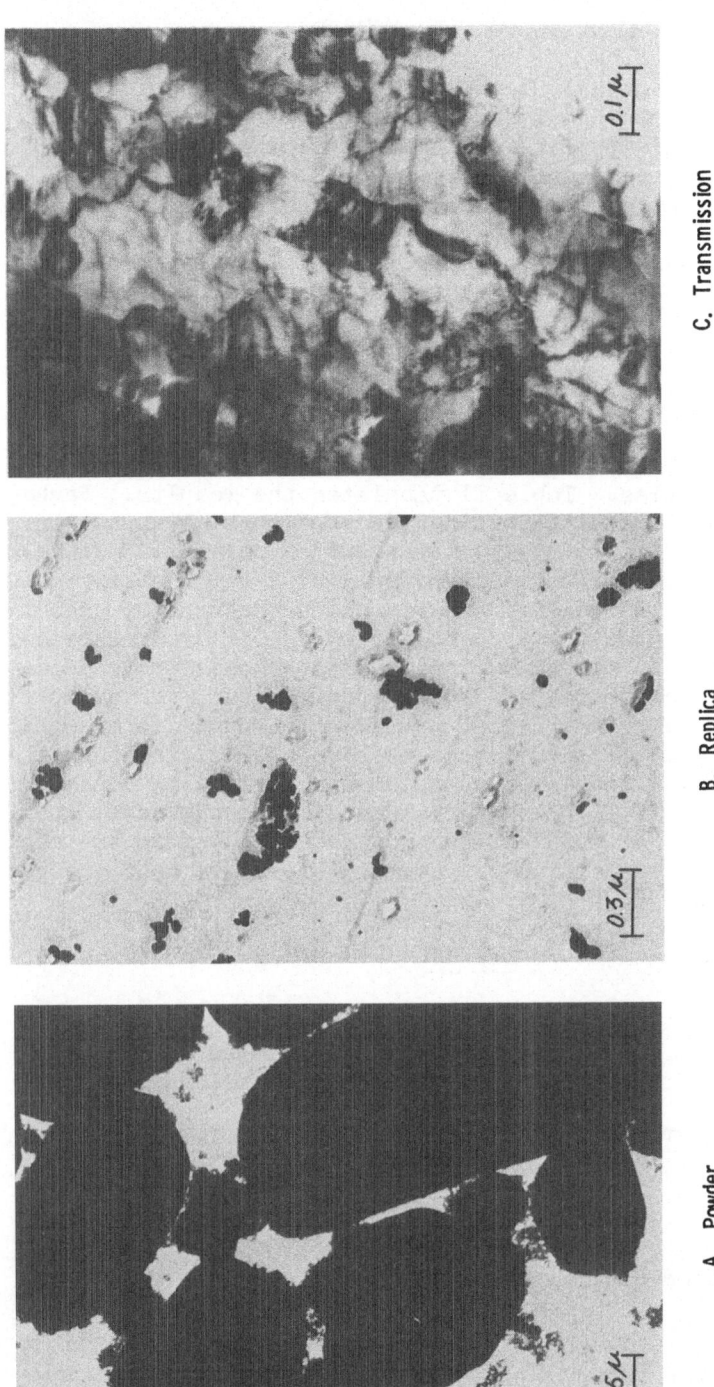

A. Powder

B. Replica

C. Transmission

FIGURE 3. ELECTRON MICROGRAPHS OF 98% NICKEL-2% THORIA: A. POWDER AT 36,000X, SMALL POWDER IS THORIA, B. SURFACE REPLICA OF PRESSED AND SINTERED HOT ROLLED SHEET 27,000X, C. TRANS-MISSION MICROGRAPH OF ELECTROPOLISHED FOIL, 90,000X.

HIGH OXIDE CONTENT POWDERS

A number of potentially interesting powders with high metal oxide content have been produced by the Microgel process. Since the stable metal oxide powder is in intimate contact with the larger metal particles, segregation does not occur in compacting steps, and the opportunity of attaining high loading of stable metal oxides in a metal matrix arises. Figure 4 shows electron photomicrographs of three Microgel powders with stable metal oxide levels over 50 vol %. All of these mixtures could be pressed at 30 ksi into uniform compacts in the MPIF standard flat tensile bar shape. When mixtures of metal and alumina powder were made of the same composition, no satisfactory compact could be pressed, even though ball milling for several days was used in an attempt to achieve homogeneity.

A range of nickel alumina Microgel compositions were examined for physical properties. Table II tabulates the results. Perhaps the most significant results obtained are the thermal expansion coefficients as function of composition and temperature. Since the thermal expansion coefficient differs only by 2×10^{-6} between 100% nickel and 50% nickel-50% alumina, a sandwich construction should be used to capitalize on the property of a high oxide surface and a high metal surface. The high oxide surface should prove excellent in oxidation resistance and sulfide corrosion. For example, a furnace tube could be fabricated with high strength inner metal core surrounded by a high alumina protective shell. In contrast to flame-sprayed alumina coatings, which when cracked will cause failure of the core material, the material made by this process should be self-healing in that more alumina would be exposed in the shell which could sinter to a refractory coating after the metal matrix surface is oxidized.

Table II

Nickel-Alumina Microgel Test Specimens
Physical Properties*

Nickel, Vol %	100	80	64
Resistance, μ ohm in.	4	5.7	22
Ultimate Tensile Strength, 1000 psi	47	35	6
Izod Impact, in. lb		3.2	2.6
Rockwell Hardness, F Scale	95	65	40
Density, g/cc	8.5	8.3	7.8
Thermal Expansion Coefficient x 10^{-5}			
160°F	1.57	1.28	1.23
500°F	1.62	1.45	1.41

*After pressing at 30 ksi and sintered four hours in hydrogen.

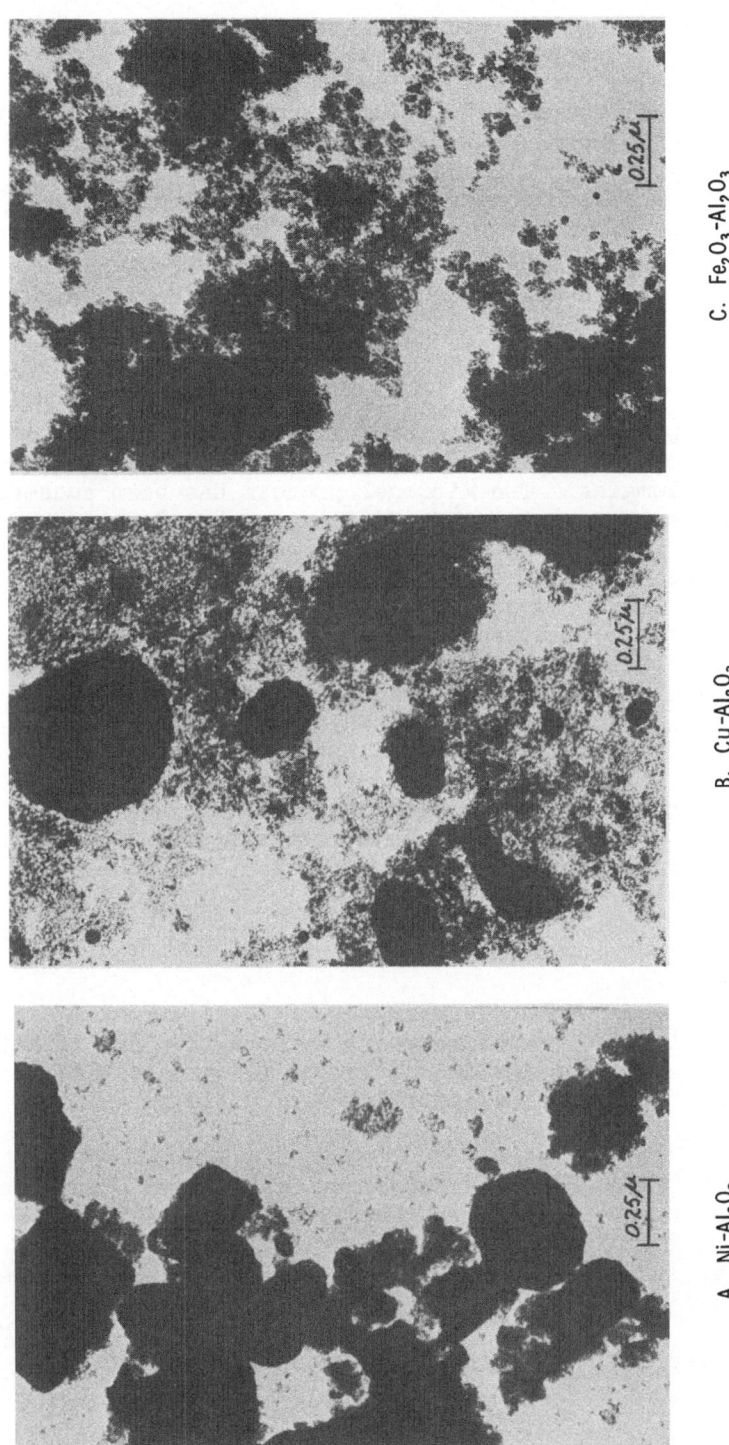

C. Fe$_2$O$_3$-Al$_2$O$_3$

B. Cu-Al$_2$O$_3$

A. Ni-Al$_2$O$_3$

FIGURE 4.   ELECTRON MICROGRAPHS AT 36,000X OF MICROGEL POWDERS: A.   30% NICKEL-70% ALUMINA,
B. 30% COPPER-70% ALUMINA, C. 50% IRON OXIDE-50% ALUMINA.

Other applications include modifying the thermal conductivity of clad metal reactors, supplying the small degree of ductility necessary in light weight composite armor,[13] diluting costly metals to reduce cost per unit volume, and producing unusual and decorative metal finishes. High alumina content produces a dark bronze luster in copper; in nickel, a high alumina content results in a pewter hue. Shrinkage of the green compact increases with increasing oxide concentration. Hot forging or recoining are desirable fabrication methods if oxide concentration is over 50 vol %. In sheet fabrication, hot rolling would be the method of choice.[14]

FERRITES AND GARNETS

For certain applications in electrical and microwave devices it is desirable to use extremely fine grain materials that behave as single magnetic domains. The Microgel process has been successfully used to make such materials.[15] Single magnetic domain particles are produced by starting with the desired metal chlorides to make a given ferrite composition, such as $Mg_{.5}Mn_{1.5}Fe_2O_4$. After making a Microgel of such composition, the gel can be broken into micron size particles in a suspended fluid consisting of aluminum chloride in methanol using a colloid mill. When the colloid solution is gelled again by addition of more propylene oxide, the resultant material is a gel-coated gel, which upon drying and calcination produces submicron size ferrite particles coated with alumina. In effect, the small ferrite particles are insulated from each other by the alumina gel. Compaction and sintering yield a fine grain material processing high speed switching advantages for use in electronic applications.

CONCLUSIONS

The Microgel process produces mixtures of metal oxides of controllable particle size in the range of 25 to 10,000 Å. An intimate combination of the metal oxides and metal powders results because of the uniformity of the initial gel of cations.

13. M. L. Wilkens, Third Progress Report on Light Armor, Lawrence Radiation Laboratory, Livermore, Rept. UCRL-50460 (1968).

14. W. G. Jaffrey, I. Davies, and R. L. S. Taylor, Science Journal, 5A, 61 (1969).

15. R. H. Lindquist and B. F. Mulaskey, U.S. Pat. 3,425,666, Feb 4, 1969.

THE DENSITY OF COEXTRUDED METAL CLADDED ROD CORE AS FUNCTION OF CARTRIDGE DENSITY AND CORE LENGTH

W. Rutkowski, W. Szteke, M. Wieczorkowski

Institute for Nuclear Research, Poland

EXTRUSION AND COEXTRUSION

Extrusion is the plastic flow of metal in the direction of the acting pressure. This method is now widely used in obtaining rods, tubes and profile shapes.
The main advantages of the extrusion processes are: good material economy, few operations needed as compared with other technologies, good dimensional tolerances and high mechanical properties, even without heat treatment.
During the extrusion process physical and mechanical properties of the material change considerably. About the extrusion, taken as a volumetric compression process, one can state that, in the case of the majority of metals, it can be considered to be the most adequate method of plastic working, utilizing the plastic properties of the metal.
The coextrusion process is, as known, one of the important technological processes used during fabrication of several types of reactor fuel elements. In this connection we have started experiments dealing with coextrusion of metal powder and ceramic dispersions cladded in alloyed aluminium.

EXPERIMENTS AIM AND APPLIED METHODS

Based on our experience with manufacture of coextruded nuclear reactor fuel elements, special investigations concerning the mechanism of material flow of the core

and canning tube during the coextrusion process.
Two experimental series were started to examine the
material flow conditions of the core and of the metal
clad.
In both cases, previously prepared cartridges were
cold extruded /520–620°K/ using steel matrices.
Fig. 1 shows the matrix assembly scheme.

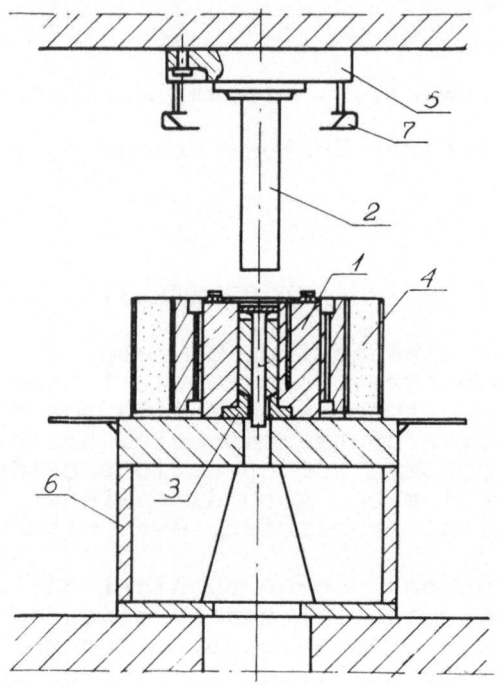

Fig.1. Matrix assembly scheme. 1 – matrix, 2 – punch,
3 – die, 4 – furnace, 5 – steel plate, 6 – table,
7 – lug.

The matrix was vertically situated in a hydraulic press
and externally heated by a special furnace. The extrusion process proceeded at a speed of 5 m/sek and an
approximate load of about 35 tons.
Special lubricant consisting of graphite, lubricating
oil /high ignition temperature/, wax and molybdenum
sulfide was used in the experiments.
The cartridge /Fig. 2/ consisted in all experiments
of a stable container from low alloyed aluminium
/Table Nr 1/. The core of the cartrigde was varying

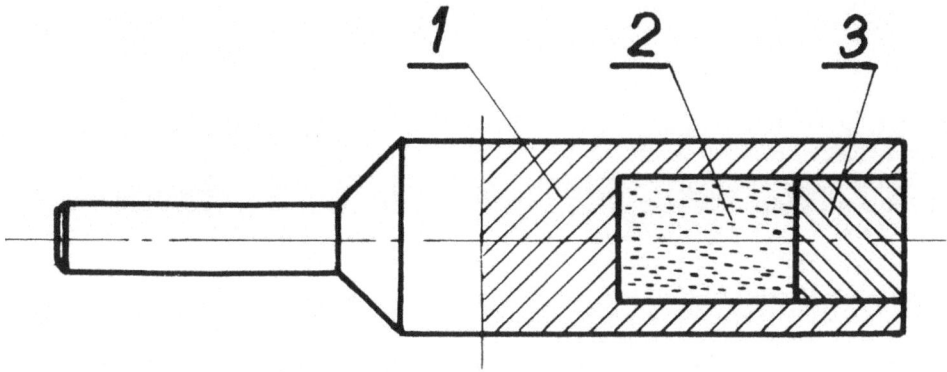

Fig. 2. The cartridge 1 - Aluminium alloy clad, 2 -
metal-oxide mixture, 3 - aluminium alloy plug.

depending on experimental series. As core compositions
following powders were used: copper, magnesium, aluminium, zinc, cobalt and their mixtures with: 5, 10,
20, 30 to 90 weight percent of alumina.
The microphotographs of the powders are shown in Fig.
3 - 8. Data on material characteristics is given
in the Table 1. The core material was pressed, before
location in the cartridge.
In the first experimental series, as constant were
taken: the cartridge core dimension /height/ and initial cartridge core density. This could be achieved
by changing the initial core constituents weight
/in uniform proportions/.
In the second experimental series, different core
densities were taken under observation, using different
pressure loads. So - the cartridge cores showed different heights and stable weights.
The coextruded rods were then tested as follons:
The core: density, hardness and microhardness distribution parallel and vertical to the extrusion direction, degree of cold work /grain size measurement/
and material texture were evaluated.
The clad tube: hardness, microhardness, degree of cold
work and material texture.

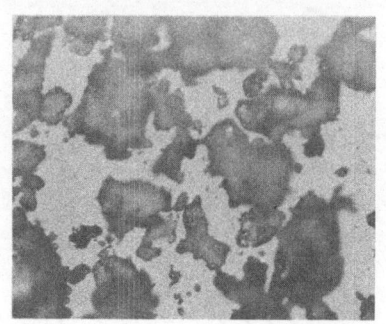

Fig. 3 Zinc powder. 100 x

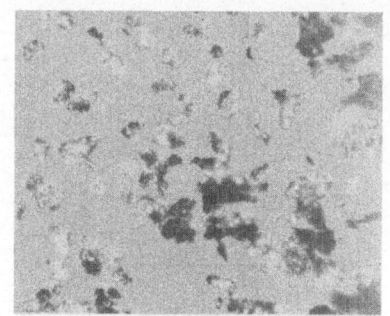

Fig. 4 Al powder. 100 x

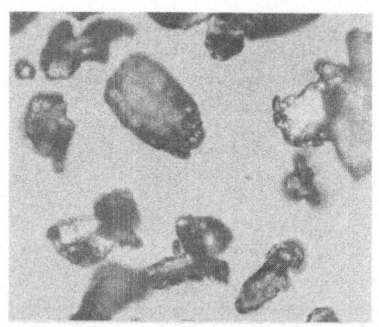

Fig. 5 Mg powder. 100 x

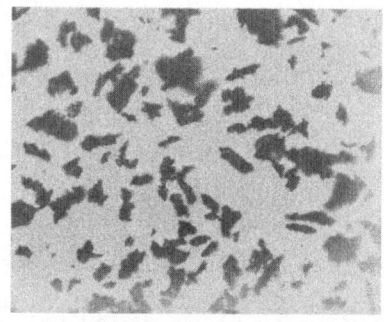

Fig. 6 Copper powder. 100 x

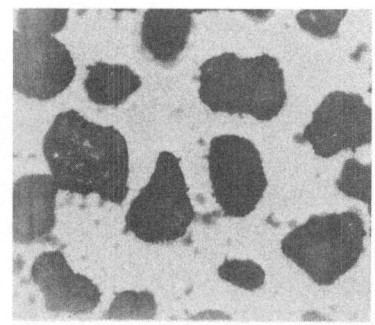

Fig. 7 Co powder. 100 x

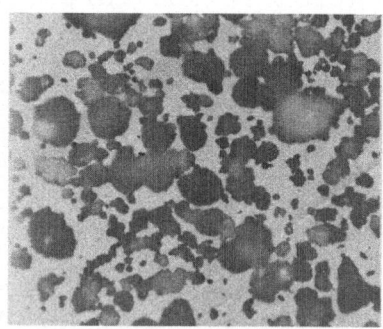

Fig. 8 Al_2O_3 powder. 100 x

Table 1.
Characteristic of materials used in our experiments.

material	Chemical composition	Specific surface /B.E.T./ m²/g	Theoretical density g/cm³	Grain size used μm	Lattice type
aluminium alloy /clad/	Si-1,03%, Fe-0,36%, Mg-0,9%, Mn-1%, Cu-0,043%, Zn-0,07%		2,69g/cm³		
aluminium powder	Cu-0,005%, Si-0,15%, Fe-0,2%, Mn-0,004% Mg-0,06%, Ti-0,01%	0,70 m²/g	2,72 g/cm³	71-125	K 12
cobalt powder	Ni-0,01%, Mg-0,1%, Al-0,05%, Fe-0,02%	0,42 m²/g	8,9 g/cm³	71-125	H 12
copper powder	Sn-0,003%, Pb-0,06%, Ag-0,01%, Fe-0,001%	0,51 m²/g	8,93g/cm³	30-60	K 12
magnesium powder	Cu-0,04%, Mn-0,005%, Si<0,01%, Fe<0,01%, Zn<0,02%, Al~0,02%	1,30 m²/g	1,74 g/cm³	71-125	H 12
zinc powder	Ca-0,01%, Cd-0,04%, Pb-1,05%, Cu-0,055%, Mg-0,004%	1,21 m²/g	7,14 g/cm³	28-200	H 6
aluminium oxide powder	Si-0,45%, Fe-0,1%, Mg-0,01%, Mn-0,01%, Ti-0,01%, Cu-0,005%, Pb-tr, Sn-tr.	32,3 m²/g	3,9 g/cm³	0-100	α H 2 β H 2 γ K 8

EXPERIMENTS RESULTS AND THEIR DISCUSSION

The rod section shapes and hardnes /density/ distributions prove what could be expected, that the clad and/
or core flow does not run identically, when using different core materials - at constant extrusion parameters.
The rod section shapes for different matrix metals
and Al_2O_3 content are demonstrated for Mg and Cu in
Fig. 9 and 10.

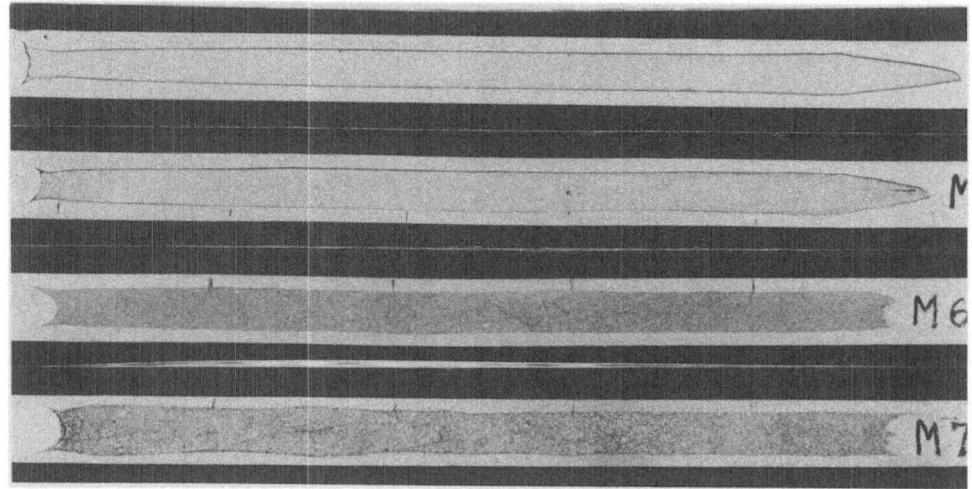

Fig. 9. Mg + Al_2O_3: 0,20,40 and 50 v.p.c. Al_2O_3

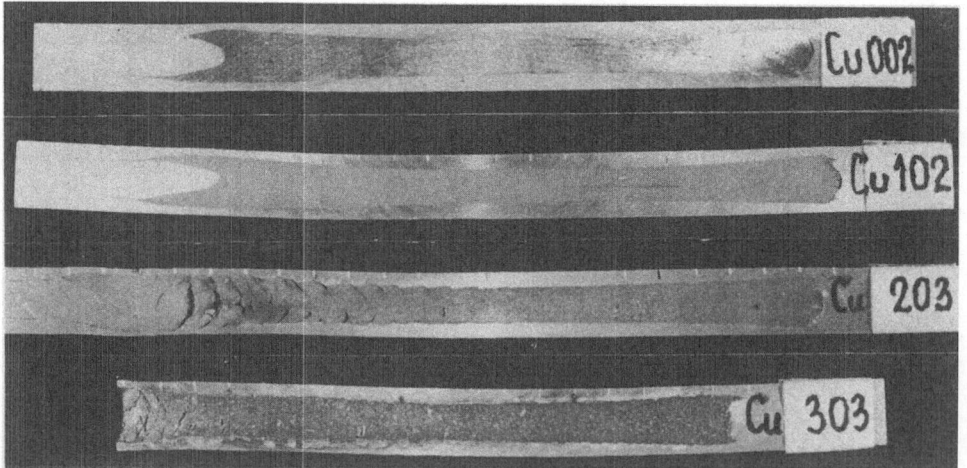

Fig. 10. Cu + Al_2O_3: 0,20,40 and 50 v.p.c. Al_2O_3

Greater amounts of Al_2O_3 destroy the rods when using
Cu as matrix metal, but do not destroy the rods containing Mg, as can be seen in Fig. 11.

Fig. 11. Mg + Al_2O_3: 65,80, v.p.c. of Al_2O_3.

But in this case, the flow of the material changes
considerably.
The measured values on the core and aluminium alloy
clad were expressed as function of the core length, tak-
en from the beginning /the origin of core formation
during extrusion/. Mean values of the core diameter
and clad thickness were evaluated, and the local deviations from this values: ΔR - deviations from the mean
rod diameter, and ΔC - deviation from the mean clad
thickness were calculated. Fig. 12 represents the ratio $\Delta R/\Delta C$ for rods of core composition Al, Mg and Cu
mixed with about 10 v.p.c. of Al_2O_3 plotted against
the core length.
From the Fig. 12 one can recognize, that this ratio
changes considerably depending on the matrix metal, even
in the case of metals with very close melting points
/Al, Mg/but showing different structure /K 12, H 12/
and different elongation at extrusion temperature.
When dividing the core length in five parts, and calculating the mean values of $\Delta R/\Delta C$ on the length of the parts
one can obtain the graph demonstrated on the Fig. 13,
showing the $\Delta R/\Delta C$ change depending from Al_2O_3 content
and its position on the core length.
It is obvious, that the Al_2O_3 content changes considerably the flow character of the rod, but it is also
interesting to show the change of the curves plotted
against the core length.
Using the ratios R_1/C_1 and R_2/C_2, where R_1, and R_2
are the core thickness values, measured on the section
from the rod axis, and the C_1 and C_2 the thicknesses
of neighbour clads, the deviations of the core central
position can be demonstrated on the core length.

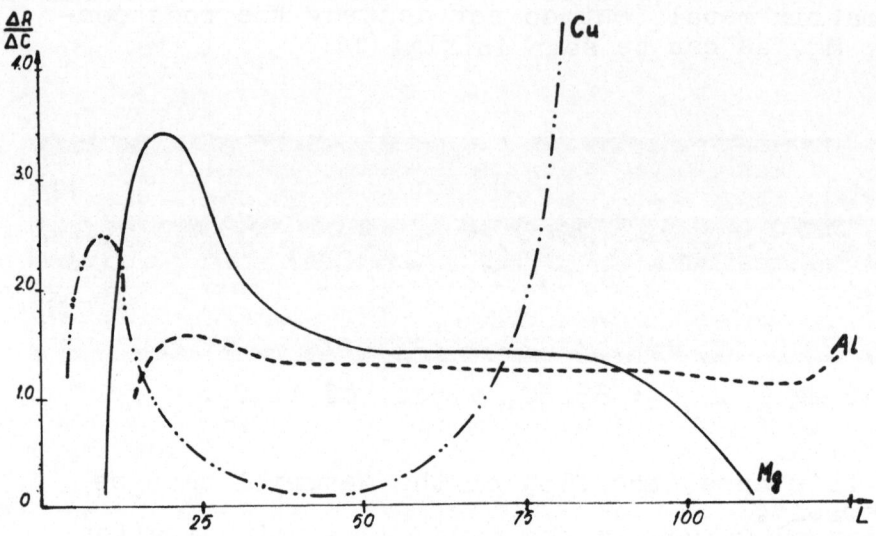

Fig. 12. ΔR/ΔC visually core length for Al, Mg and Cu
+ 10 v.p.c. of Al₂O₃.

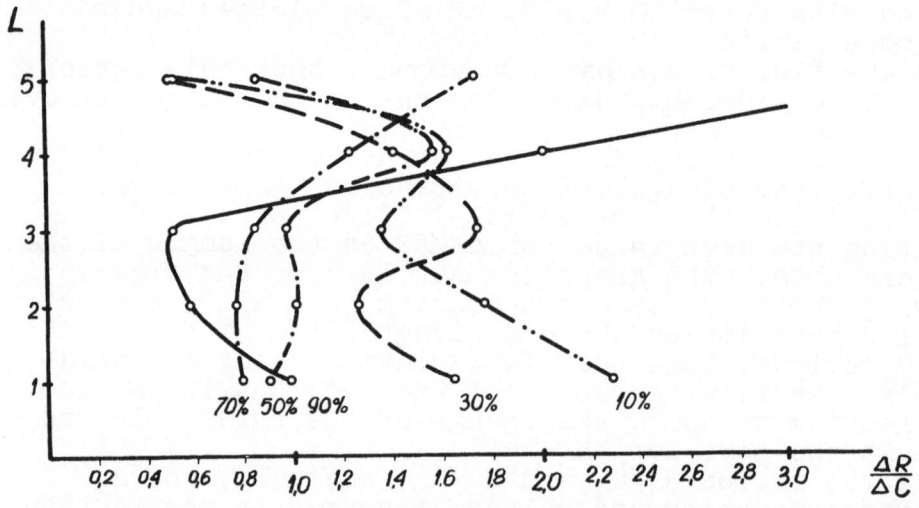

Fig. 13. ΔR/ΔC plotted against the Al₂O₃ content in
Mg matrix and the position on the core length.

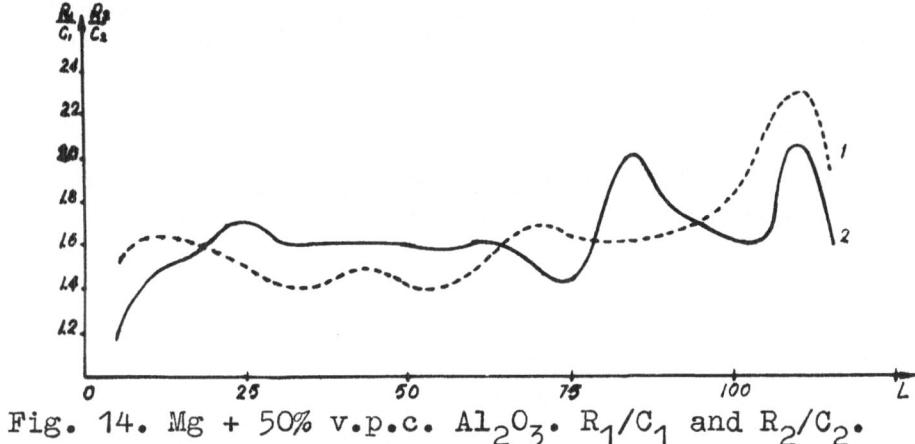

Fig. 14. Mg + 50% v.p.c. Al_2O_3. R_1/C_1 and R_2/C_2.

In Fig. 14 one can observe these deviations. It can
be seen that the core position in the rod changes with
the core length during coextrusion.
The microhardness of the core and the clads measured
on the rod length is shown in Fig. 15.

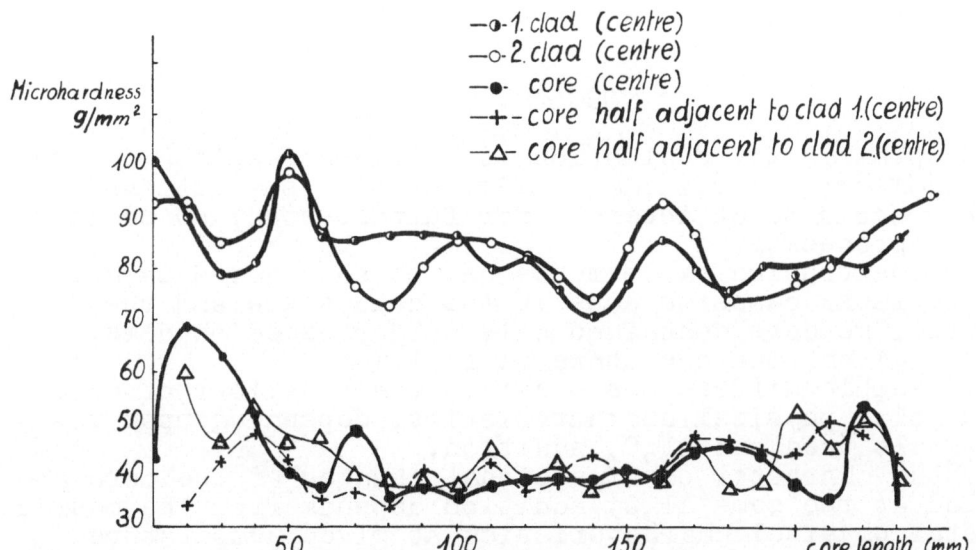

Fig. 15. The microhardness of the core and clad visually
core length for pure magnesium core.

Using the Rockwell hardness measurement one can see
that the diagonals of the diamond pyramide impressions
differ considerably in the direction of the extrusion
and transverse to this direction.
Fig. 16 shows the change of the ratio d_1/d_+, d_1 being
the longitudinal diagonal and d_+ the transverse diagonal in μm, for both clad sides on the rod section,
plotted against the core rod length.

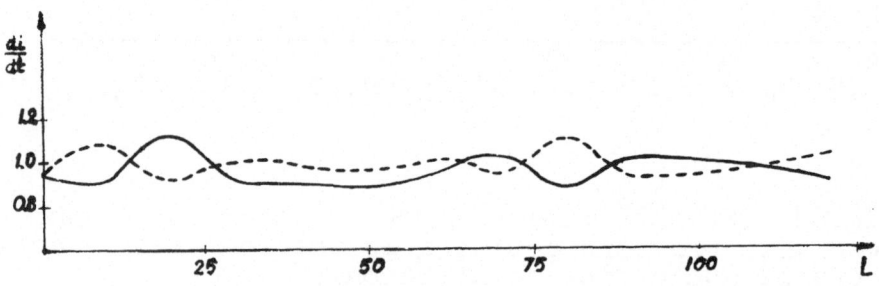

Fig. 16. Ratio d_1/d_+ visually core length /Mg: Al_2O_3 =
         1 : 1 volume/.

It can be seen, that these ratios change in a reverse
manner, their mean value being about 1.

CONCLUSIONS

The preliminary evaluation of obtained results in many
 continuing experiments, indicates some suggestions
that could be of interest for further study on coextrusion processes.
The coextrusion experiments can be considered in all
experiment cases as done in the cold state and the
cartridge core contained only cold pressed powders.
This given, one can state as follows:
1. visible differences occur in the behaviour of core
and clad physical characteristics, depending upon materials used and Al_2O_3 addition.
2. the character of dimensional changes of the core and
clad at the same Al_2O_3 addition depends from the matrix
metal physical characteristics at given temperature
and the core length /Fig. 12/.
3. the Al_2O_3 content in the core acts in its volumetric

ratio and changes the character of dimensional deviations of the core and clad on the core length very distinctly /Fig. 13 and 14/.
4. the microhardness changes according to the core
length and dimensional deviations /Fig. 15 and 14/.
The more specified evaluation of correlations existing between the physical properties of materials used
and results partly presented in this paper can give
more accurate views concerning the coextrusion process.

REFERENCES

1. Fuel Element Fabrication with Special Emphasis on
   Cladding Materials, Vienna /1961/.
2. J. Belle – Uranium Dioxide /1958/.
3. New Nuclear Materials Including Nonmetallic Fuels,
   Prague /1963/.
4. E.A. Evans, Staft – Novel Ceramic Fuel Fabrication
   Processes Fuel Element Fabrication with Special
   Emphasis of Cladding Materials /1960/.
5. G. Cogliati, R. De Leone, S. Ferrari, M. Gabaglio,
   A. Liscia – Lo estrusione Dell'ossido Mistro di Torio
   E Uranio, Roma, CNEN, /1963/.
6. H.H. Hausner – Powder Metallurgy in Nuclear Reactor
   Construction, IAEA, Vienna, 1961.
7. C. Sanve – Lubrication Problems in the Extrusion
   Process, Journ. Inst. Met., 156–65, Vol.
8. L. Meny, M. Champingny, P. Guyader – Eléments com-
   bustibles UO_2 – inox cylindriques et tubulaires –
   Fabrication ět propriétés, New Nuclear Mat. Incl.
   Non-Metallic Fuels, Vienna, 1963.
9. W. Rutkowski, W. Szteke, M. Wieczorkowski – Swaging
   Metallic and Nonmetallic Powders in Metallic Tubes.
   Abh. d. DAW, 1966, 1.
10. W. Rutkowski, W. Szteke, M. Wieczorkowski – EK–10
    Dispersion Fuel Element for Experimental and Univer-
    sity Reactors, Report INR Nr 585/1964/.
11. W. Rutkowski, W. Szteke, M. Wieczorkowski – Some
    Coextrusion Experiments of Aluminium Cladded Cerme-
    tallic Core Rods, Report INR No 728/XIV/R /1966/.
12. S. Stolarz, J. Kurzeja – Inf. Bull. of the Inst.
    of Nonferrous Metals, 1969, 4.

THE INFLUENCE OF POWDER CHARACTERISTICS ON PROCESS
AND PRODUCT PARAMETERS OF URANIUM DIOXIDE

Ulf Runfors

AB Atomenergi

Stockholm, Sweden

INTRODUCTION

Uranium dioxide is the predominant fuel in nuclear reactors, particularly in water cooled reactors. It is an oxide ceramic which implies that the pelletizing technology is more closely related to the technics used in powder metallurgy than those in traditional ceramics.

It was early realized that powder characterization was necessary in order to make predictions concerning the outcome of a pelletizing process. An impressive number of powder characterization methods have consequently been developed over the years. However, these methods have proved to be of limited value in the sense that they were able to give some guidance, when applied to an established manufacturing process, but were of less value in establishing more logically conclusive correlations of basic nature.

This paper is divided into two parts. The first one is concerned with Swedish UO_2 with the object of creating a better understanding of an already established production technic. It was at the outset assumed that the individual processes in pelletizing: powder treatment, pressing, dewaxing and sintering offer so many variables that predictions of the properties of the final product by means of powder characteristics would be rather futile. The efforts were consequently concentrated on finding

311

correlations between the powder properties on the one hand and on the
other hand the variables of the individual or partial processes such as
pressing, sintering etc. or intermediate product properties, which often
introduce boundary limitations in the process.

The second section deals with six commercial UO_2-powders from UK, USA
and Canada. The original object in the latter investigation was to
establish suitable pelletizing technics. Data obtained in the course
of this piece of development work have been extracted and treated in
the same way as in the case of the Swedish powder in order to test the
generality of the findings in the first section.

REQUIREMENT ON SINTERED UO_2

Reactor performance requirements on sintered UO_2 can be classified
into the following categories:

1. Neutron economy

2. Compatability with respect to
    a. the canning material
    b. the coolant

3. Dimensional stability

4. Fission product release

5. Thermal performance

These requirements govern product specifications which in turn govern
production methods and parameters. The principle product properties
are sintered density, stoichiometry, impurity content and dimensions.
In this context it suffices to say that densities can vary between
85-98% of theoretical density depending on the actual reactor construction. The composition shall be stoichiometric which generally is
expressed as $O/U = 2.00$. The powder should further be very pure
in particular with respect to carbon, fluorine and neutron absorbing
elements.

PART I

SWEDISH UO_2 POWDER

The essential feature characterizing the Swedish powder is the particular precipitation process used in the manufacture of the mother
substance, ammonium uranate from a pure uranyl nitrate solution.
The precipitating substance is carbamide $(NH_2)_2CO$ which hydrolyzes
when the solution is brought to boiling. The nucleation of ammonia
starts a homogeneous precipitation which by suitable choice of variables produces a freeflowing powder of almost spherical uranate
agglomerates (fig. 1). Isomorphus UO_2 agglomerates are produced
by air calcination and hydrogen reduction (fig. 2). The single
parameter varied in the manufacture of the powders for this investigation was the temperature of reduction. The powder was produced
in the ordinary production equipment of AB Atomenergi (Atomic Energy
Co), Stockholm, Sweden. Approximately 3 tons of UO_3 were homogenized
in a large cone mixer. Nearly 400 kg were then reduced in a continuous
rotary furnace using a 95 per cent hydrogen, 5 per cent nitrogen gas
mixture representing 100 per cent surplus over the theoretically
required amount of hydrogen. The residence time in the hot zone of
the furnace was approximately 20 minutes. The reduction was carried
out at 550, 585, 630, 700, 800 and $900°C$. The powders were stored
for more than two months, during which time provision was made for
slow oxidation by air, in order to prevent rapid pyrophoric oxidation.

CHARACTERIZATION OF THE POWDERS

The following determinations were carried out on the powder:

1. Apparent powder density
   a. according to the American standard method MPA-4-45
   b. in the pressing dye
2. Flow rate (American standard MPA-4-45)
3. Particle size distribution (turbidimetric method, sieve analysis)
4. Specific surface area (nitrogen adsorption, BET)
5. Microscopy
   a. optical
   b. electron
6. Chemical analysis (O/U-ratio, carbon)

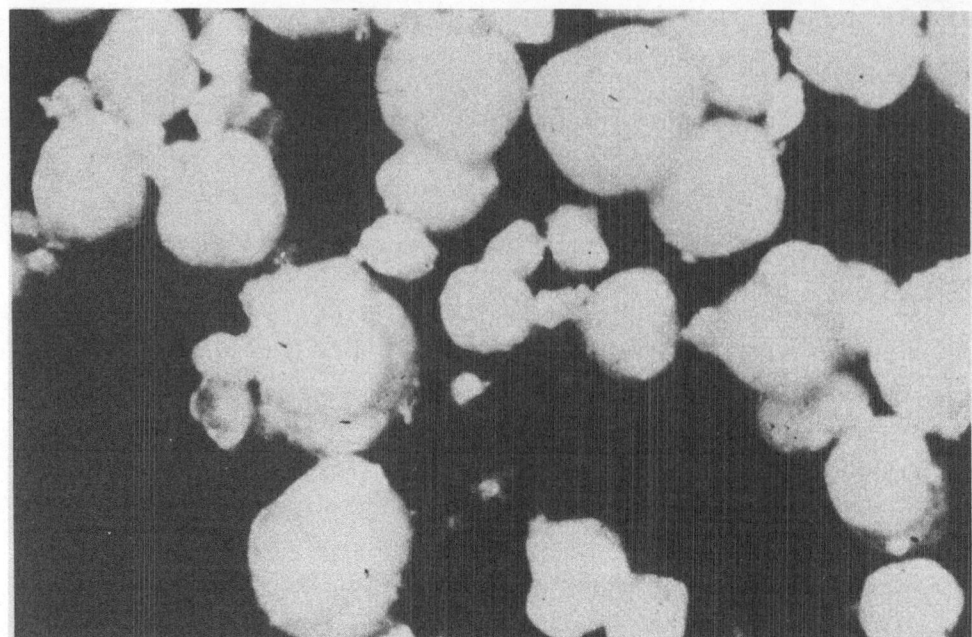

Fig. 1. Carbamide precipitated ammonium uranate powder. Mean
         particle size ∼48 μm.

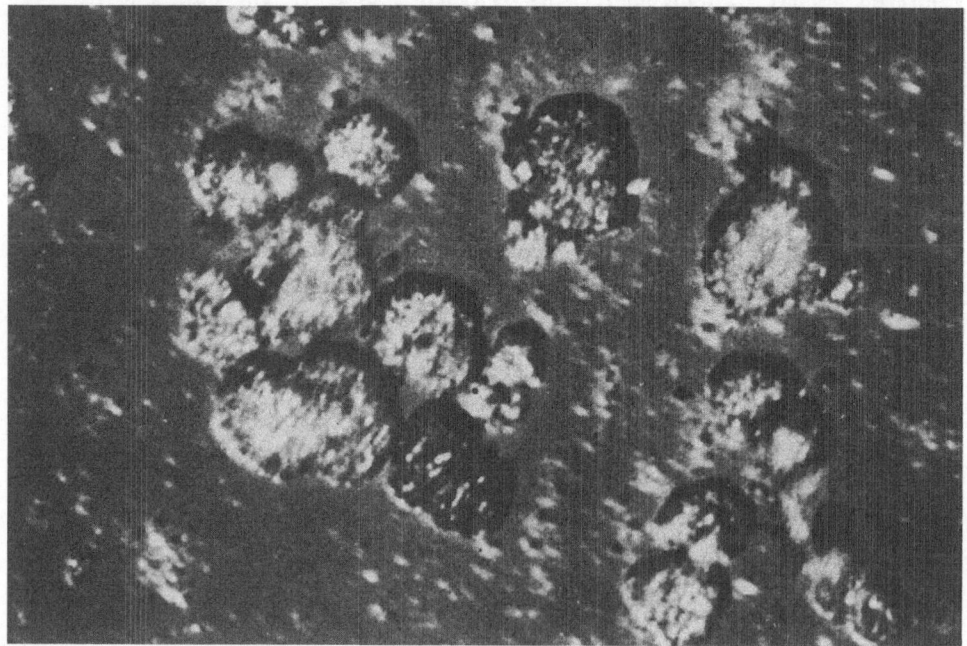

Fig. 2. Uranium dioxide powder made from carbamide precipitated
         ammonium uranate. Mean particle size ∼ 46 μm.

PELLETIZING METHODS

The free-flowing nature of the UO_2 eliminates powder granulation.
For lubrication purposes 0,6% zinc stearate was added in a dry mixing
operation (10 minutes) prior to pressing which was performed over
a wide range of pressures in a hydraulic press. Dewaxing and sintering
were performed in hydrogen at 800°C and 1700°C respectively. Soaking
time during sintering was six hours. Non-isothermal hydrogen sintering was studied in a heating microscope using a linear heating rate
of 390°C/h.

PRODUCT CHARACTERIZATION

Pressed pellets: dimensions, density, abrasion strength. In the
abrasion test three pellets were rotated in a mesh wire cylinder for
15 minutes at a fixed speed. The weight of the abraded material is
a measure of the pellet strength. These abrasion values can be
correlated with the percentage of scrap produced in mechanical
handling in an established production line.

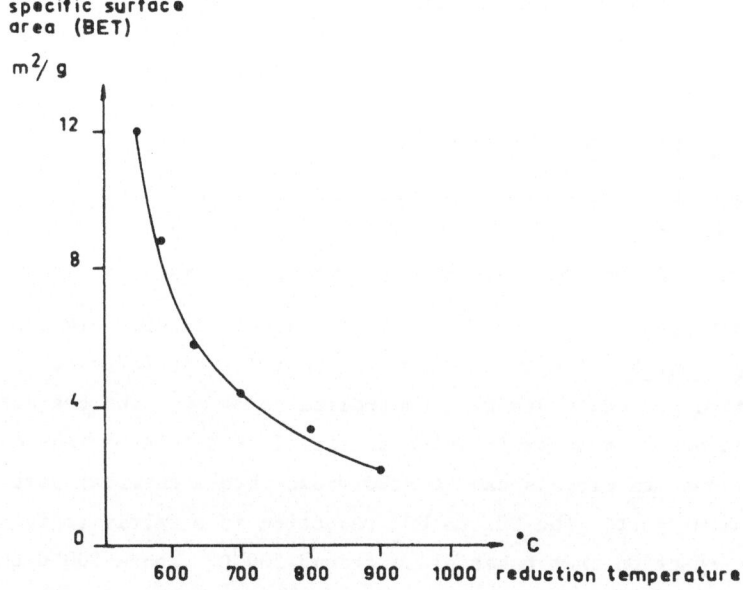

Fig 3   Specific surface area of UO_2-powders
         reduced at various temperatures and
         constant flow of hydrogen.

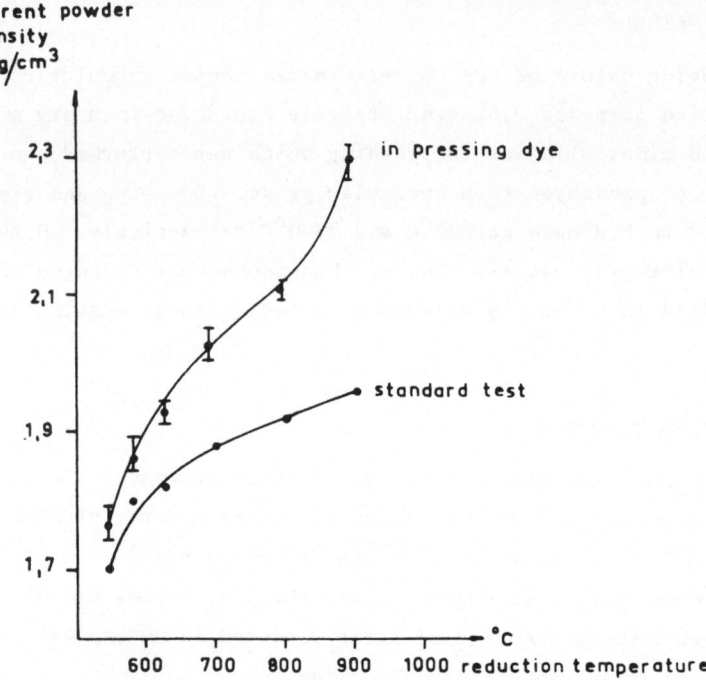

Fig 4   Apparent powder densities of UO$_2$-powders
          reduced at various temperatures.

Dewaxed pellets: carbon content.

Sintered pellets: dimensions, density and carbon content.

CORRELATIONS: POWDER MANUFACTURE - POWDER PROPERTIES

Powder properties as a function of reduction temperatures are shown
in fig. 3, 4 and 5. The regularity of the relations indicates that
the reduction procedure was well controlled as were   the characterization methods. As an explanation to fig. 3 it should be mentioned
that the reduction process can be subdivided into a chemical part
and a physical part. The UO$_3$ to UO$_2$ reduction is a fairly rapid
exothermic reaction that takes place around 500°C. Above 500°C the
powder is then heat treated which reduces the specific surface area
or sinters the agglomerates. This is revealed by the shrinkage of
the agglomerates as determined by sieve analysis, fig. 5.

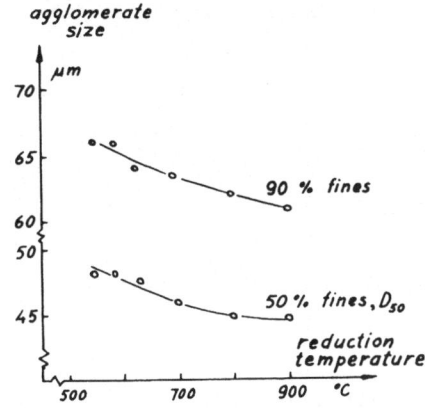

Fig 5

Agglomerate size of powders reduced at various temperatures.

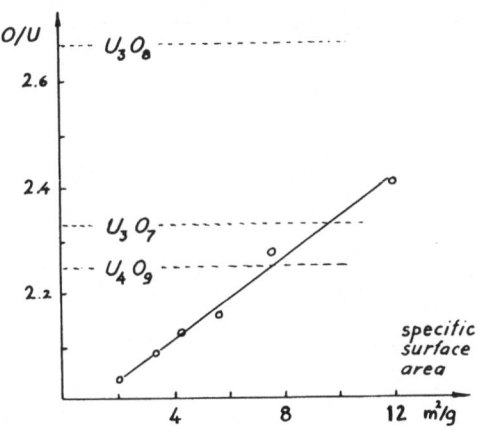

Fig 6

Composition of uranium oxide powders expressed as O/U-ratio as a function of specific surface area.

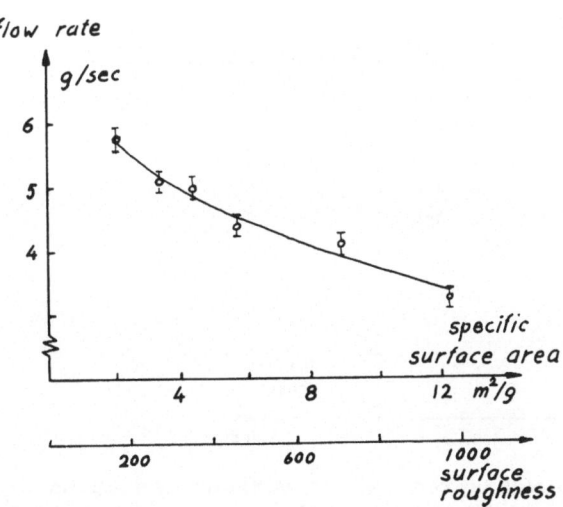

Fig 7

Flow rate, as measured in a Hall's flow meter, for uranium oxide powders reduced at various temperatures. The flow rate is plotted both as a function of specific surface area and of surface roughness.

After the final controlled oxidation of the powder, which has the
object of making it stable in air, the hyperstoichiometry of the
powder shows a linear relationship with the specific surface area,
fig. 6.

The flow rate of the powder increases with increasing reduction
temperature. The results are given as function of specific surface
area or surface roughness in order to illustrate the discussion in
the following chapter, fig. 7.

CORRELATIONS: POWDER PROPERTIES - PRESSING VARIABLES

The green densities achieved in compacting all the powders over a wide
range of specific pressures are given in fig. 8. The green density

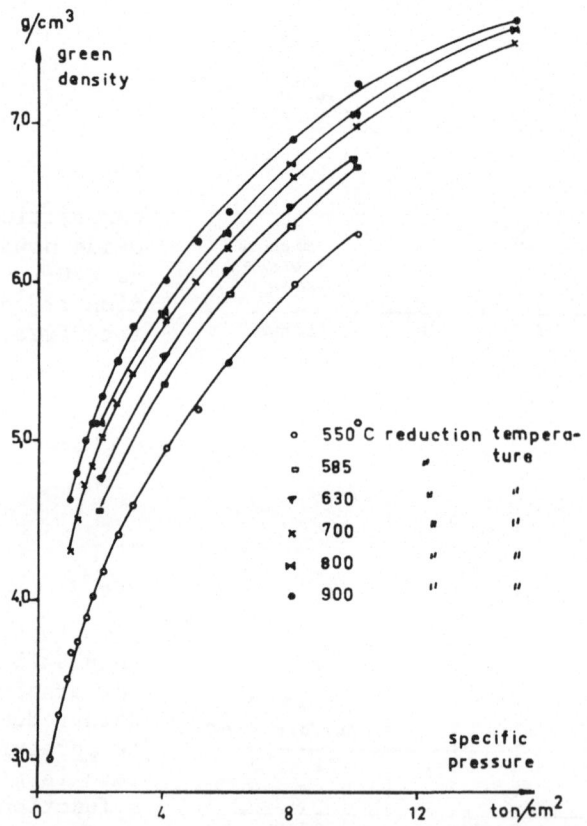

Fig 8   Green density for powders reduced at various reduction
         temperatures as a function of specific pressure.

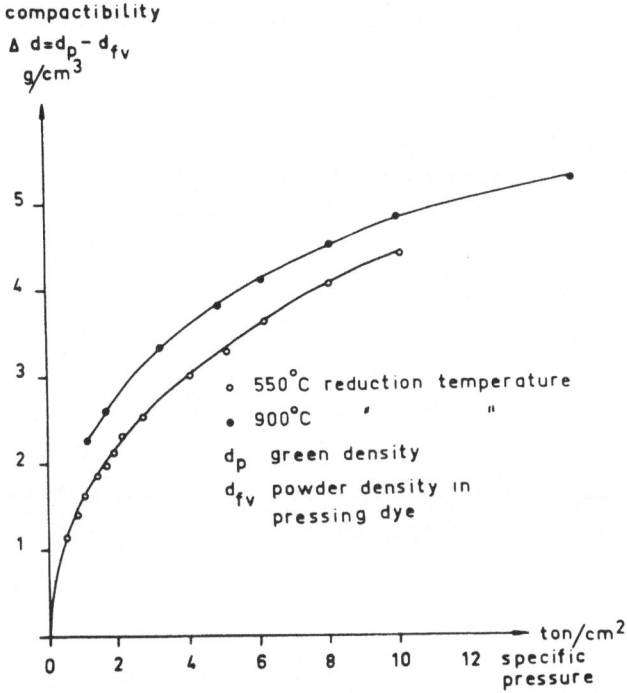

Fig 9   Compactibility of two powders reduced
         at 550 and 900°C.

is no direct measure of the compactibility as the starting density
before pressing also varies with the specific surface area or the
reduction temperature. A better measure of the compactibility of
a ceramic powder is the densification during pressing which is shown
in fig. 9. In general it can be said that densification is improved
by lower specific surface areas (fig. 10). A look at the electron
microscope photos in fig. 11 helps to understand why this is so.
It is obvious that the powder with a high specific surface area has
a much rougher surface than one with a low specific surface area.
As a consequence the smooth particles can move more easily past one
another, either by the force of gravity as in the case of powder
packing in powder density or flow rate (fig. 7) measurements, or
under the pressure applied in pressing, than do the high specific
surface particles with plenty of asperities. In order to quantify
surface roughness, the following relationship has been used:

$A_M = R_s \cdot A_C$ where A_M = specific surface area, measured (BET)

R_s = surface roughness

A_C = specific surface area, calculated = $\dfrac{6}{D_{50} \cdot d_{th}}$

D_{50} = mean particle size

d_{th} = theoretical density

The surface area can be calculated from a particle size distribution
if it follows a normal distribution. As this is not the case with
any of the UO_2-powders, the calculations were very much simplified
by assuming that the powder particles are spherical, smooth and mono-
dispersed with a size equal to the D_{50} value in the particle size
distribution curves. This coarse approach was deemed justified as
the widths of the distributions were closely equal. As the particle
size distribution is virtually constant in the Swedish powders studied,
the calculated surface area becomes equally constant and hence the

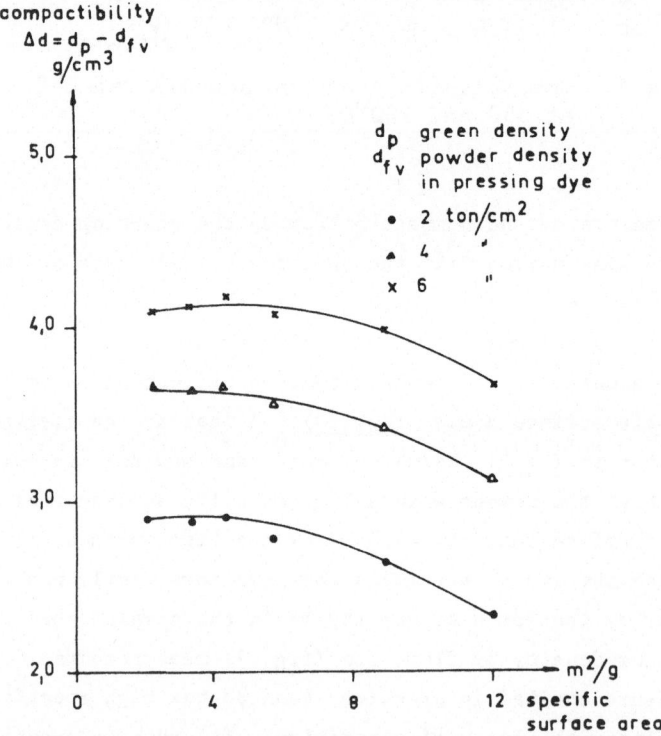

Fig 10   Compactibility for powders of various
surface areas at different pressures.

550°C

Fig 11   Particles of UO_2 manufactured
at various reduction temperatures. Magnification 25 000 x.

630°C

700°C

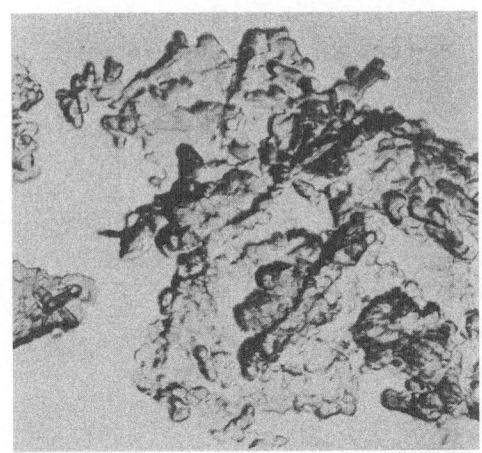

800°C

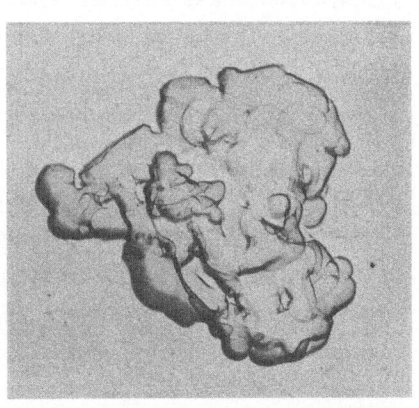

900°C

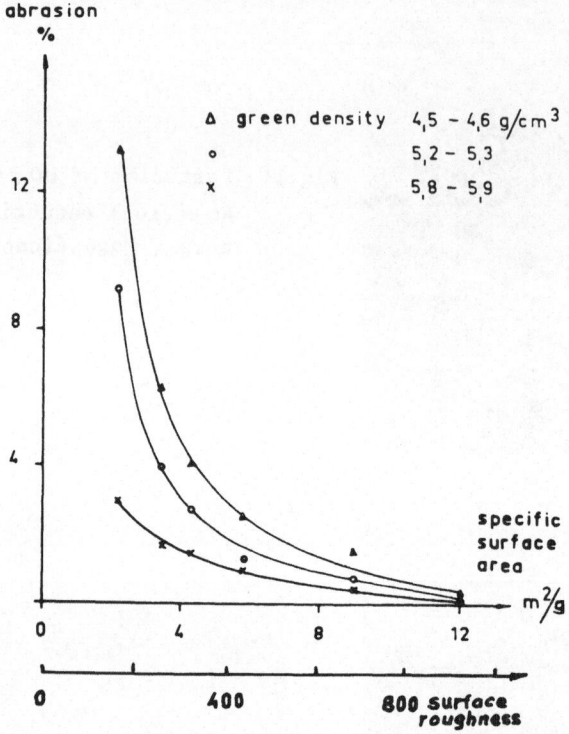

Fig 12   Abrasion strength of pressed pellets
         as function of specific surface area
         or surface roughness at various
         green densities.

surface roughness becomes almost linearly proportional to the measured
BET-surface. The value of the surface roughness concept becomes more
obvious when the particle size distribution varies which is the case
with the UO_2 powders of foreign origin (Part II).

The surface roughness explains quite clearly why the abrasion strength
of pressed pellets increases with the specific surface area (fig. 12).
Interlocking of particles is improved when there are more asperities
on the surface of the particles and when the asperities are brought
into increasingly closer contact, i.e. when pressed to higher green
densities.

CORRELATIONS: POWDER PROPERTIES - DEWAXING VARIABLES

Dewaxing of zinc stearate in UO_2 can be divided in three sub-processes:

1. Decomposition of zinc stearate and evaporization of the decomposition products at about 300°C.
2. Reduction of ZnO and evaporization of Zn at about 650°C.
3. Elimination of residual carbon above 750°C.

The carbon elimination is the difficult operation for several reasons:

1. The equilibrium of the reaction $C + 2H_2 \rightarrow CH_4$ is displaced towards the left with increasing temperature.
2. The reaction rate is slow at lower temperatures, say 800°C, when the equilibrium circumstances still are favorable.
3. In order to counteract the equilibrium disadvantage the gas flow of hydrogen has to be such that the ventilation of the intricate pore channel system in the pellet is good. High pressed densities and pore closure by sintering are hence detrimental to ventilation and carbon removal.

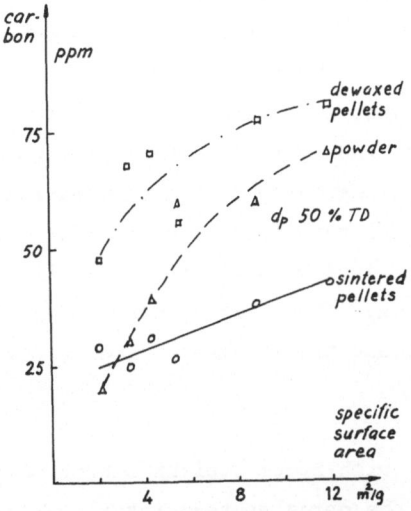

Fig 13   Carbon content of powders, dewaxed pellets and sintered pellets as a function of specific surfaces of the uranium oxide powders.

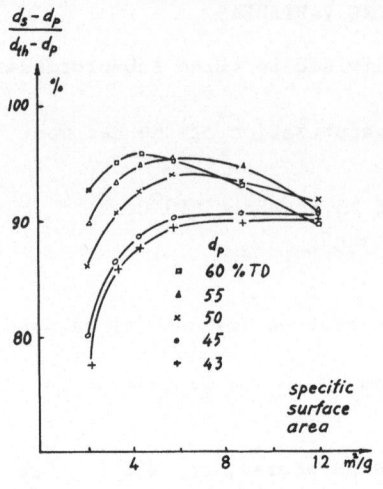

Fig 14   Sinterability of uranium
oxide powders of various
specific surfaces pressed
to various levels of
pressed or green density.
Sinterability is defined
as $d_s - d_p / d_{th} - d_p$.

With this background it is quite understandable that the carbon content
in the powder decreases from 70 to 20 ppm as the reduction temperatures
increase from 550 to 900°C, fig. 13.

The zinc stearate addition increases the carbon content by about
4000 ppm.   After dewaxing the carbon content ranges from 50 to 80 ppm
and after sintering from 25 to 40 ppm.   There is a trend to find more
residual carbon the higher the specific area of the powder is.   The
explanation closest at hand is that sintering or pore closure makes
pore ventilation and consequently carbon elimination difficult.
This difficulty increases the higher the specific surface area is.
The higher the specific surface area, the lower is the starting
temperature of sintering.   (See next chapter.)   In other words high
surface area powders close their pores earlier and leave more carbon
in closed porosity.

Further, carbon reacts with hyper-stoichiometric oxygen in UO_2 to
form CO which dilates the pores at temperatures when UO_2 is sufficiently plastic (1500-1700°C).   This phenomenon is implicitely indicated in fig. 14 by the falling sintered density of the powders which
were reduced at 550, 585 and 630°C (12, 8.9, 5.7 m^2/g) and pressed
to high pressed densities.

As volume increase of UO_2 pellets during reactor operation is undesirable carbon elimination must be almost complete during dewaxing
and sintering. The carbon content hence constitutes one example of
boundary conditions in pelletizing.

CORRELATIONS: POWDER PROPERTIES - SINTERING VARIABLES

Sinterability is often expressed as

$$\frac{d_s - d_p}{d_{th} - d_p} \text{ , where } d_s = \text{sintered density}$$

$$d_p = \text{pressed } "$$

$$d_{th} = \text{theoretical } "$$

The results in <u>fig. 14</u> show that the sinterability increases with
increasing specific surface area as theory predicts only when pressed
pellets of low density are sintered. When pellets of higher pressed
density are used, maximum sintered densities are found in the intermediate range of specific surface areas. The decrease that can be
observed with high surface area powders is explained by insufficient

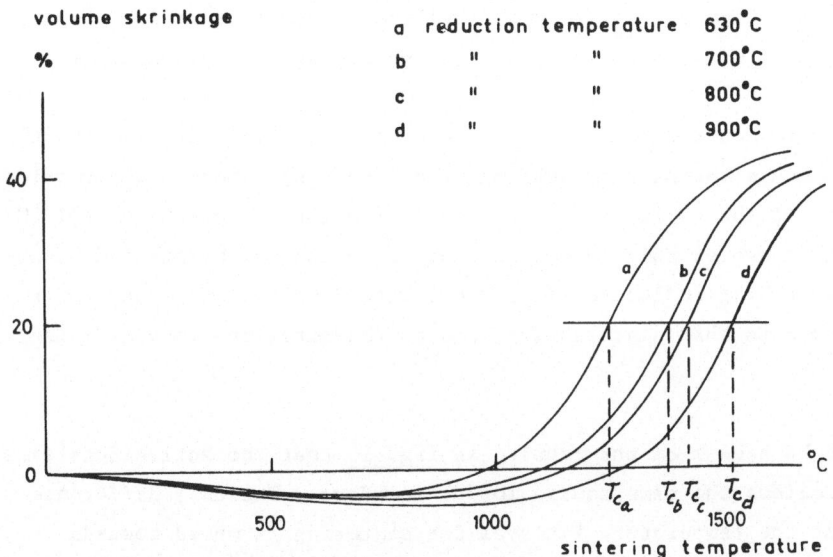

Fig 15   Shrinkage of UO_2-pellets sintered in a heating microscope. Heating rate: $390°C/h$. T_c is the temperature
used to characterize the entire shrinkage curve.

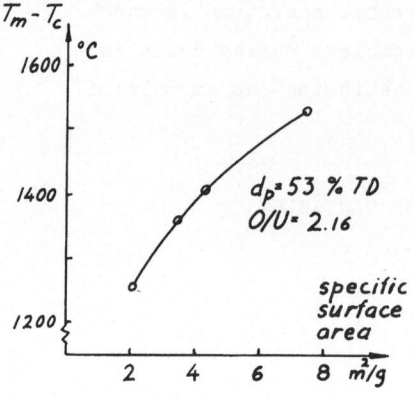

Fig 16

Sinterability of uranium oxide powders, defined as $T_m - T_c$ where T_m = melting temperature and T_c = the characteristic temperature shown in fig 15. Sinterability is shown as a function of specific area.

carbon elimination and volume expansion by CO trapped in closed porosity, fig. 18.

Hydrogen sintering of UO_2 is complex because the powder is hyper-stoichiometric and the final product stoichiometric. Sintering in UO_2 can be considered to use not only surface energy and thermal energy but also chemical energy. The influence of chemical energy was standardized in one series of experiments performed on the powders reduced at 630, 700, 800 and 900°C which were oxidized to the composition $UO_{2.16}$. Further pellets of constant pressed density, 53 per cent of theoretical density, containing no lubricant, were used.

The entire non-isothermal sintering process was followed in a heating microscope with constant heating rate (390°C/h), where the dimensional changes of the pellet were registered from room temperature to 1700°C. One advantage of this method is that sintering can be studied before the disturbing influence of carbon induced swelling appears. Another advantage is that sinterability can be characterized in a different way.

It can be seen from the results in fig. 15 that the entire densification follows the same course for all powders. The only difference is that the temperature interval for sintering is moved towards successively lower temperatures, the higher the specific surface area becomes. In a simplified manner one could say that the more surface energy is build into the powder, the less thermal energy is

required to reach a certain degree of densification. In order to
represent the S-shaped curves by a one figure value, a characteristic
temperature, T_c, is used in order to express the displacement of the
curves in the direction of the abscissa (temperature axis). If T_c is
taken as a measure of the thermal energy required for sintering to
a certain degree (20 per cent volume shrinkage), $T_m - T_c$ (T_m = melting
temperature) might serve as a measure of the energy built into the
powder facilitating sintering. However crude, this definition has
the advantage of taking the entire densification into account and
the correlation between sinterability and specific surface area
follows the expectations from theory, fig. 16.

The linear shrinkage of pellets pressed at constant pressure is
shown as a function of specific surface area in fig. 17. Shrinkage
is defined as

$$\frac{\phi_p - \phi_s}{\phi_p} \cdot 100, \text{ where } \phi_p = \text{diameter of pressed pellets}$$

$$\phi_s = \text{" " sintered "}$$

ASPECTS ON THE ENTIRE PELLETIZING PROCEDURE

Looking at the pelletizing procedure as a whole the final result is
best characterized by the sintered density as a function of specific

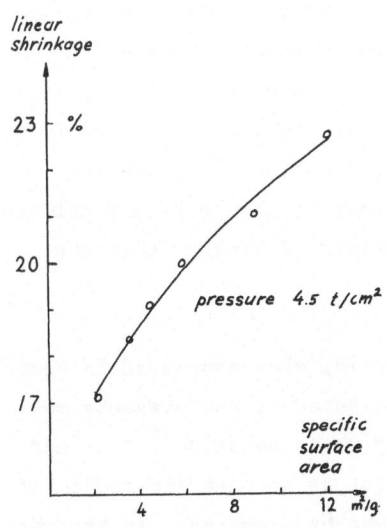

Fig 17

Linear diameter shrinkage
of uranium dioxide pellets
pressed from powders of
various specific surfaces.

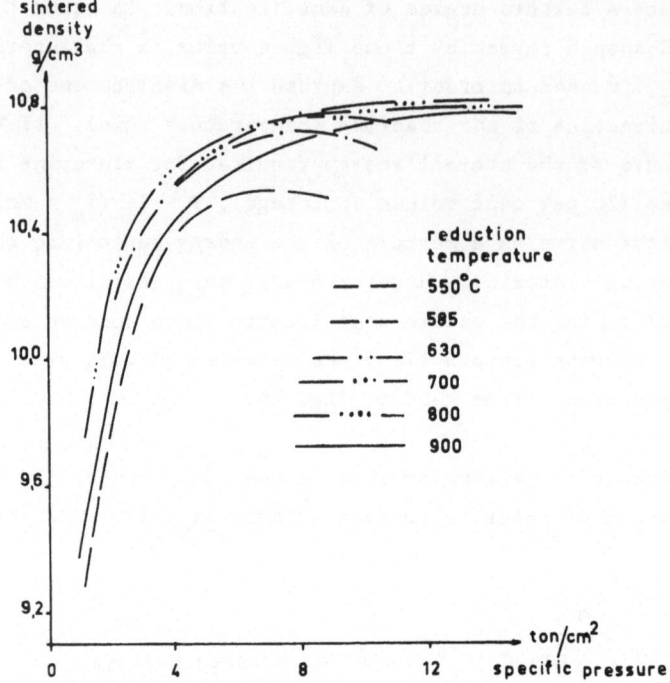

Fig 18   Sintered density as function of specific pressure
and powder reduction temperature.

pressure (<u>fig. 18</u>). The highest sintered densities are achieved
with the powders in the intermediate range of reduction temperatures
and consequently specific surface areas. From the foregoing one can
conclude that a high specific surface area improves sinterability
and green strength but hampers compactibility and dewaxing. Pelleti-
zation hence presents an optimization problem to the powder metallurgist where he has to balance pressing, dewaxing and sintering
variables against powder properties in order to fabricate a specified
product with a minimum of effort or a maximum of yield within the
framework of his equipment.

Optimization becomes particularly interesting when expressed in economic   terms. Pressing costs can be represented by the pressure required to reach a certain pressed density while abrasion losses and
the carbon content of dewaxed pellets ought to give an indication of
the costs caused by mechanical handling and by dewaxing. As the charac-

teristic temperature, T_c, was used as a measure of the thermal energy
needed for sintering, this temperature might also serve as a measure
of the sintering costs. Increasing shrinkage gives rise to a greater
spread in sintered diameter, which in turn causes more grinding losses
in the final adjustment of the pellet dimensions. Grinding costs are
consequently assumed to be proportional to the shrinkage. Fig. 19.

The total costs derived in this simplified manner appear to reach
minimum around 3 to 5 m^2/g specific surface area of the powder. It is
interesting to note that this range of surface areas was adopted
by trial and error development work and by manufacturing experience on
Swedish UO_2 years before this investigation disclosed the reasons.

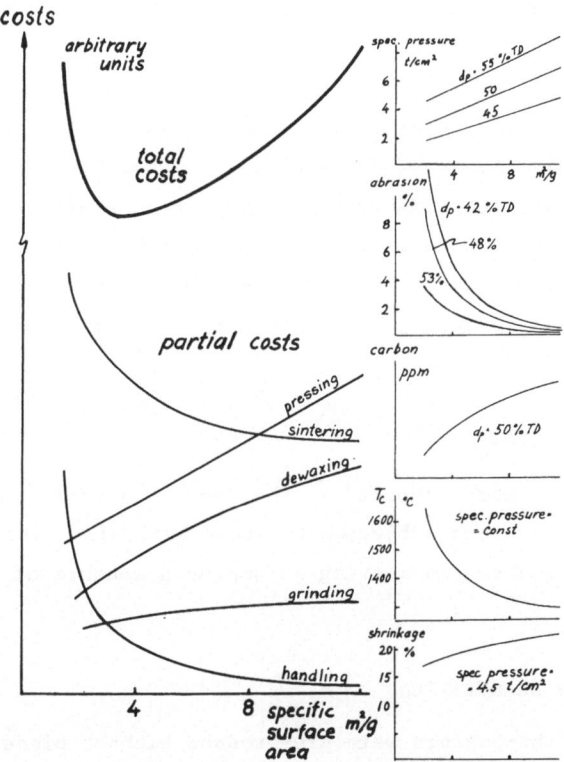

Fig 19   Schematic diagram outlining the tendencies for the costs
         of various partial processes in pelletizing as well as
         the costs for the entire fabrication of pellets from
         uranium oxide powders of various specific surfaces.

An observation concerning UO_2-powders available on the international
market points in the same direction. Formerly the powders covered
a range from 2,5 to approximately 10 m^2/g. In later years this
interval seems to have shrunk to 3-6 m^2/g which is somewhat higher
than for the Swedish UO_2. It should be kept in mind that virtually
all powders require granulation which, of course, will have an effect
on the optimization. This is particularly true when granulation by
means of pre-pressing and crushing is used. A certain surface roughness of the powder then becomes necessary to give the granules
sufficient strength, which sometimes leads to specific surface areas
in upper part of the interval.

PART II

SIX UO_2 POWDERS OF FOREIGN ORIGIN

MANUFACTURE

Fairly little knowledge concerning the manufacture of the foreign
powders is available. Some of the powders are known to be precipitated
by ammonia. The rest is believed to be ammonia precipitated as well.
Calcination and reduction conditions are with one exception unknown.
Most powders have probably been mildly comminuted. One powder (I) has
definitely been very intensely milled in a micronizer (fluid energy
mill).

POWDER CHARACTERIZATION

In addition to the methods applied to the Swedish powders the granulated foreign powders were subjected to sieve analysis. The amount of
fines (<0.12 mm) produced in sieving served as a measure of granule
strength.

POWDER TREATMENT - GRANULATION

Prior to pressing the powders were pre-pressed without binder or
lubricant at a pressure generally substantially lower than that of
the final pressing. The pellets or slugs were crushed and sieved.
The fines below 0.12 mm were recycled. The granulated powders were
moderately free-flowing.

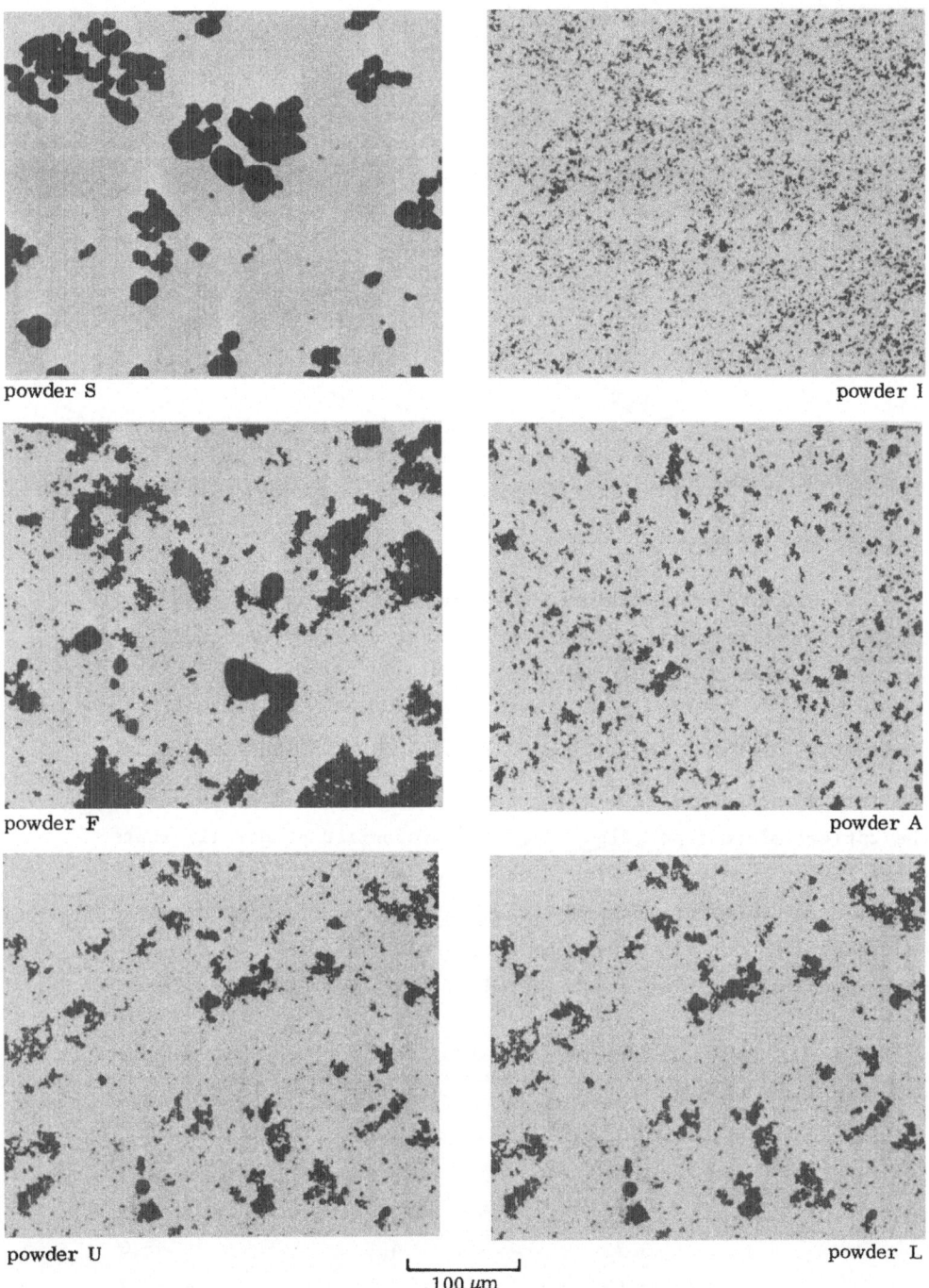

powder S

powder I

powder F

powder A

powder U

powder L

100 μm

Fig 20   Particles of six UO₂-powders.
Magnification: 200 x

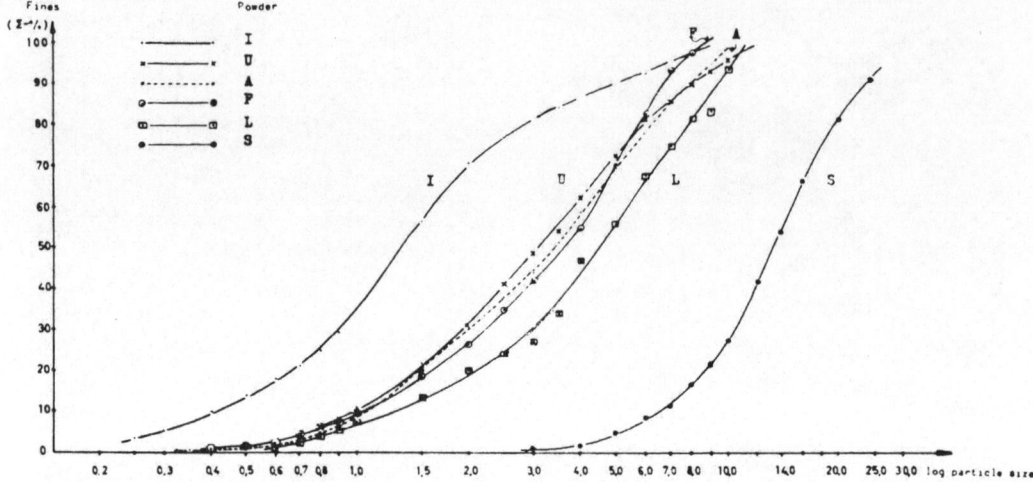

Fig 21   Particle size distribution as turbidimetrically determined.

PRESSING AND SINTERING

Pressing and sintering followed the same lines as in the case of
the Swedish powders with the exception that the lubricant addition
was reduced to 0.2% zinc stearate.

COMPARISON OF POWDER PROPERTIES OF FOREIGN AND SWEDISH POWDERS

The most eye-catching difference appears already in a study under
the optical microscope (fig. 20). The micronized powder (I) represents a very fine powder with a D_{50}-value of only 1.3 μm while
powder S is ten times coarser (fig. 21). The specific surfaces
on the other hand differ only to a rather small extent as can be
seen in table 1. It. should be noted in table 1 that the surface

TABLE 1.   POWDER CHARACTERISTICS OF SIX FOREIGN UO_2 POWDERS

Powder	Spec. surface area (m^2/g)	D_{50} (μm)	R_s (-)
I	3.76	1.3	8.9
S	3.28	13.0	77.9
L	5.73	4.3	45.0
A	5.19	3.4	32.2
F	3.54	3.6	23.2
U	3.26	3.0	17.9

roughness varies over a much wider range than the specific surface
areas which is due to the great variation in average particle size.

COMPARISON OF VARIOUS CORRELATIONS

While no correlation can be found between specific surface area
and compactibility or green strength, there is fairly good correlation between compactibility and the surface roughness of the powder
(fig. 22) as in the case of the Swedish powder. Green strength
first plotted versus surface area (fig. 23) and then versus surface
roughness (fig. 24) illustrates an interesting transition from
obscurity to insight merely by the choice of a parameter better
suited for describing a certain phenomenon. The green strength-
surface roughness correlation shows fairly good similarity with the
results for the Swedish powders, provided the results from powder A
are discounted. It is, however, obvious that the powders have to be
better described in order to improve the correlation. The deviation
of powder A can possibly be explained by the fact that in compacting
fairly strong prepressed granules, uneven green density distribution
is obtained in the pellets when insufficient amount of fines are present

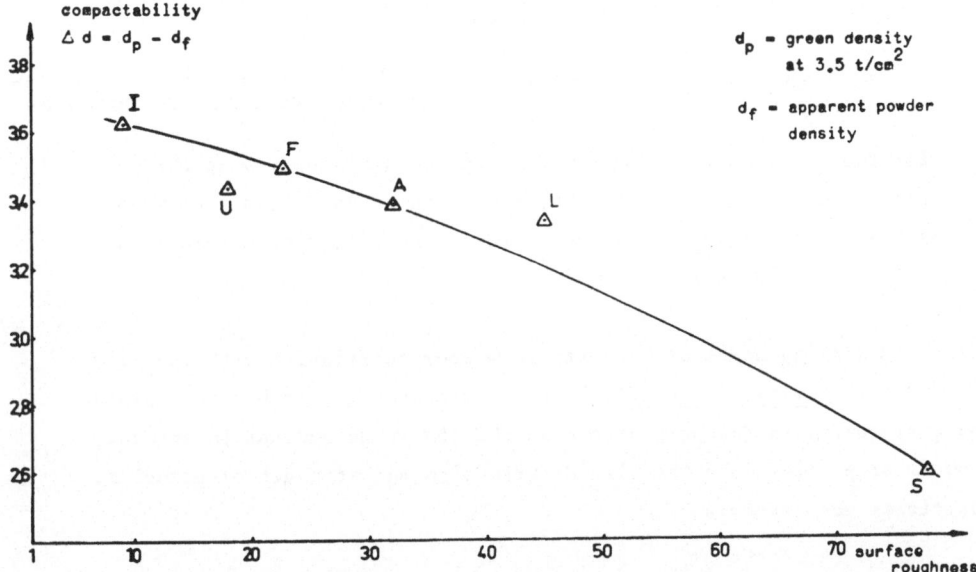

Fig 22   Compactibility of six UO_2-powders as function of particle
surface roughness.

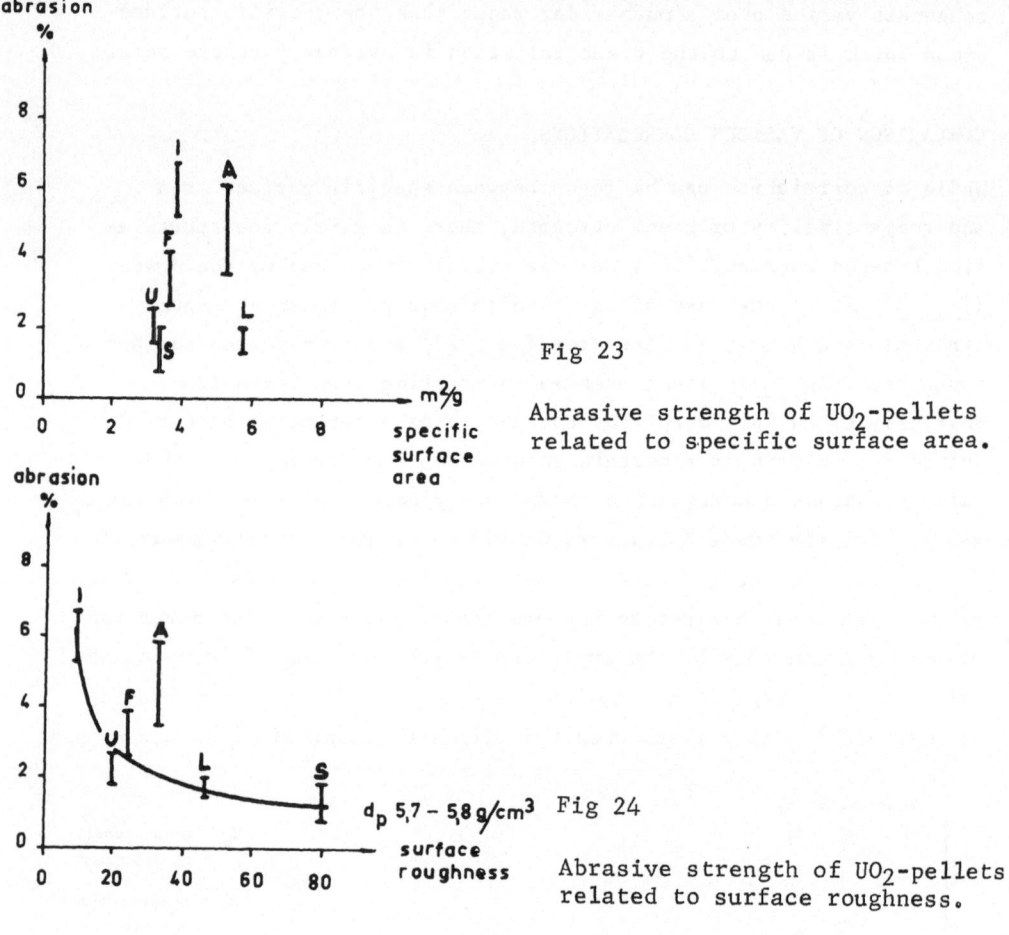

Fig 23

Abrasive strength of UO_2-pellets related to specific surface area.

Fig 24

Abrasive strength of UO_2-pellets related to surface roughness.

to fill the voids between the granules to the same density as the granules. The fair correlation between green strength and the amount of granule fines (fig. 25), which in itself seems rather meaningless, supports this hypothesis.

Sintered density and sinterability show poor correlation with specific surface area (fig. 26 and 27). The deviations can probably be explained by differences in the pore size distribution which depends on several powder properties like the size distribution and strength of granules, particles and crystals.

Densification can be regarded as pore elimination. Small pores are more easily eliminated during sintering than big ones. An obvious

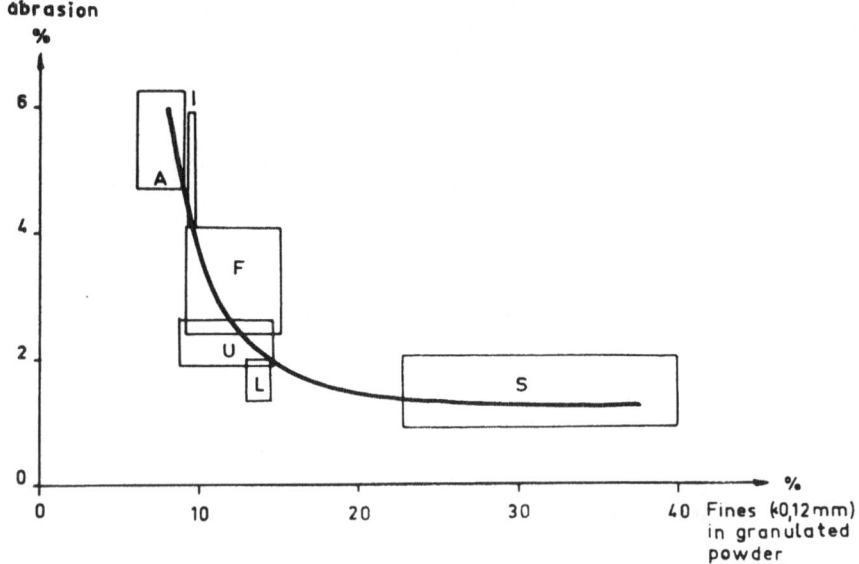

Fig 25   Abrasion strength correlated to the percentage
          of fines among granules.

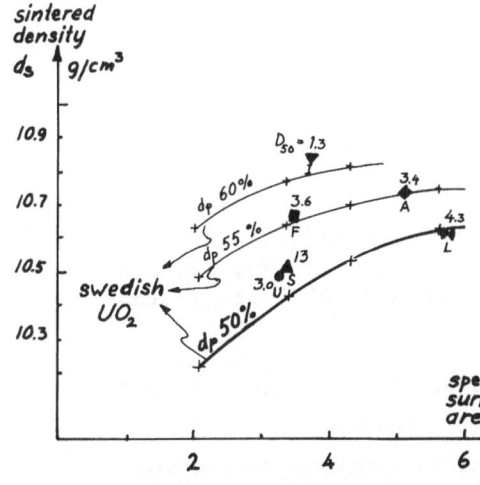

Fig 26

Sintered density as a function of
specific surface area for six UO_2-
powders of various origin compared
with corresponding data for Swedish
UO_2. Green density, d_p: 50% TD.
Mean particle size of the powders
is given near the points.

Fig 27

Sinterability, defined as $T_m - T_c$,
as a function of specific surface
area for six UO_2-powders compared
with corresponding data for
Swedish UO_2.

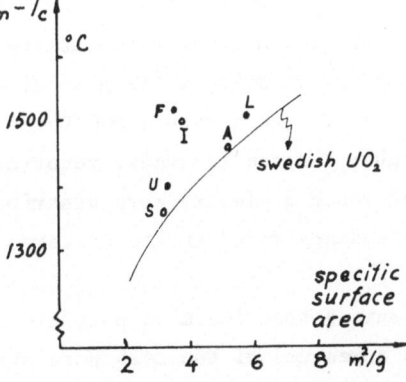

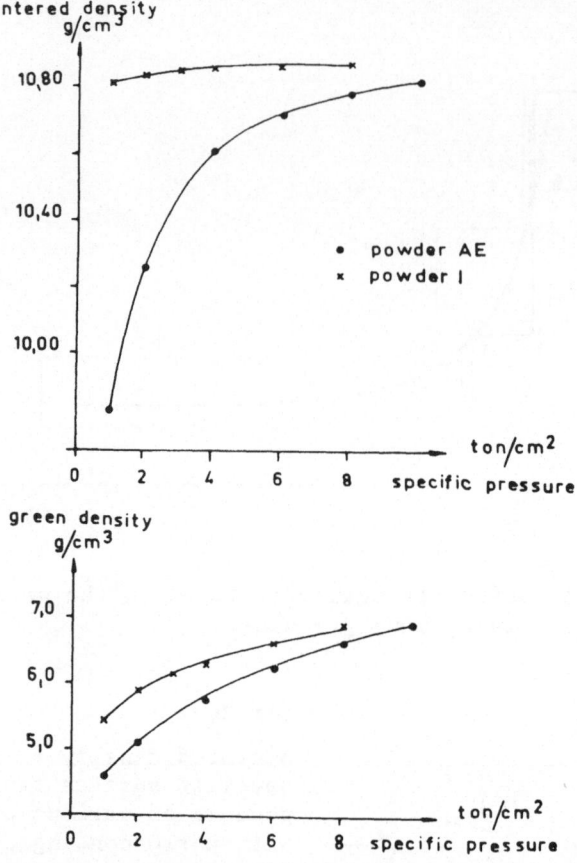

Fig 28   Sintered density of two UO$_2$-powders with nearly
         the same specific surface areas (I:3.76; AE,3.29)
         but with differing particle and pore size distri-
         butions.

example of the beneficial effect of small pores on sintering is given
in fig. 28, where the intensely comminuted I-powder with the lowest
surface roughness already at low pressures is compacted to contain
small pores only, while a Swedish powder of corresponding surface area
(reduction temperature 800°C) due to its fairly strong granules and
higher surface roughness requires high pressure to eliminate big pores
and reach a similar pore distribution. This explains the strong pressure
dependency found in the sintered density of Swedish UO$_2$.

Assuming that the mean particle size of the powders given in fig. 26
is a measure of the mean pore size in the pressed compacts and assuming

that the sintered density is dependent only on pore size at constant
specific surface area, the results become somewhat more consistent.
The expected tendency to reach higher sintered densities with powders
of smaller particle size is obvious but not without exception. Even
if there exists a very clear relation between pore size distribution
and sintered density, the problem still remains to establish precise
correlations between suitable powder properties and pore size distribution.

DISCUSSION AND CONCLUSIONS

The results have demonstrated that good correlations between powder
properties on the one hand and process variables and product properties
on the other can be established, provided

1. one can keep a sufficient number of powder properties constant
    leaving only one important parameter to be varied. This is, of
    course, a scientific truism. The problem is rather how to achieve
    this desirable state with powders. The experiments with the
    Swedish powder indicate that this is possible.

2. process variables, powder and product properties have to be care-
    fully chosen and defined in order to describe a phenomenon clearly.

3. powder properties are correlated to the various partial processes
    making up pelletization. The entire processes can only be under-
    stood as the sum of the parts.

The work on the foreign powders has been deemed to fulfil provision
number 1 and 2 insufficiently. Not very precise but still valuable correlations can be found, especially if projected against the background
of the results from the Swedish powders. Additional meaningful characteristics have to be introduced in order to improve the accuracy when
application is made on arbitrary powders. This becomes particularly
true if ductile powders are studied.

The major conclusion to be drawn from this work is that powder metallurgy
or pelletizing can be regarded as an optimization problem which could
become quite valuable to the production man if supplemented by econo-

mic parameters.   Conversely one can say that once a pelletizing process
has been firmly established in an optimal fashion the powder cannot be
allowed to vary more than marginally.   Reliable powder manufacture and
powder control are consequently vital to secure reproducibility.

ACKNOWLEDGEMENT

This paper is compiled from the work of many people, most of whom
work or have worked in the laboratory for ceramic fuels and powder
metallurgy, AB Atomenergi, Stockholm.   They are:

Håkan Vangbo, Bertil Jansson, Sven Eriksson, Berndt Pettersson,
powder metallurgy on Swedish and foreign UO_2;

Branislav Živanovič and Dimitrij Sušnik, sintering studies;

Albrecht Fickel, foreign UO_2-powders;

Ivan Stamenkovič, Boris Kidrič, Institute of Nuclear Sciences,
Beograd-Vinča, Yugoslavia, electron microscopy;

Gunnar Lindelöf, Jernkontorets Laboratorium för Pulvermetallurgi,
Stockholm, particle size analysis.

MATRIX HARDENING IN DISPERSION STRENGTHENED POWDER PRODUCTS

Niels Hansen and Hans Lilholt

Metallurgy Department, Danish Atomic
Energy Commission,
Research Establishment Risö, Roskilde, Denmark

ABSTRACT

Dispersion hardening is one of the methods of strengthening a soft matrix. Other methods such as grain boundary hardening and solid solution hardening are known. Experimental data for various dispersion hardened systems have been examined in order to investigate whether it is possible to add the stress contributions from these three different strengthening processes. The effect of dispersed particles has been accounted for by the Orowan-expression $\sigma_p = 2 \dfrac{G \cdot b}{\lambda}$ The effect of grain boundaries has been described by the Petch-expression $\sigma_{gb} = k D^{-1/2}$. The contribution from solid solution is entered as the empirically found value. Considering the flow stress values (0.2% offset) it has been determined to what extent the various contributions can be added independently, i.e. whether

$$\sigma = \sigma_o + \sigma_p + \sigma_{gb} + \sigma_{sol}$$

is valid when inserting the stress values expected if the individual strengthening agents had occurred alone in the matrix. Data have been obtained from the following systems: $Al-Al_2O_3$, $Ni-ThO_2$, $Zr-Y_2O_3$, and Fe-carbides. The upper limit for the amount of dispersed phase has been about 10 vol. pct. The grain sizes have been in the range from 0.3 μ to the order of 1 mm (i.e. extruded, cold worked, and recrystallized materials). Within some limitations the strength data at room temperature indicate that the contributions from the various strengthening agents can be added, i. e. that each mechanism acts independently of the others. At higher temperatures in the range 0. 5 to 0. 8 T_m, the three strengthening processes may all contribute to the tensile strength and to the creep strength, but the use of the simple additive rule was inconclusive. At temperatures above 0. 8 T_m dispersion hardening is predominant.

INTRODUCTION

In dispersion hardened products advantage is taken of the strengthening effect of hard particles on a soft ma-trix. The particles impede the motion of dislocations through the matrix by forcing them to bend between the particles. Experimental evidence as well as theoretical considerations indicate that the important strength determining parameter is the distance between the particles; a smaller spacing makes the dislocation movement more difficult and thus in-creases the strength. Practical limitations make it difficult to achieve high strengths by reducing the spacings to very small values. Therefore it becomes of interest to consider other ways af strengthening the matrix. In this paper we shall discuss methods like grain boundary hardening and solid solution hardening, and investigate to what extent they can be used simultaneously with dispersion hardening.

HARDENING MECHANISMS

Dispersion Hardening

Several investigations of dispersion hardened systems have indicated that the yield stress is determined by the force to bend a dislocation between two particles (Orowan (1)). This situation is described by

$$\tau_a = \frac{T}{b R}$$

(τ_a is the shear stress, T the line tension, b the Burgers vector, and R the radius of curvature). Modifications to this stress are brought in by assigning different values to the line tension T. However, the differences between the various values are not great. and the experimental diffi-culties in distinguishing between them makes it reasonable to consider a very simple estimate for T

$$T = \frac{1}{2} Gb^2$$

(G is the shear modulus).

Thus the yield stress is given by

$$\tau = \tau_o + \frac{G \cdot b}{\Lambda} \quad \text{and} \quad \sigma = 2\tau = \sigma_o + 2\frac{G \cdot b}{\Lambda}$$

($R = \frac{1}{2}\Lambda$, Λ is the surface to surface spacing, τ is the shear yield stress, and σ is the tensile yield stress). We shall use this approximate value to account for the effect of particles. The spacing Λ is calculated on the assumption that the particles intersecting a plane form a square array:

$$\Lambda = L - \sqrt{\frac{2}{3}}\,d = \left[\sqrt{\frac{\pi}{6f}} - \sqrt{\frac{2}{3}}\right]d$$

(L is the centre to centre distance, f is the volume fraction and d is the particle diameter).

Grain Boundary Hardening

It has been established empirically that the effect of grain boundaries is described by the Petch-relation (2)

$$\sigma = \sigma_o + k \cdot D^{-1/2}$$

(D is the grain size, and σ_o and k are constants). We shall apply this equation to account for the effect of grain boundaries and use the empirically determined values of k. We shall not try to include any of the theoretical models, which have been presented to explain the Petch-relation.

Grain boundaries of different nature, i. e. large angle grain boundaries and small angle boundaries composed of dislocations, may not behave in the same way during the movement of dislocations in the matrix. It has been found (e. g. Al (3, 4))that the values of k do not depend significantly on the type of grain boundary present. We shall therefore use an average value of k, when more than one type of grain boundary is present.

Solid Solution Hardening

The effect of solute atoms in the matrix will only be treated experimentally, i. e. we shall compare the stresses sustained by the alloy and by the pure metal directly, without relating the stress increase to the amount of solute.

Additive Contributions

This study will investigate to what extent the various contributions can be added independently, i. e. whether

$$\sigma = \sigma_o + \sigma_p + \sigma_{gb} + \sigma_{sol} \qquad (1)$$

is valid when inserting the stress values expected if the individual strengthening agents had occurred alone in the matrix.

This expression will be an approximation for several reasons. Firstly, the formulas adopted for the individual contributions are inaccurate, and the parameters (Λ and D) entering the formulas can, at best, only be average values for the material. Secondly, we have neglected the effects of the various strengthening agents on each other. Such interactions could be the effect of solute atoms on the bending of a dislocation between two dispersed particles, as well as the direct effect of solute atoms on the movement of a dislocation in the matrix. Also, particles may sit preferentially on grain boundaries and possibly cause an interaction contribution (positive or negative), as well as the individual effects of particles and grain boundaries.

It should be noted that Kocks (5) has recently remarked on the superposition rule when several strengthening mechanisms operate. The only non-trivial cases occur either when many weak and few strong obstacles are present, or when about equal numbers of equally strong obstacles are present. In the first case Kocks showed that a simple linear addition, as in eq. (1), is operative. In the second case, applied to point obstacles (particles), it was found (5), when summing the obstacles in a plane, that

$$\tau = \sqrt{\tau_1^2 + \tau_2^2}$$

which is less than $\tau_1 + \tau_2$. This relation describes approximately the resulting stress calculated by Foreman and Makin (6) for a system of two sets of obstacles.

This uncertainty regarding the superposition rule is another reason why eq. (1) is only a first approximation. Furthermore, the experimental data are sometimes insufficiently accurate to allow a distinction to be made between these possible ways of adding the strength contributions.

It has been very difficult to treat different alloy systems in the same manner, mainly because each investigation has had its own objective. Therefore we shall discuss the systems individually, and in each case point out how they demonstrate our idea. We will first consider dispersion hardening and grain boundary hardening, and after-

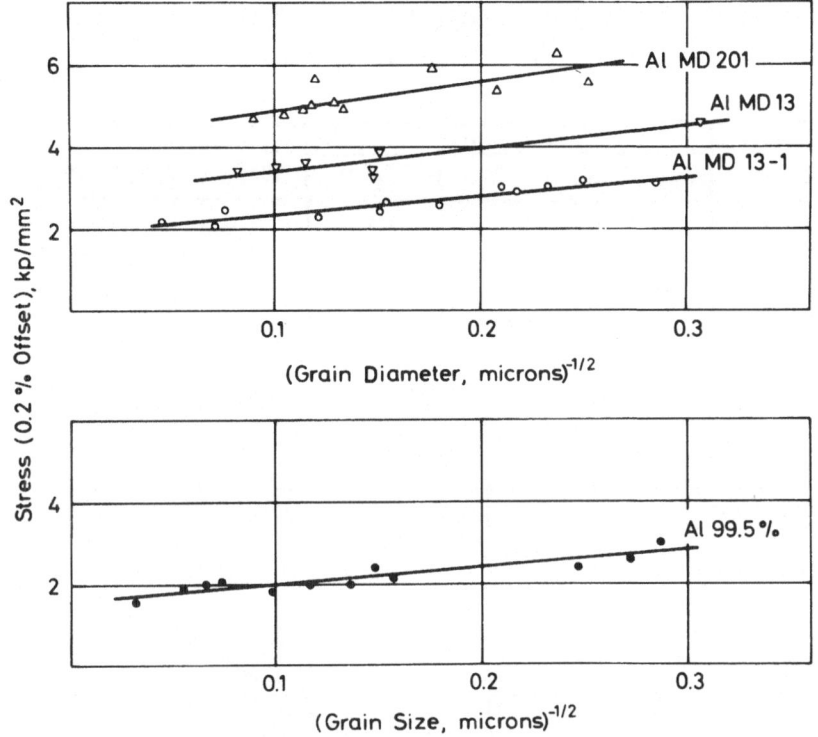

Figs. 1 + 2. Petch-plots at room temperature for Al and Al$_2$O$_3$ products (3).

wards dispersion hardening combined with solid solution hardening.

The first part of the paper will deal with the room temperature strength (measured as the stress at 0.2% off-set, $\sigma_{0.2}$). This treatment will be quantitative, as sufficient data are available. The second part covers high temperature strength ($>$ 0.5 T_m), where the scarcity of suitable experimental data has allowed only a qualitative treatment.

STRENGTH AT LOW TEMPERATURES

Dispersion Hardening and Grain Boundary Hardening

Al-Al$_2$O$_3$. This system includes alloys with different amounts of oxide and having different grain sizes. The data are presented as a Petch-plot for each alloy of a given oxide content, figs. 1, 2 and 3 (3,4). The data on the oxide phase are given in table 1.

Table 1

Al-Al$_2$O$_3$ products (7)

Material	Oxide plates			Equivalent sphere (Å)	Spacing (Å)
	(vol pct)	Diam. (Å)	Thickn. (Å)		
MD 13	0.19	450	350	470	7400
MD 201	0.45	460	150	360	3600
MD 105	0.79	520	83	320	2300
SAP 960	3.8	800	100	460	1300

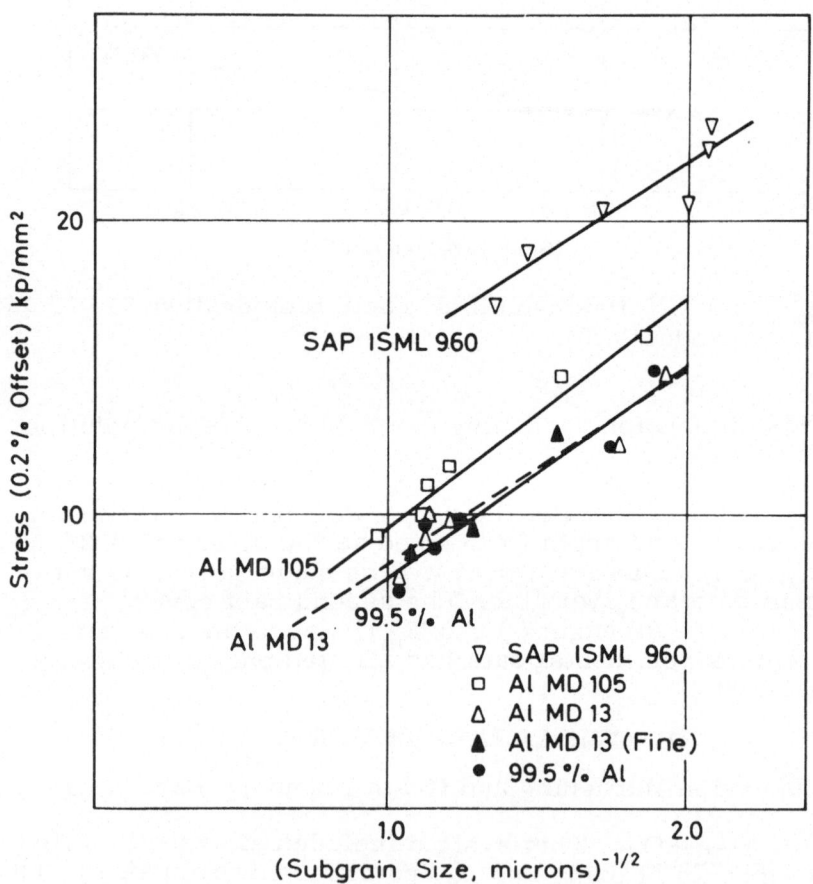

Fig. 3. Petch-plots at room temperature for Al-Al$_2$O$_3$ products, drawn and heat treated (4).

The straight lines of figs. 1, 2 and 3 show that the nature of grain boundary hardening is the same in all alloys. The slopes (k-values) being nearly the same, show that the oxide does not have a great influence on grain boundary hardening.

The (nearly) parallel shift of the lines indicates the effect of the oxide dispersion. We have used the values of σ'_0 at $D^{-1/2} = 0$ (figs. 1 and 2), and of σ_1 at $D = 1\,\mu$ (fig. 3) to evaluate the contribution from the oxide, which is found by subtracting the stress for pure Al from the stress of the appropriate alloy.

We should compare the Δ-values with the calculated contributions from the oxide. The experimental values lie

Table 2

Petch-slopes for Al-Al$_2$O$_3$ products

Material	k (figs. 1 and 2) $(kp/mm^2\ \mu^{\frac{1}{2}})$	k (fig. 3) $(kp/mm^2\ \mu^{\frac{1}{2}})$
Al	4.0	7.2
MD 13-1*	4.5	
MD 13	5.3	6.3
MD 201	7.0	
MD 105		7.8
SAP 960		6.5

*MD 13-1 has a less uniform particle distribution than MD 13

Table 3

Strength contributions in Al-Al$_2$O$_3$ products

Material	$\tau_p = \frac{Gb}{\Lambda}$	$\sigma_p = 2\tau_p$	Figs. 1 & 2		Fig. 3			
			σ'_0	$\Delta\sigma'_0$	σ'_0	$\Delta\sigma'_0$	σ_1	$\Delta\sigma_1$
	(kp/mm^2)		(kp/mm^2)		(kp/mm^2)			
Al	-	-	1.6	-	0.7	-	7.7	-
MD 13	1.0	2.0	2.9	1.3	2.1	1.4	8.4	0.7
MD 201	2.2	4.4	4.2	2.6				
MD 105	3.4	6.8			1.9	1.2	9.7	2.0
SAP 960	5.8	11.6			9.0	8.3	15.5	7.8

$G = 2700\ kp/mm^2$, $b = 2.87$Å

Table 4

Ni-ThO$_2$ products

Material	Grain size, D (μ)	$D^{-\frac{1}{2}}$ ($\mu^{-\frac{1}{2}}$)	$\sigma_{0.2}$ (kp/mm^2)	$\sigma_{0.1}$	Ref.
Recr.		0.1-0.3	24		8
As-received	1.6*		38.5		8
Recr.	5-11			28	9
Non-recr.	0.4			63	9

*Cylindrical grains of diameter 1.1 μ; grain dimension on a plane at 45° to the tensile axis is 1.6 μ.

between the calculated shear stress and tensile stress, except for the MD 105 material. We may conclude that in the Al-Al$_2$O$_3$ system the stress contributions from dispersion and from grain boundaries can be added with reasonable accuracy.

Ni-ThO$_2$. No single investigation on this sytem exists, but we have combined information from two sources, both of which used the commercial TD-Ni. From Wilcox and Jaffee (8) we have taken a recrystallized material with $D^{-1/2}$ in the range 0. 1-0. 3 $\mu^{-1/2}$, and an as-received specimen of grain size about 1. 6 μ. Heimendahl and Thomas (9) have listed $\sigma_{0.1}$-stresses for a recrystallized specimen of grain size 5-11 μ , and for a non-recrystallized specimen. No grain size is quoted for the latter, which had been rolled 90% and annealed at 500°C for 1 hr. It is assumed that this anneal did not change the grain size appreciably, thus we have estimated the grain size from their fig. 10, which shows the microstructure of the rolled material. We have also included the Petch-plot for pure Ni, taking the data from (9).

The straight line of fig. 4 for Ni-ThO$_2$ indicates that grain boundary strengthening is of the same nature as in pure Ni. The k-values are nearly the same, i. e. the oxide is not affecting the grain boundary hardening. The parallel shift of the two lines is a measure of the contribution from the dispersion.

We find that the experimental value lies between the calculated shear and tensile stresses, and may conclude that the contributions from the dispersion and

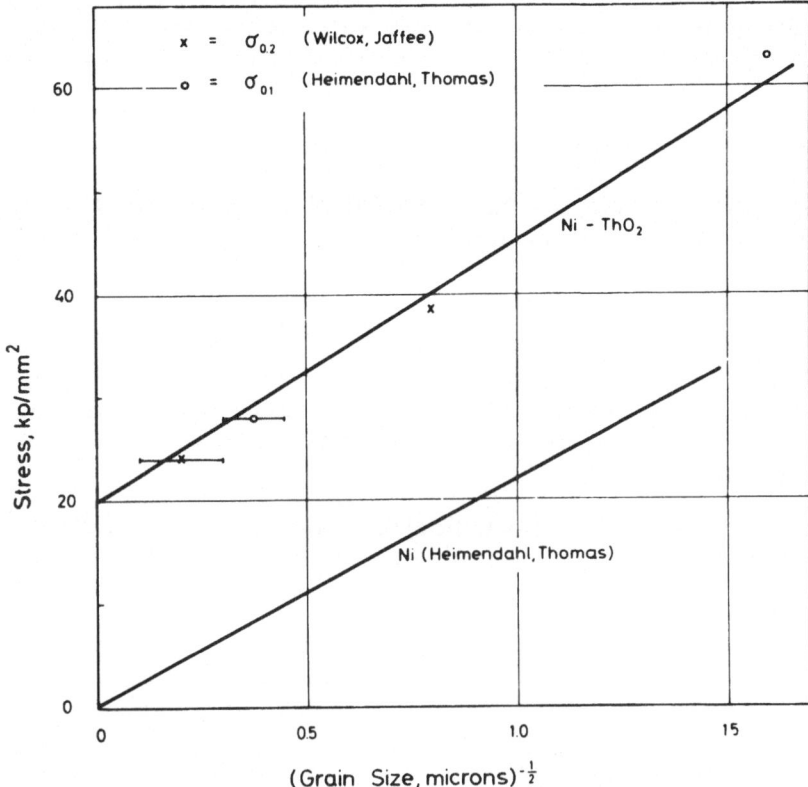

Fig. 4. Petch-plots at room temperature for Ni and Ni-ThO$_2$ products; Wilcox and Jaffee (8); Heimendahl and Thomas (9).

from grain boundaries can be added with a reasonable accuracy.

      Fe-carbides. A set of data from Embury, Keh and Fisher (10) can be applied, as presented in fig. 5. The

Table 5

Strength contributions in Ni-ThO$_2$ products

Oxide (vol pct)	Av.part. size (Å)	Av. part. spacing (Å)	$\tau_p = \dfrac{Gb}{\lambda}$ (kp/mm2)	$\sigma_p = 2\tau_p$	$\Delta\sigma$ (fig.4) (kp/mm^2)
2.3	370	1460	13.5	27	~20

G = 7900 kp/mm^2, b = 2.49 Å

lower curve shows the grain size dependence for pure Fe
(Ferrovac E) and the upper curve corresponds to Fe with
precipitated carbides. The straight and parallel lines
again indicate that the presence of particles is not affecting the strengthening by grain boundaries.

To account for the parallel shift we must calculate
the particle spacing. Embury et al. (10) have not indicated
the type of carbides present. We have assumed these to be
Mo_2C, formed during the annealing at 720°C (11, 12), and
ε-carbide Fe_2C, formed during the annealing at 60°C (13)
from the carbon in solid solution in iron at 720°C (14). The
excess of carbon is assumed to form large particles of
Fe_3C during the annealing at 720°C (11), and to have no significant effect on the strengthening due to the fine carbides,
Mo_2C and Fe_2C.

We find that the experimental value lies between the
calculated shear and tensile stresses. The agreement may

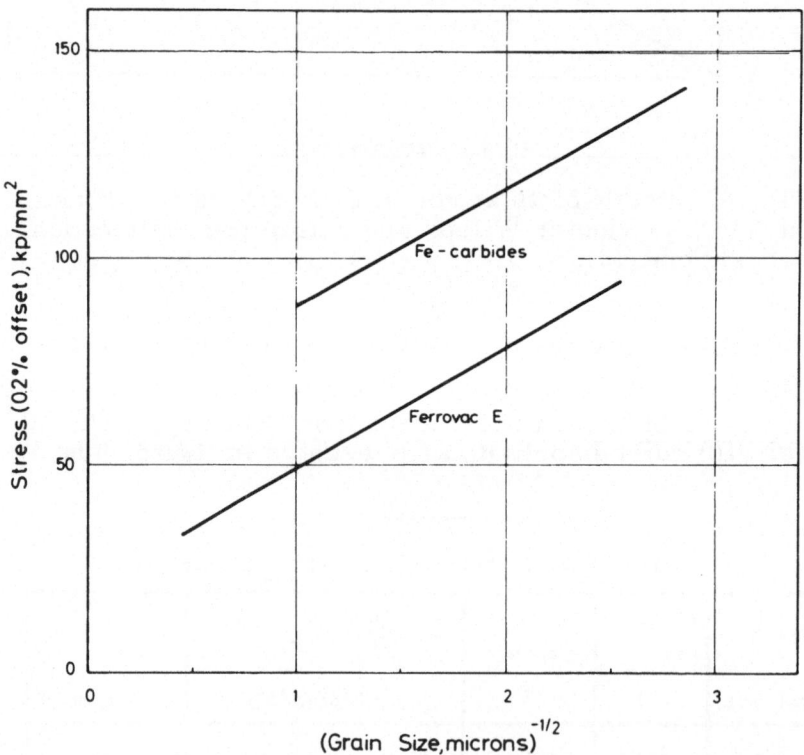

Fig. 5. Petch-plots at room temperature for pure Fe and
Fe with carbides; taken from Embury et al. (10).

Table 6

Strength contributions in Fe-carbides products

C	Mo	Mo₂C		Fe₃C		Carbides			$\tau_p = \frac{Gb}{\lambda}$ $\sigma_p = 2\tau_p$	$\Delta\sigma$
Weight pct.	Weight pct.	Weight pct.	Density (g/cm³)1)	Weight pct.	Density (g/cm³)2)	Vol. pct.	Size (Å)	Spac. (Å)	(kp/mm²)	(fig.5) (kp/mm²)
0.10	0.47	0.50	9.18	0.20	7.18	0.64	100	825	25.5 51	42

1) Density from (15)

2) Density from (13)

   G = 8500 kp/mm², b = 2.48 Å

be coincidental because of the assumptions made regarding
the carbides. But we conclude tentatively that the contributions from dispersion and grain boundaries can be added
with reasonable accuracy.

Dispersion Hardening and Solid Solution Hardening

Al-Mg-Al₂O₃. In Al-Al₂O₃ products the strength is
increased appreciably by solid solution hardening with Mg
(16), see table 7. However, in addition to solid solution
hardening there is also a reaction between Al₂O₃ and Mg
to form particles of a spinel, Al₂O₃· MgO and MgO (18),
and thus the contributions to the strength from addition of
Mg cannot be identified individually.

Ni-Mo-ThO₂. This system has been investigated
(19) for a Mo-content of approximately 12 pct. (of matrix).
In fig. 6 we have plotted some data for Ni-ThO₂ (experimental alloy) and for NiMo-ThO₂ versus the reciprocal
particle spacing. Even with a rather large scatter the
NiMo-alloys clearly have greater stress-values than the

Table 7

Effect of Mg on the tensile strength at
room temperature of an Al-8 vol pct Al₂O₃ product

Material	Flow stress 0.2% offset (kp/mm²)	Ultimate tensile strength (kp/mm²)	Ref.
Al-8 vol pct Al₂O₃	22.2	30.7	17
Al-4.4 pct Mg-8 vol pct Al₂O₃	38.5	44.5	16

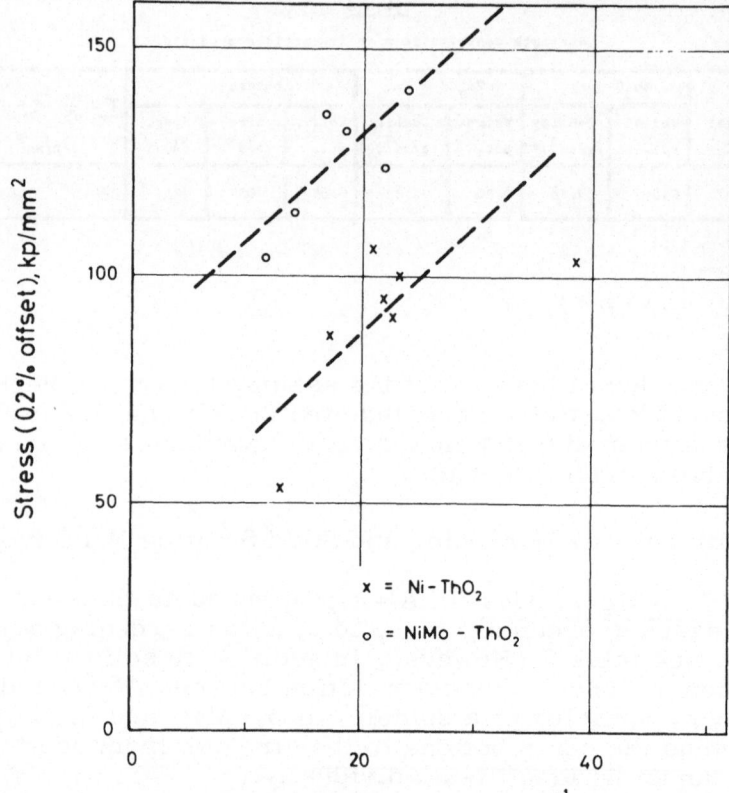

Fig. 6. Orowan-plots at room temperature for Ni-ThO$_2$ and NiMo-ThO$_2$ products (19).

Ni-alloys, demonstrating the effect of the solid solution. The tentative straight and parallel lines indicate that to a first approximation the solid solution effect is independent of the dispersion hardening.

Ni-Cr-ThO$_2$. A limited amount of information on this system can be found in Webster (20). The data include grain sizes, and therefore an additional comparison can be made with the Ni-ThO$_2$ system. The data in fig. 7 on NiCr and NiCr-ThO$_2$ have been taken from Webster's fig. 7. It should be noted that not all points are mentioned in the text of (20), thus there is some ambiguity regarding the position of the lines in fig. 7. The tentative parallel lines (Webster has drawn parallel lines) indicate that the grain boundary contribution is the same in NiCr-ThO$_2$ as in NiCr; this is equivalent to the situation for Ni-ThO$_2$.

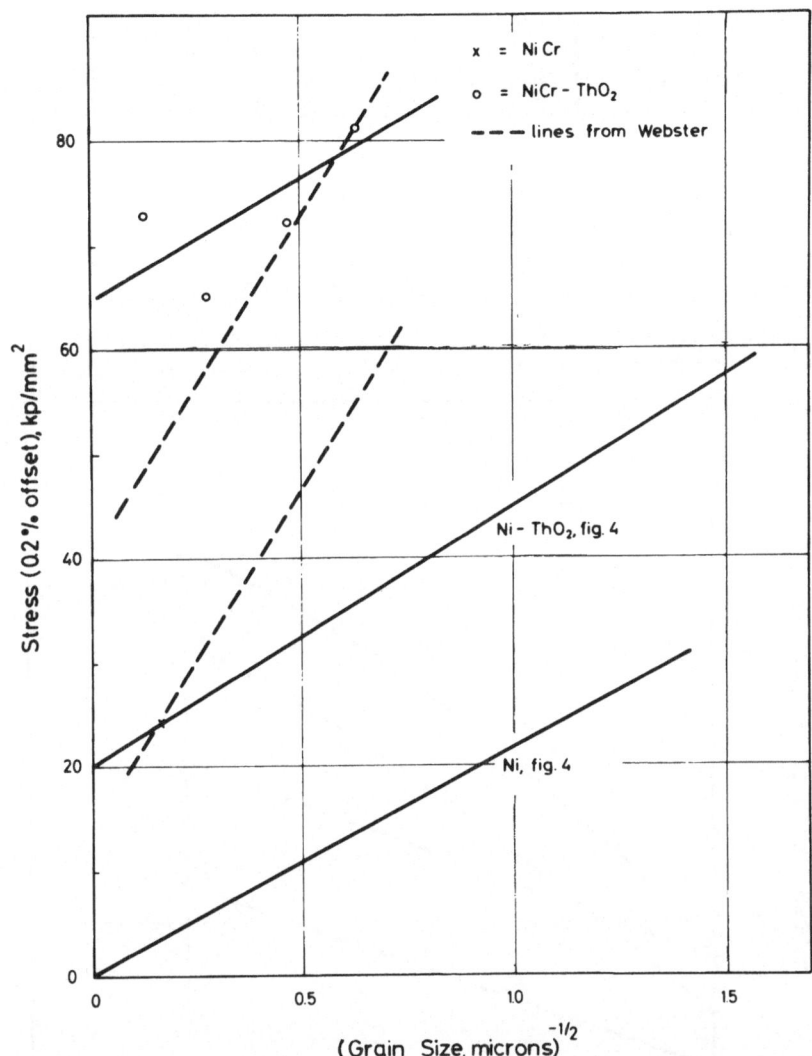

Fig. 7. Petch-plots at room temperature for NiCr and NiCr-ThO$_2$ products. The points are taken from Webster (20), the heavy line is drawn by us, the dotted lines are taken from Webster.

Table 8

Strength contributions in NiCr-ThO$_2$ products

Oxide vol. pct.	Avg. particle size (Å)	Avg. particle spacing' (Å)	$\tau_p = \dfrac{Gb}{\Lambda}$ (kp/mm^2)	$\sigma_p = 2\tau_p$ (kp/mm^2)	$\Delta\sigma$ (fig. 7) (kp/mm^2)
2.3	232	917	21.5	43	25-40

G = 7900 kp/mm^3, b = 2.49 Å

The parallel shift is a measure of the dispersion hardening contribution, (table 8).

We thus find an agreement, which may be coincidental due to the uncertainty of the position of the lines.

By comparison with the lines for Ni and Ni-ThO$_2$ in fig. 7 we find a clear increase in strength due to solid

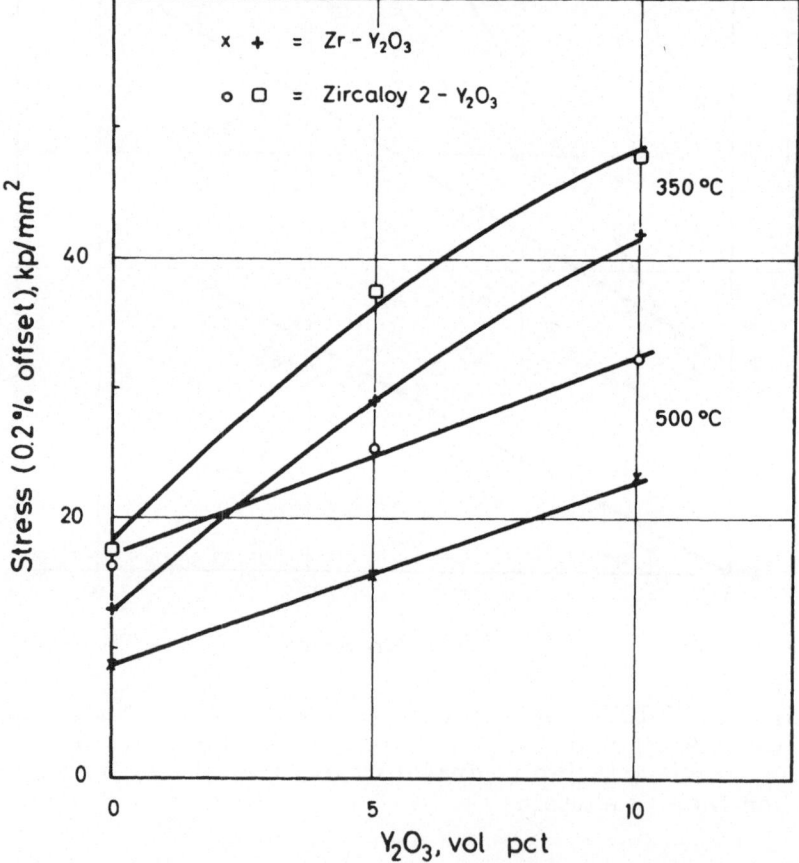

Fig. 8.   Flow stress at 350°C and 500°C for Zr-Y$_2$O$_3$ products and Zircaloy 2-Y$_2$O$_3$ products (21).

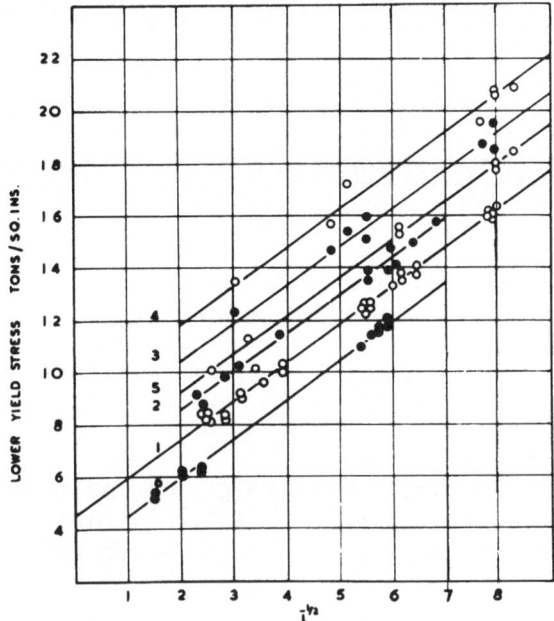

Fig. 9. Petch-plot at room temperature for the lower yield stress of irons and steels. Code: 1 = annealed mild steel En 2; 2 = En 2 nitrided; 3 = En 2 quenched from 650°C; 4 = En 2, quenched, aged 1 hr at 150°C; 5 = En 2, quenched, aged 100 hrs at 200°C; 6 = annealed Swedish iron. Copied from Cracknell and Petch (22).

solution. As the contributions from dispersion hardening agree with the calculated values for both systems we can conclude that the solid solution effect is the same in the $Ni-ThO_2$ alloy as it is in the pure Ni alloy (the extra increase for the $NiCr-ThO_2$ alloy is caused by a smaller particle spacing, see tables 5 and 8).

The influence of solid solution on grain boundary hardening indicated by the larger slopes of the lines for NiCr and for $NiCr-ThO_2$ will not be discussed.

Zircaloy $2-Y_2O_3$. The $Zr-Y_2O_3$ system can only be treated semi-quantitatively, because the parameters are not known in detail. Tensile data (21) on $Zr-Y_2O_3$ and on Zircaloy $2-Y_2O_3$ at 350°C and 500°C have been plotted versus the oxide content in fig. 8. (All specimens have grain sizes of the order 1 µ). The data at room temperature are not included, because the amount of oxygen and nitrogen in solid solution causes some scatter in the data. The parallel lines, both at 350°C and at 500°C, indicate that the contributions from the solutes Sn, Fe and Cr in Zircaloy 2 are independent of the dispersion hardening effect of the oxide in the range of 0 to 10 vol. pct.

Fe(C, N)-carbides and nitrides. This system has been investigated by Cracknell and Petch (22). Various amounts of carbon and nitrogen were introduced into solid solution in a pure α-Fe matrix by selected annealing, quenching and nitriding treatments. Carbon and nitrogen also give strengthening contributions due to dispersion hardening by carbides and nitrides, but as no separate data are given either for solid solution or for dispersion hardening no quantitative comparison can be made.

However, from the straight and parallel lines of fig. 9 we can conclude that the solid solution effect and/or the dispersion hardening effect does not influence the grain boundary strengthening.

STRENGTH AT HIGH TEMPERATURES ($>0.5\,T_m$)

Dispersion Hardening and Grain Boundary Hardening

Grain boundary strengthening at high temperatures is still a matter for discussion in pure metals, and this is also the case for dispersion strengthened products. Quali-

Table 9

Tensile properties at room temperature of extruded Al-Al$_2$O$_3$ products (3)

Material	Oxide (vol pct)	Heat treatment	Subgrain size (μ)	Flow stress 0.2% offset (kp/mm^2)	Ultimate tensile strength (kp/mm^2)
Al MD 13	0.2	6 hr at 500°C	3.8	6.1	9.1
		12 hr at 635°C	1)	2.6	7.0
Al MD 201	0.5	6 hr at 500°C	1.9	7.9	12.5
		12 hr at 635°C	1.9	7.6	12.0
Al R 400	0.9	6 hr at 500°C	1.60	10.4	14.7
		12 hr at 635°C	1.41	10.8	14.6
SAP-ISML 960	3.8	6 hr at 500°C	0.78	14.0	20.9
		12 hr at 635°C	0.90	13.2	20.7

1)Recrystallized grains, diameter 0.2 mm, length 0.1-2 mm.

Table 10

Hardness (VPN) of extruded Ni-2.5 vol pct ThO_2 product (23)

Heat treatment	Hardness (VPN)
none[x]	185
1 hr at 800° C	175
1 hr at 1000° C	160
1 hr at 1200° C	145

[x]After annealing $\frac{1}{2}$ hr at 1000° C, the subgrain size was ~1 μ

tatively, grain boundaries may act as barriers to moving dislocations, giving increased strength with increasing grain boundary area (decreasing grain size). In contrast to this, grain boundary shearing may also take place to give decreased strength with increasing grain boundary area (decreasing grain size). In addition to the grain size, the grain shape may be of importance.

The dispersed particles may strongly influence the size and the shape of the grains in the matrix, thus giving indirectly the observed effect of the grain structure. Important in this context is the effect of particles in stabilizing the structure against recovery, recrystallization and grain growth. This effect is illustrated in table 9, for extruded $Al-Al_2O_3$ products heat treated at temperatures near the melting point of aluminium. The stable structure is reflected in the tensile properties, which are unaffected by the heat treatment. The same stability is not observed in $Ni-ThO_2$ products as illustrated in table 10. The observed decrease in hardness is related to a change in the matrix structure, e.g. grain growth, since the ThO_2 particles are stable (8).

The resistance to recovery and recrystallization after cold work is illustrated in table 11, showing that recrystallization may be incomplete even after heavy deformation followed by a heat treatment at high temperature. For deformations, which do not lead to recrystallization, recovery by subgrain growth may take place. During the recovery the effect of the cold work on the strength may be gradually annealed out (see table 12).

Tensile Strength. Grain boundary strengthening has been found in $Al-Al_2O_3$ (27) tested between room temperature and 426°C (0.75 T_m), and in $Ni-ThO_2$ (8) tested

Table 11

Recrystallization of Al-Al$_2$O$_3$ and Ni-ThO$_2$ products

Material	Oxide (vol pct)	Initial state	Deformation	Heat treatment	Degree of re-crystallization (%)	Ref.
Al-Al$_2$O$_3$	0.5	Recryst.	Cold drawing 30%	24 hr at 600°C	Incomplete	-
"	0.5	"	" 60%	24 hr at 600°C	100	-
"	0.9	Extruded	" 73%	1 hr at 454°C	100	24
"	2.4	"	" 86%	24 hr at 593°C	90	24
"	3.8	"	" 95%	12 hr at 600°C	100	-
"	4.2	"	Cold rolling 98.8%	½ hr at 620°C	Incomplete	25
Ni-ThO$_2$	2.3	Rolled	none	200 hr at 900°C	44	26
"	"	Rolled + 9000C/200 hr.	Cold rolling 50%	1 hr at 1300°C	59	26
"	"	"	" 80%	1 hr at 1300°C	83	26

Table 12

Effect of heat treatment on cold worked Al-Al$_2$O$_3$ and Ni-ThO$_2$ products

Material	Oxide (vol pct)	State of deformation	Heat treatment	Ultimate tensile strength (kp/mm^2)	Hardness (VPN)	Ref.
Al-Al$_2$O$_3$	5.9	Extruded	none	28.6		27
"	"	Extruded + cold rolled 50%	none	32.8		"
"	"	"	8 hr at 538°C	29.6		"
Ni-ThO$_2$	2.5	Extruded	none		185	23
"	"	Extruded + cold drawn 50%	none		230	"
"	"	"	1 hr at 1000°C		183	"
"	"	"	1 hr at 1200°C		173	"
"	"	Extruded + cold drawn 83%	none		250	"
"	"	"	1 hr at 1000°C		200	"
"	"	"	1 hr at 1200°C		180	"

between room temperature and 1200°C (0.85 T$_m$). This effect can be observed in figs. 10 and 11, where a drop in strength follows an increase in grain size in extruded and recrystallized Al-Al$_2$O$_3$ and as-received and recrystallized Ni-ThO$_2$ products[*].

---

[*] In disagreement with fig. 11 it has been found (30) by tensile testing at 1093°C that as-received and recrystallized TD-Ni has practically the same strength. The recrystallization treatment consisted of cold rolling 50% followed by a heat treatment 1 hr at 1204°C. With reference to table 11, it could, however, be possible that the absence of a strength decrease was due to an incomplete recrystallization.

Further evidence for grain boundary strengthening
was found in Al-Al$_2$O$_3$ products containing 0. 19 vol. pct.
oxide; the flow stress at 400°C is plotted in fig. 12 in
accordance with the Petch equation. A straight line was
fitted by the method of least squares, and the k-value was
calculated to 1. 4 kp/mm^2 μ$^{1/2}$ with a standard deviation
of about 30 pct. For 99. 5 pct. Al, also plotted in fig. 12,
the k-value was 2. 9 kp/mm^2 μ$^{1/2}$; the standard deviation
is about the same as for MD 13-1, and it is proposed that
additive strengthening takes place.

A simple model of grain boundary strengthening
does not explain the behaviour of products having very
small subgrains, e. g. introduced by cold work. This is
illustrated in fig. 10 and fig. 11, showing stress-tem-
perature curves for cold drawn Al-Al$_2$O$_3$ products (23)
and cold rolled Ni-ThO$_2$ products (29). At temperatures
below 175°C (∼ 0. 5 T$_m$) for Al-Al$_2$O$_3$ and below 1000°C
(∼ 0. 75 T$_m$) for Ni-ThO$_2$ the strength increases with de-

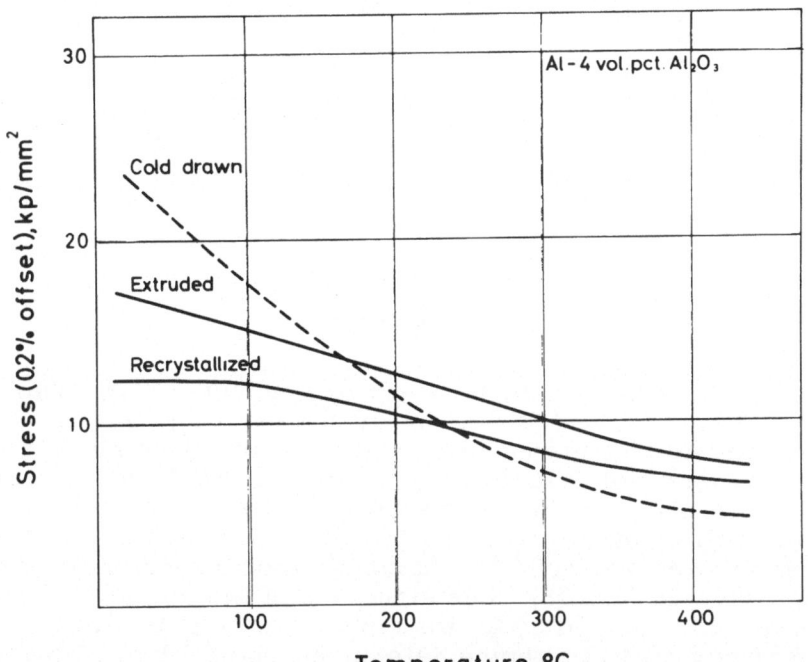

Fig. 10. Flow stress of an Al-Al$_2$O$_3$ product containing
about 4 vol. pct. Al$_2$O$_3$. The extruded material consists
of ∼ 1 μ equiaxed subgrains, which by cold drawing (∼98%)
are reduced to ∼ 0. 25 μ subgrains. Recrystallization (1 hr
at 593°C) after cold drawing gives grains of millimeter
size. Copied from Towner (27).

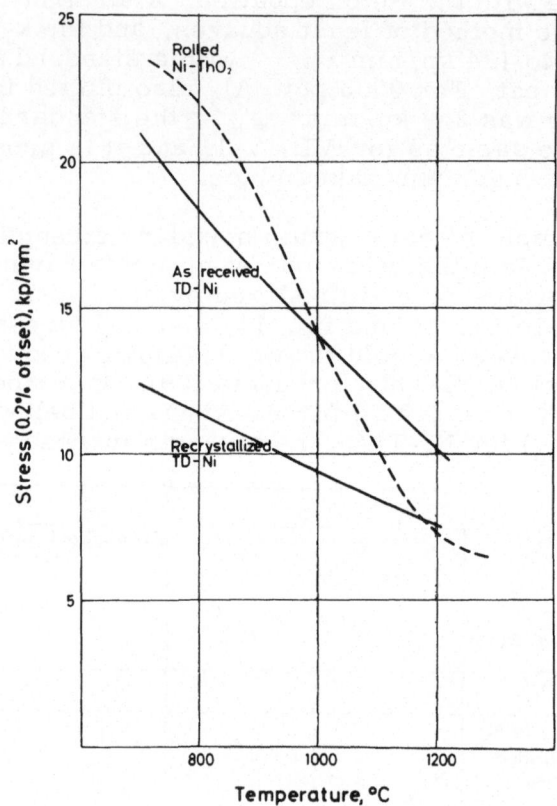

Fig. 11. Flow stress of Ni-ThO$_2$ products. The structure of the as-received TD-Ni is formed by swaging a hot extruded bar 95% in reduction and annealing 1 hr at 1000°C, and consists of cylindrical grains about 10-15 x 1 μ (28). The structure of the recrystallized TD-Ni is formed by cold rolling an as-received bar 94% in reduction and annealing 3 hr at 1300°C, and consists of irregular grains 10-100 μ in size (8). The structure of the rolled Ni-ThO$_2$ product is formed by 21 working cycles (cold rolling 10% in reduction and intermediate annealing 0.5 hr at 1204°C), and consists of 0.5-1.5 μ subgrains (29). The volume fraction and the size of the ThO$_2$ particles in this product is approximately the same as in TD-Ni.

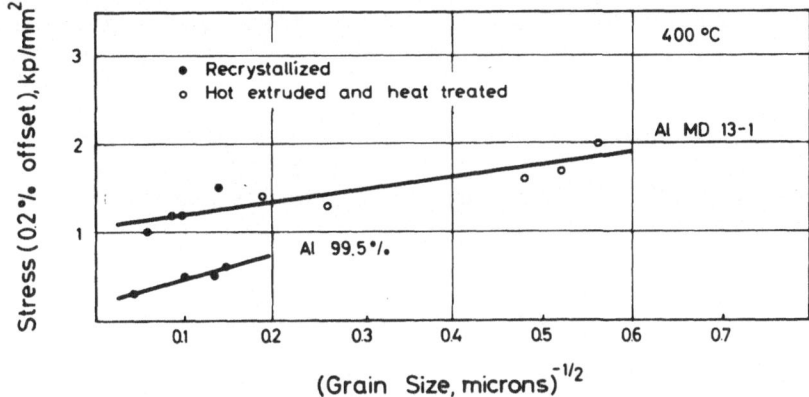

Fig. 12. Flow stress of recrystallized and hot extruded Al-materials at 400°C. Al MD 13-1 is an Al-Al$_2$O$_3$ product containing 0.2 vol. pct. Al$_2$O$_3$. (For the recrystallized Al MD 13-1 showing cylindrical grains, the grain size is the dimension on a plane at 45° to the tensile axis).

creasing grain size, whereas at higher temperatures the products with a medium grain size are the strongest. At temperatures above 240°C (~ 0.55 T$_m$) for Al -Al$_2$O$_3$ and above 1175°C (~ 0.85 T$_m$) for Ni-ThO$_2$, the products with the smallest grain size are the weakest. The heat treatment prior to testing was 0.5 hr at the testing temperature for cold drawn Al-Al$_2$O$_3$, and 0.5 hr at 1204°C for cold rolled Ni-ThO$_2$, thus it is possible that the structures were not stable during testing. Recrystallization has probably not occurred, but the weakening observed might be due to recovery processes taking place in the cold worked structures. Another possibility is that the weakening of the products with the fine grains is due to processes such as grain boundary shearing and the generation of dislocations from grain boundaries. Taking into account such mechanisms, the strength depends probably not only on the grain size but also on the grain shape and the nature of the grain boundaries. Therefore the suggestion of grain boundary weakening processes being strength controlling agrees with the observation that the products with a medium grain size are the strongest at high temperatures.

Creep Strength. Generally the experimental data give no support to a grain boundary strengthening mechanism, e. g. it has been shown (31) for an Al-Al$_2$O$_3$ product creep tested in the temperature range 400 to 600°C (~ 0.7-0.95 T$_m$) that a coarse-grained product has a strength superior to that of a fine-grained product. More detailed results (32) for $Al-Al_2O_3$ products are given in table 13, showing for coarse-grained (recrystallized) products that the creep rate at 300 and 350°C (0.61-0.67 T_m) is approximately the same as that found in fine-grained (extruded) products, whereas at higher temperatures the creep rate is much smaller in the coarse-grained products. Qualitatively these results indicate that grain boundary strengthening and grain boundary weakening processes counterbalance each other at 300 and 350°C, whereas at higher temperatures grain boundary weakening predominates. The structural changes (32) taking place during testing were the formation of subgrain boundaries and dislocation networks in the fine-grained products; such features were practically absent in the coarse-grained products (except for MD 13). Calculations of the activation energy gave values in the range 40 to 200 kcal/mol, which is about one to five times the activation energy for self-diffusion in aluminium; similar values has been found (33) for fine-grained (extruded) SAP products, containing from 4 to 14 wt pct. of Al_2O_3, creep tested in the temperature range 350 to 600°C. These findings do not

Table 13

Creep rates of coarse-grained and fine-grained
$Al-Al_2O_3$ products (32)

Material	Temperature °C	Stress (kp/mm²)	$\dot{\varepsilon}_{min}(h^{-1})$ coarse-grained[1]	$\dot{\varepsilon}_{min}(h^{-1})$ fine grained[2]
MD 13	300	2	$4.5 \cdot 10^{-4}$	$5.6 \cdot 10^{-4}$
"	350	2	$1 \cdot 10^{-2}$	$9.4 \cdot 10^{-3}$
"	400	1	$<5 \cdot 10^{-7}$	$2.7 \cdot 10^{-6}$
"	450	1	$<5 \cdot 10^{-7}$	$3.4 \cdot 10^{-4}$
MD 201	300	3	$6 \cdot 10^{-3}$	$2.7 \cdot 10^{-3}$
"	350	1	$6 \cdot 10^{-7}$	$5 \cdot 10^{-7}$
"	400	2	$3.9 \cdot 10^{-5}$	$6 \cdot 10^{-2}$
"	500	1	$<5 \cdot 10^{-7}$	$1.4 \cdot 10^{-4}$
R 400	400	4	$4.7 \cdot 10^{-3}$	$4 \cdot 10^{-1}$
"	500	2	$<5 \cdot 10^{-7}$	$1.9 \cdot 10^{-6}$

[1] Recrystallized at 600° C for 24 hours. Grain diameter 0.1-0.5 mm, grain length 0.5-12 mm.

[2] Heat treated before testing at 600° C for 24 hours. Grain size 1-5 µ.

lead to any firm conclusion about the creep processes taking place. The results do not, however, exclude the possibility that grain boundary weakening may be due to processes such as grain boundary shearing, as found in as-
received TD-Ni (28), or the generation of dislocations from
grain boundary sources, as proposed in ref. (31).

In the temperature range 675 to 1050°C ($\sim$ 0. 55-0. 8
T_m), the detrimental effect of small subgrains has also
been found in Ni-ThO$_2$ products (34) by comparing a product having a subgrain size of 0. 5 to 2 μ (and superim-
posed cylindrical grains 65 x 2 μ) with as-received TD-Ni
having a larger grain size (cylindrical grains 10-15 x 1 μ).
The creep processes have not been identified with certain-
ty, but based on a structural examination and a calculation
of the activation energy it has been concluded (34) that
grain boundary shearing was probably the most important
creep process in as-received TD-Ni, whereas for the fine-
grained product the rate controlling process might be
climb of dislocations over particles, as originally proposed
by Ansell and Weertman (31) for creep in dispersion hardened alloys.

On the basis of the results for Al-Al$_2$O$_3$ and Ni-ThO$_2$
products it may be concluded that small equiaxed subgrains
of micron size are more beneficial to the tensile strength
than to the creep strength. The best creep strength at high
temperatures has been obtained in products having a larger
grain size, e. g. coarse-grained Al-Al$_2$O$_3$ products and as-
received TD-Ni products. Considering these products, a
common structural feature is the cylindrical grains with
the cylinder axis parallel to the loading direction. It is
possible therefore that the good creep strength is due to
the grain shape rather than to the grain size. The importance of grain shape has been indicated for as-received
TD-Ni by creep testing specimens with the stress axis
respectively parallel to and at an angle of $\sim$ 20 deg. to
the axis of the cylindrical grains (28). Testing at 950°C
($\sim$ 0. 7 T_m), where grain boundary shearing took place,
gave rupture lives of 3. 4 hr and 6 min, respectively.

Based on the experimental findings it is tentatively
concluded that the structure giving the best creep strength
at high temperatures (i. e. above 0. 75 T_m), should consist of large grains, possibly shaped as cylinders with
their longitudinal axis in the loading direction. Based on
this conclusion it is suggested that TD-Ni with a grain
size larger than in the as-received state (10-15 x 1 μ)
might show improved creep properties. Preparation of

Table 14

Stress (kp/mm^2) to give rupture in 100 hours (37)

Material	Temperature	
	982° C	1093° C
TD-Ni	8.4	6.7
TD-Ni - 8 pct Mo	11.2	8.8

such a coarse-grained product may be difficult, as Ni-ThO$_2$ products tend to recrystallize with rather small grains of uneven size (8, 30).

Experimental data on the effect of cold work on the creep strength and the tensile strength are in general agreement. By rolling (35), drawing (17) and hammer swaging (35) extruded Al-Al$_2$O$_3$ products, and then creep testing them in the temperature range 400 to 538°C (~ 0.7–0.9 T_m) a detrimental effect of cold working has been observed, whereas for a Ni-2.5 vol. pct. ThO$_2$ product tested at a lower temperature 815°C ($\sim 0.65\,T_m$) the rupture life has been increased by cold drawing (23).

Dispersion Hardening and Solid Solution Hardening

Tensile Strength. Solid solution hardening is for many dispersion strengthened products only a low temperature phenomena; e.g. for NiCr-ThO$_2$ (36) and for AlMg-Al$_2$O$_3$ (16) no solid solution strengthening has been observed at 871°C and 316°C ($\sim 0.65\,T_m$), respectively.

Creep Strength. No solid solution strengthening was found in AlMg-Al$_2$O$_3$ (18) by creep testing at 450°C ($\sim 0.8\,T_m$). Improvement in creep properties have been reported for NiMo-ThO$_2$ alloys (37) tested in the temperature range 0.7 to 0.8 T_m, see table 14.

CONCLUSION

For materials strengthened by dispersed particles, grain boundaries and elements in solid solution, the room temperature flow stress at 0.2% offset is approximately expressed by the equation

$$\sigma = \sigma_o + 2 \cdot \frac{G\,b}{\Lambda} + k \cdot D^{-1/2} + \sigma_{sol}$$

where the different stress contributions correspond to those found in alloys where the strengthening mechanisms operate individually.

A simple additive rule has not been identified at high temperatures. In the temperature range 0.5 to 0.8 T_m the three strengthening processes may contribute to the tensile strength and to the creep strength, while at higher temperatures dispersion hardening is predominant.

REFERENCES

1. E. Orowan, Symp. Internal Stresses in Metals and Alloys, Oct. 15-16, 1947, London, Inst. Metals, 1948, 451.

2. N. J. Petch, J. Iron Steel Inst., 174 (1953), 25.

3. N. Hansen, Trans. TMS-AIME, 245 (1969), 1305.

4. N. Hansen, Trans. TMS-AIME, 245 (1969), 2061.

5. U. F. Kocks, In: A. S. Argon (ed.): Physics of Strength and Plasticity, M. I. T. Press, 1969; chap. 12.

6. A. J. E. Foreman and M. J. Makin, Can. J. Phys., 45 (1967), 511.

7. N. Hansen, Acta Met., 18 (1970), 137.

8. B. A. Wilcox and R. I. Jaffee, Suppl. to Trans. Japan Inst. Metals, 9 (1968), 575.

9. M. von Heimendahl and G. Thomas, Trans. TMS-AIME, 230 (1964), 1520.

10. J. D. Embury, A. S. Keh and R. M. Fisher, Trans. TMS-AIME, 236 (1966), 1252.

11. E. Smith and J. Nutting, J. Iron Steel Inst., 187 (1957), 314.

12. K. Kuo, J. Iron Steel Inst., 173 (1953), 363; espec. p. 372.

13. E. W. Langer, Dr. Techn. Thesis, Copenhagen 1967, espec. pp. 19 and 27.

14. M. Hansen: Constitution of Binary Alloys, 2. ed., 1958, p. 362.

15. Handbook of Chemistry and Physics, Chemical Rubber Publishing Co., 1958, pp. 610-611.

16. A. S. Bufferd and N. J. Grant, ASM Trans. Quart., 60 (1967), 305.

17. D. J. Boerman, M. Grin and M. Veaux, Euratom (EUR 4074 e), part 1, 1969.

18. D. Gualandi, D. Gelli, P. Jehenson, L. Mori and M. Paganelli, Euratom (EUR 4035 e), 1968.

19. J. G. Rasmussen, Sc. D. Thesis, M. I. T., 1964.

20. D. Webster, AD 669 208, 1968; also ASM Trans. Quart., 62 (1969), 936.

21. D. J. Marsh, J. C. Balling Jensen and J. Kjøller, Danish AEC, Metallurgy Department, Report B-107, Sept. 1969.

22. A. Cracknell and N. J. Petch, Acta Met., 3 (1955), 186.

23. V. A. Tracey and D. K. Worn, Powder Met., 1 (1962), 34.

24. E. J. Westerman and F. V. Level, Trans. TMS-AIME, 218 (1960), 1010.

25. D. Nobili and R. De Maria, J. Nucl. Mat., 17 (1965), 5.

26. E. R. Kimmel and M. C. Inman, ASM Trans. Quart., 62 (1969), 390.

27. R. J. Towner, Trans. TMS-AIME, 230 (1964), 505.

28. B. A. Wilcox and A. H. Clauer, Trans. TMS-AIME, 236 (1966), 570.

29. R. W. Fraser, B. Meddings, D. J. I. Evans and V. N. Mackiw, Modern Developments in Powder Metallurgy, vol. 2, Plenum Press, New York, 1966, 87.

30. G. S. Dobble and R. J. Quigg, Trans. TMS-AIME, 233 (1965), 410.

31. G. S. Ansell and J. Weertman, Trans. TMS-AIME, 215 (1959), 838.

32. N. Hansen and B. Cech, To be presented at the 3rd International Powder Metallurgy Conference in Czechslovakia, Karloy Vary, Sep. 29 - Oct. 2, 1970.

33. K. Milicka, J. Cadek and P. Rys, Creep of Aluminium Strengthened by Aluminium Particles. To be published in Acta Met.

34. B. A. Wilcox and A. H. Clauer, In: Oxide Dispersion
    Strengthening, Second Bolton Landing Conference.
    Bolton Landing, New York, June 27-29, 1966. Gordon
    and Breach, Science Publishers, New York, 1968,
    323.

35. W. A. Friske, Atomics International (NAA-SR-4233),
    1960.

36. L. P. Rice, DMIC - Memo-210, 1965.

37. N. J. Grant, H. J. Siegel, and R. W. Hall, Oxide
    Dispersion Strengthened Alloys, NASA, SP-143, 1967.

P/M PROPERTIES AND TESTING

EXPERIMENTS AND TESTING

CHARACTERIZATION OF COMMERCIAL METAL POWDERS WITH SCANNING ELECTRON MICROSCOPE

Hung-Chi Chao

Applied Research Laboratory

United States Steel Corporation

INTRODUCTION

The scanning electron microscope, in conjunction with conventional optical microscopy, has been used to study the transparticle microstructure and the particle-surface morphology of commercially available iron powders. The powders had been produced by five different processes: (1) by atomization from molten metal, (2) by direct reduction of oxides, (3) by direct reduction followed by other processing steps, (4) by electrolytic deposition, and (5) by chemical reduction. In addition, various metallurgical properties of the powders were determined under controlled laboratory conditions.

MATERIALS AND EXPERIMENTAL WORK

Twenty samples of iron powder made by various processes were obtained. Most samples were obtained from manufacturers of powder-metallurgy (P/M) parts, and the rest were obtained directly from the powder makers. Code numbers were assigned to the iron powders evaluated. These are given in Table I, together with the manufacturing process used, as described in the literature.* Because this investigation was conducted mainly to study the particle morphology with the scanning electron microscope, only relatively small quantities of powder were obtained; thus, that

---

* S. A. Gregory and A. V. Bridgwater, "Iron Powder: Prices, Costs, and Possibilities," Powder Metallurgy, Vol. 11, No. 22, pp. 233-260 (1968).

Table I

Iron Powders Evaluated

Sample Code	Manufacturing Process
A1	Atomization (water)
A2	Atomization (water)
A3	Atomization (water)
A4	Atomization
A5	Atomization
A6	Atomization (air)
A7	Atomization (centrifuged in inert gas)
R1	Direct reduction
R2	Direct reduction
R3	Hydrogen reduction of iron ore
RP1	Direct reduction followed by mechanical densification and recarburization
RP2	Direct reduction followed by resulfurization
RP3	Direct reduction followed by resulfurization
RP4	Direct reduction followed by resulfurization and recarburization
RP5	Direct reduction followed by mechanical densification
RP6	Direct reduction followed by blending with electrolytically reduced powder
E1	Electrolytic reduction
C1	Hydrogen reduction of iron chloride
C2	Hydrogen reduction of iron chloride followed by mechanical densification
C3	Carbonyl iron powder

Table II

Composition of Iron Powders Evaluated—Percent

Sample Code	C	Mn	P	S	Si	O	N	H_2 Loss
Atomized Powders								
A1	0.022	0.20	0.007	0.015	<0.007	0.188	0.001	0.19
A2	0.029	0.32	0.009	0.017	0.069	0.450	0.002	0.32
A3	0.007	0.19	0.007	0.021	0.028	0.235	0.004	0.22
A4	0.028	0.010	0.209	0.007	0.008	0.227	0.006	0.21
A5	0.013	<0.03	0.026	0.006	0.014	0.151	0.003	0.10
A6	0.15	0.35	0.009	0.031	0.026	0.471	0.003	0.60
A7	0.16	0.74	0.006	0.025	0.029	0.02	0.005	0.10
Powders Produced by Direct Reduction of Oxides								
R1	0.041	<0.032	0.005	0.010	0.050	0.642	0.008	0.25
R2	0.029	0.55	0.014	0.004	0.093	1.468	0.001	1.0
R3	0.015	0.025	0.010	0.005	0.003	0.697	0.001	<0.31
Powders Produced by Direct Reduction Followed by Other Steps								
RP1	0.24	0.032	0.005	0.009	0.056	0.658	0.011	0.47
RP2	0.027	0.029	0.005	0.58	0.060	0.973	0.009	1.10
RP3	0.030	0.030	0.005	0.81	0.061	0.98	0.008	1.40
RP4	0.98	0.049	0.004	0.52	0.067	1.080	0.022	2.30
RP5	0.051	0.010	0.005	0.008	0.011	0.497	N.D.	0.18
RP6	0.045	0.47	0.007	0.013	0.14	0.863	0.003	0.59
Powder Produced by Electrolytic Deposition								
E1	0.005	0.010	0.008	0.004	0.003	N.D.	N.D.	0.08
Powders Produced by Chemical Reduction								
C1	0.012	0.27	0.006	0.005	0.006	N.D.	N.D.	0.22
C2	0.02	0.19	N.D.	0.007	0.006	N.D.	N.D.	0.25

ND—Not determined because of limited quantity of sample.

part of the study concerned with the metallurgical properties of
the powders was necessarily limited.

A screen analysis was conducted on each sample and its
chemical composition and the hydrogen loss was determined. APMI
standard procedures were used to obtain the flow rate and the
apparent and tap densities. Where the quantity of sample permitted,
the green densities and the green strength of the samples were also
determined as a function of compacting pressure. In these studies,
no additives were mixed with the powder, but the die walls were
lubricated by zinc stearate. A 25-gram-load microhardness
indenter was used to determine the particle hardness on polished
cross-sectional surface of the particles. Metallographic studies
were conducted on all samples. Both light and scanning electron
microscopes were used in these studies.

RESULTS AND DISCUSSION

Chemical Composition

Shown in Table II are the results of chemical analyses on the
19 samples of iron powder evaluated. The results are grouped into
five categories on the basis of the manufacturing process:
(1) atomization, (2) direct reduction of oxides, (3) direct
reduction followed by other processing steps, (4) electrolytic
deposition, and (5) chemical reduction.

Except for the A6 powder, the atomized powders generally had
low carbon contents (0.007 to 0.029%), relatively low oxygen
contents (0.15 to 0.50%), and low hydrogen-loss values (0.19 to
0.32%). Because of the relatively low oxygen values, these
powders would be expected to contain only a small amount of oxides
and therefore would be relatively clean. The A6 powder is a rela-
tively-high-carbon powder made by atomizing very-high-carbon
(about 4%) liquid metal with air. The powder is then annealed;
the reaction between carbon and oyxgen during annealing lowers the
carbon and oxygen levels to about 0.15 and 0.5 percent, respectively.

The powder produced by the direct-reduction processes had
higher oxygen contents than the atomized powder; as a result,
the R2 powder had a higher hydrogen-loss value (1.0%) than the
atomized powders. The low hydrogen-loss value of R1 and R3 powder
in comparison with its oxygen content is due to the nonreducible
oxides from the ore, which are present within the powder particles
as a residue.

When directly reduced powders were given resulfurization or
recarburization treatments or both (RP1, RP2, RP3, and RP4), the

hydrogen-loss values of these treated powders were relatively high due to the tendency of these powders to decarburize and desulfurize during hydrogen-loss testing.

The electrolytic powder, E1, was a high-purity powder with a very low carbon content (0.005%) and a low hydrogen-loss value (0.08%). The compositions of the chemically reduced powders (C1 and C2) were similar to those of the atomized powders.

Metallurgical Parameters

Loose-Powder Properties. The screen analysis, flow rate, apparent and tap density, and hardness of the iron powders evaluated are given in Table III. Generally, the powders had a bimodal distribution of screen analysis.

Inspection of Table III shows that except for the A6 powder, the atomized powders as a class generally had better flow rates than powders produced by other methods.

Although there was some overlap, the atomized powders (except for the A6 powder) generally had slightly better apparent densities than powders made by other methods. The hardness of the powders varied widely, depending on composition and processing treatment; the powders produced by chemical reduction were the softest, and the mechanically densified and recarburized powders were the hardest.

Green Density. The green density is plotted as a function of compacting pressure in Figure 1. In general, the atomized powders, such as the A1 molding-grade powder, had the higher densities for a given compacting pressure, and the direct-reduced powders, such as the R2 molding-grade powder, had the lower densities. The variation in density was about 0.8 g/cc.

Green Strength. Relatively large variations in green strength among the powders investigated were observed, Figure 2. Also, the differences in green strength became greater as the compacting pressure was increased. The high-carbon, high-sulfur, high-oxygen special powder (RP4) had the lowest green strength, whereas the directly reduced iron powders (R1 and R2) and the resulfurized reduced powder (RP3) were among the powders with the highest green strength. The recarburized reduced powder (RP1) and the mechanically densified reduced powder (RP5) had relatively low green strengths.

Among the various atomized powders, the A4 and A5 powders had the lowest green strength. The A3 powder had a slightly higher green strength, and the A1 powder had a green strength lower than

Table III

Screen Analysis, Flow Rate, Apparent and
Tap Density, and Hardness of Iron Powders Evaluated

Sample Code	Screen Analysis, percent ASTM Screen No.							Flow Rate, sec	Apparent Density, g/cc	Tap Density, g/cc	Hardness,* DPH25
	+80	+100	+140	+200	+230	+325	Pan				
Atomized Powders											
A1	3.73	4.16	15.55	18.35	19.10	31.26	7.84	26.2	3.10	3.92	111
A2	0	0.11	7.30	13.15	19.37	44.42	15.93	25.1	2.93	3.77	97
A3	0.09	0.33	21.39	23.02	25.01	26.50	7.88	26.7	3.03	3.88	159
A4	0	0.26	13.52	22.10	22.11	31.32	9.97	28.0	2.86	3.70	115
A5	0.60	0.75	14.8	28.2	20.15	26.9	7.3	29.4	2.85	3.58	N.D.
A6	0.07	2.04	15.46	17.17	22.76	30.68	11.01	32.3	2.32	2.99	N.D.
A7	0.03	0.03	0.11	0.06	60.94	36.44	1.48	13.5	4.53	4.98	N.D.
Powders Produced by Direct Reduction of Oxides											
R1	0	0.05	11.75	18.15	20.15	20.27	32.89	32.2	2.38	3.22	N.D.
R2	0.5	0.38	20.77	27.25	19.63	24.62	7.13	31.7	2.58	3.29	94
Powders Produced by Direct Reduction Followed by Other Steps											
RP1	0	0.19	23.06	29.06	20.69	20.87	5.95	32.4	2.46	3.18	201
RP2	0	0.23	13.72	20.36	24.49	29.71	11.15	30.5	2.72	3.51	145
RP3	0	0.51	18.22	21.72	21.25	26.33	11.63	31.3	2.66	3.40	198
RP4	0.32	0.37	19.68	25.76	19.90	24.30	8.89	36.2	2.91	3.53	159
RP5	0.29	0.60	18.44	31.12	25.48	20.62	3.00	25.9	2.95	3.76	179
RP6	0.08	0.83	10.27	20.83	27.12	30.97	9.2	33.8	2.46	3.25	124
Powder Produced by Electrolytic Deposition											
E1	0	0.5	14.8	25.8	28.0	13.9	17.1	28.2	2.95	N.D.	N.D.
Powders Produced by Chemical Reduction											
C1	0	1.5	15.8	23.1	12.5	18.0	28.6	33.0	2.41	N.D.	76
C2	0	2.5	13.4	21.4	22.5	22.5	28.1	29.0	2.79	N.D.	87

* DPH25 is a diamond pyramid hardness reading under a 25-gram load on powder particle surface.

ND Not determined.

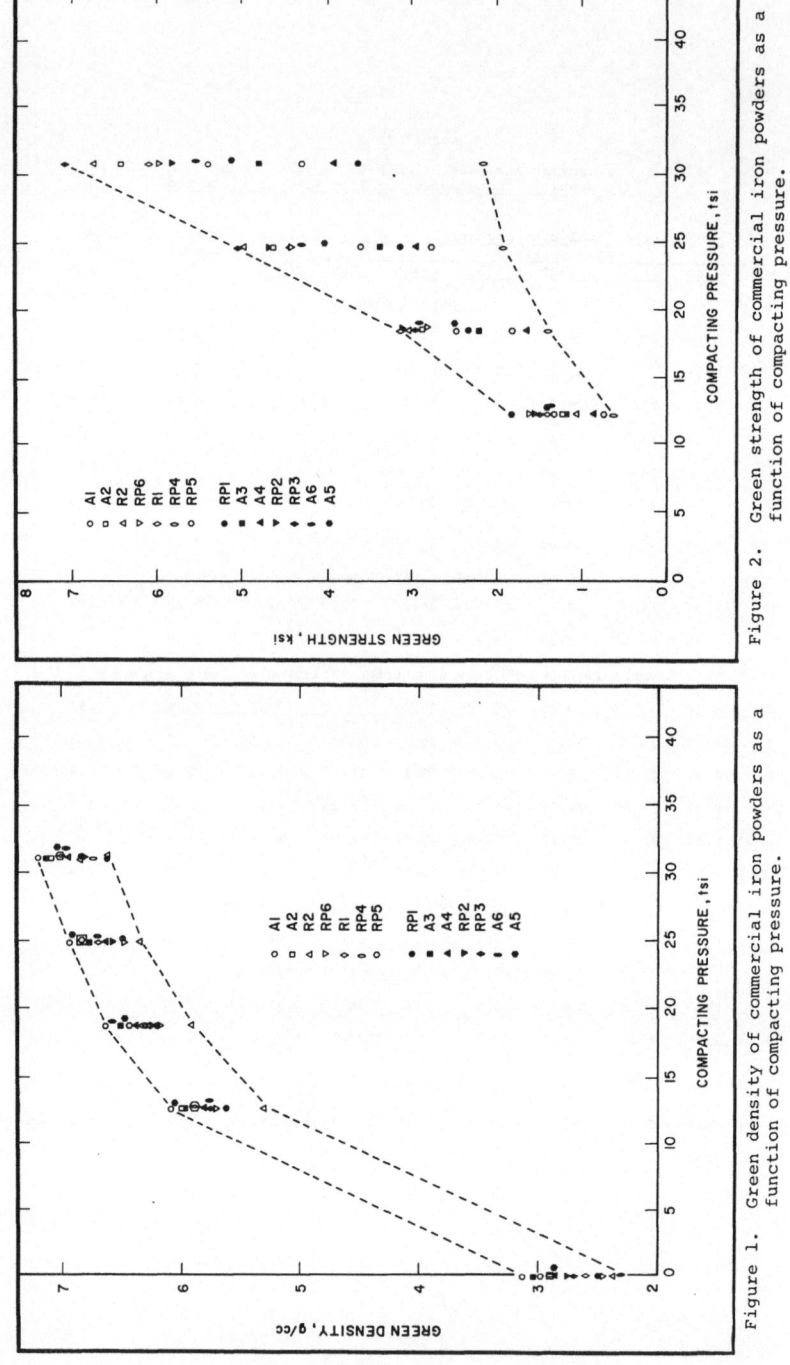

Figure 2. Green strength of commercial iron powders as a function of compacting pressure.

Figure 1. Green density of commercial iron powders as a function of compacting pressure.

that of most reduced iron powders but occupied about the middle position of the commercial powders evaluated. The high green strength of the atomized A2 powder may be attributed to the fact that this powder is produced by decarburizing the high-carbon iron powder by annealing it with mill scale; thus, the resultant powder has certain similarities to reduced iron powder, and therefore, has properties similar to those of reduced iron powder.

Metallographic Studies

Atomized Powders. Atomized iron powders are presently produced by atomization with inert gas, air or water. In gaseous atomization the cooling rate is relatively slow so that the liquid droplets have time to spheroidize before solidification. Thus powder produced by inert-gas atomization is generally characterized by its nearly spherical configuration, as is illustrated by the sample of inert-gas-atomized steel powder (A7) shown in Figure 3. Consequently, inert-gas-atomized powder has excellent flowing properties, high loose-packing density, but low green strength after processing. As shown in Figure 3B, the spherical particles exhibit a fine as-cast dendritic structure.

Figure 4 shows the surface and cross section of a sample of air-atomized high-carbon iron powder (A6). This powder is produced by a process in which high-carbon liquid metal is atomized by air. Atomization in such a highly oxidizing atmosphere results in the production of many fine oxide particles and large oxide-covered particles. Subsequent annealing in a reducing atmosphere results in sintering of reduced fine particles to reduced large particles. Thus, after grinding, very irregular particles are produced, Figure 4. The ultimate result is an improvement in green strength. Since voids are inevitable in this type of atomization and agglomeration, the apparent density of this powder is relatively low.

Atomization by a liquid medium such as water results in the freezing of the liquid-metal droplets during the early stage of atomization before spheroidization occurs. Thus, irregular-shaped particles are produced. Furthermore, the much higher pressure that can be used in water atomization results in relatively high water velocity; thus, disintegration of the liquid metal is more effective than in gas atomization. Also, collision and agglomeration of particles occur frequently during the atomization process, and, as a result, very irregular powders are produced.

Figure 5 shows that one of the newer types of water-atomization processes produces a powder (A1) with a relatively large number of elongated, irregular "sweet-potato-shaped" particles; the finished powder contains relatively few aggregates. At high magnification, Figure 5B, the particles have an irregular shape with smooth, clean

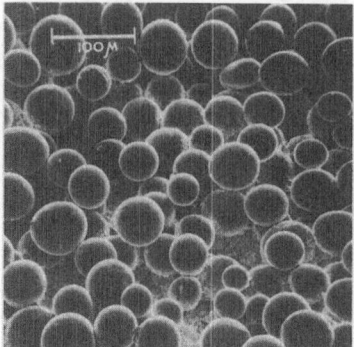

Figure 3A. Surface.

Figure 4A. Surface.

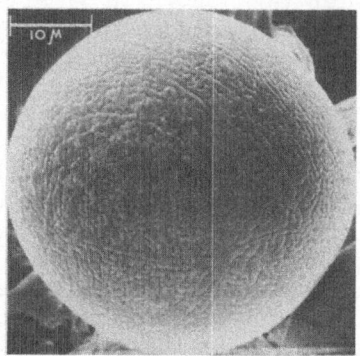

Figure 3B. Surface.

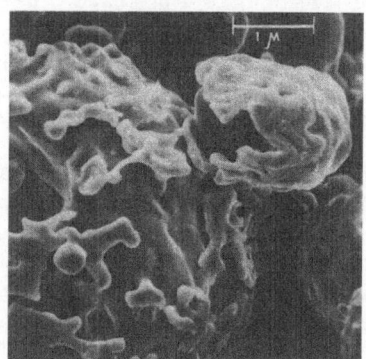

Figure 4B. Surface.

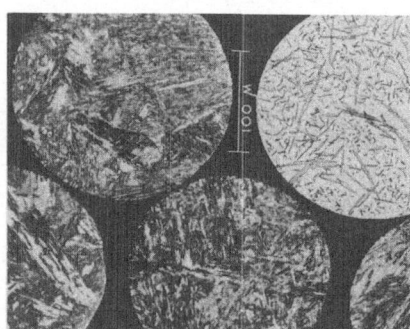

Figure 3C. Cross section. 2 percent
           nital etch.

Figure 3. Surface and cross section
           of a sample of A7 steel
           powder.

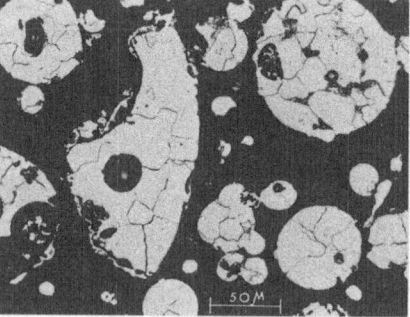

Figure 4C. Cross section. 2 percent
           nital etch.

Figure 4. Surface and cross section
           of a sample of A6 iron
           powder.

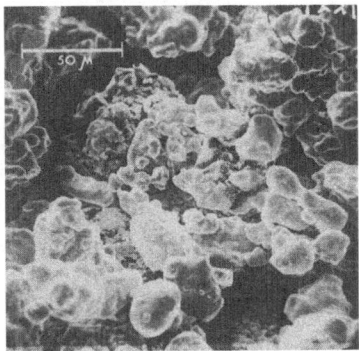

Figure 5A. Surface.

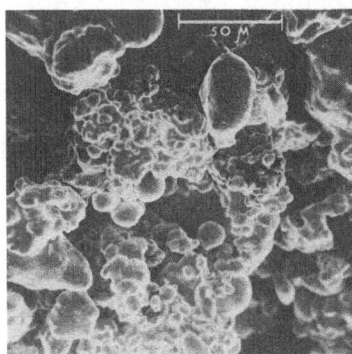

Figure 6A. Surface.

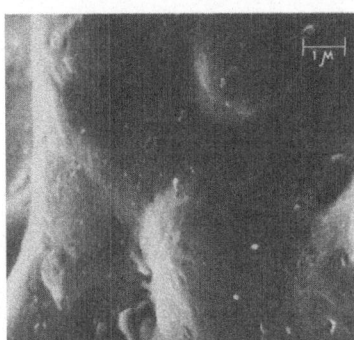

Figure 5B. Surface.

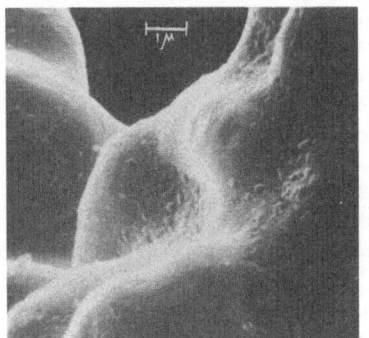

Figure 6B. Surface.

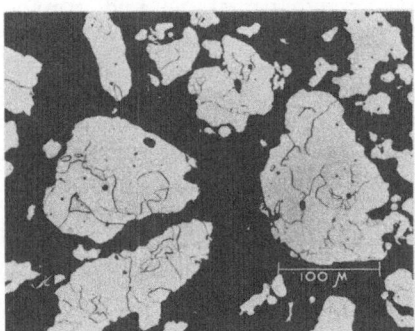

Figure 5C. Cross section. 2 percent
            nital etch.

Figure 5. Surface and cross section
           of a sample of A1 iron
           powder.

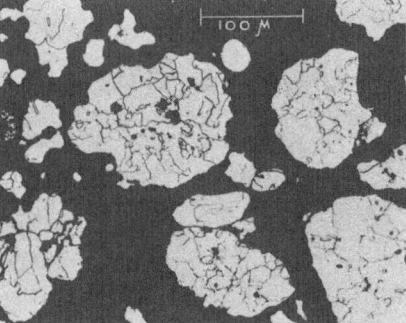

Figure 6C. Cross section. 2 percent
            nital etch.

Figure 6. Surface and cross section
           of a sample of A3 iron
           powder.

surfaces. The irregular shape, the small number of aggregates and
the cleanliness of this powder are the main reasons for its high
ductility and its high compact density. Thus, this powder should
have sufficient green strength so that the agglomeration effect
in annealing would not be necessary. In addition, the low-
temperature-annealed fine grain size should contribute to high-
efficiency in carbon pickup during sintering with graphite additives.
Another powder (A3, Figure 6) appears to be similar to that of A1
powder; at least the appearance and properties of these two powders
are very similar.

Some atomized-iron-powder producers depend on a post-
atomization agglomeration technique to produce irregular-shaped
molding-grade powder. With this technique, partly sintered
irregular aggregates are produced through a combination of annealing
and grinding. A typical example of a powder made by this technique
is the A2 powder shown in Figure 7. This powder is an atomized
high-carbon powder, which is mixed with mill scale, annealed in a
reducing atmosphere, and then ground. A4 and A5 powders, Figure 8,
may also have been produced by a similar technique. Note the
similarity in the surfaces and in the cross sections of the powders
illustrated in Figures 7 and 8.

Powders Produced by Direct Reduction of Oxides. The surface
and cross section of a sample of R1 molding-grade powder are shown
in Figure 9. This powder was made by mixing relatively high-
purity iron ore with a solid carbonous reducing agent. At high
magnification, the powder is seen to consist of irregular, smooth-
cornered, spongy particles with numerous tiny, pipe-like randomly
dispersed voids with average spacing of 5 microns. Large patches
of inclusions were observed to be mingled with irregular shaped
internal voids. This type of powder does not give a very high
apparent density because of the very irregular shape and contour of
its particles and because of numerous internal voids.

Figures 10 and 11 show the surfaces and cross sections of
samples of R2 and R3 powders, respectively. These powders were
produced by the hydrogen reduction of mill scale and iron oxide,
respectively, and thus are similar to the R1 powder, Figure 9.
Because the oxides used to produce the R2 and R3 powders may not be
as pure as the ore used to produce the R1 powder, the R2 and R3
powders are not as clean as the R1 powder. This is clearly shown
in Figures 10C and 11C.

A variety of aftertreatments to impart certain special properties and to improve the density of the direct-reduced powders are
being used by powder producers. A cross section of a mechanically
densified powder, RP5, which was probably made from R1 powder, is
shown in Figure 12. Figure 13 shows a cross section of RP1 powder,
a mechanically densified and recarburized powder. A resulfurized

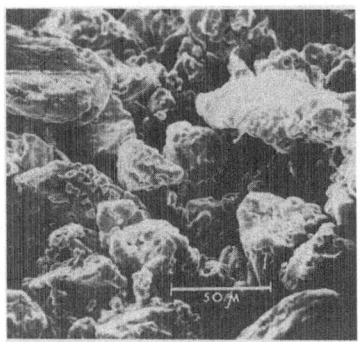

Figure 7A.   Surface.

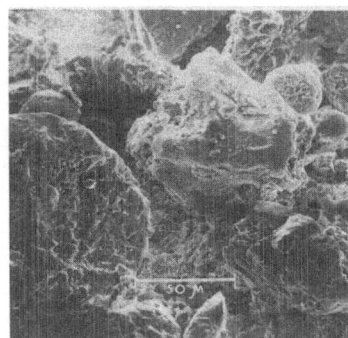

Figure 8A.   Surface.

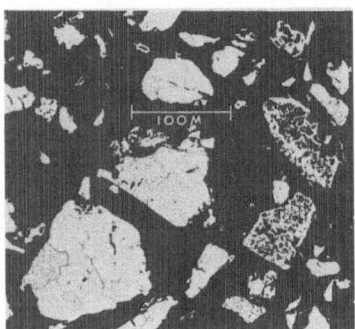

Figure 7B.   Cross section.   2 percent
nital etch.

Figure 7.   Surface and cross section
of a sample of A2 iron
powder.

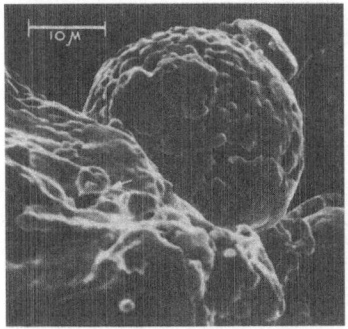

Figure 8B.   Surface.

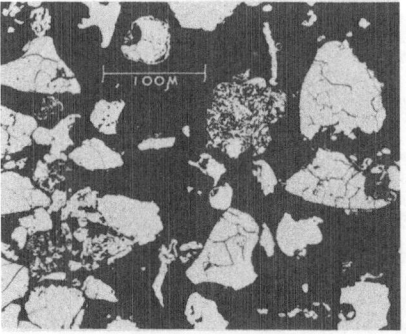

Figure 8C.   Cross section.   2 percent
nital etch.

Figure 8.   Surface and cross section
of a sample of A4 and A5
iron powders.

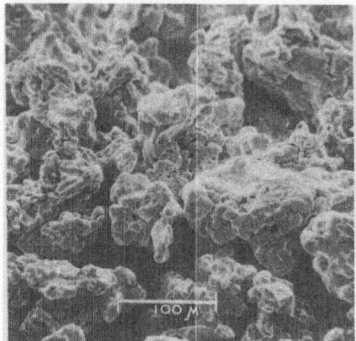

Figure 9A. Surface.

Figure 10A. Surface.

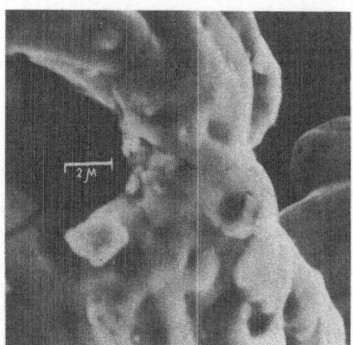

Figure 9B. Surface.

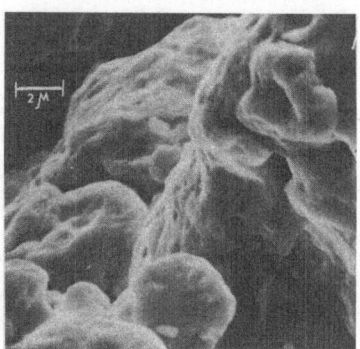

Figure 10B. Surface.

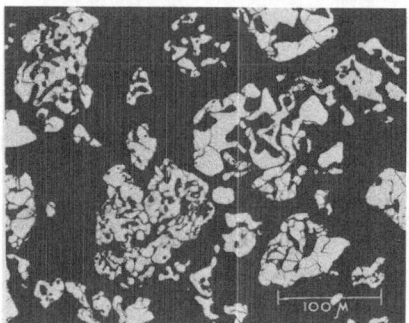

Figure 9C. Cross section. 2 percent
             nital etch.

Figure 9.   Surface and cross section
             of a sample of R1 iron
             powder.

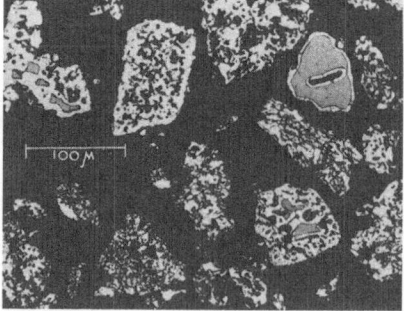

Figure 10C. Cross section. 2 percent
              nital etch.

Figure 10. Surface and cross section
             of a sample of R2 iron
             powder.

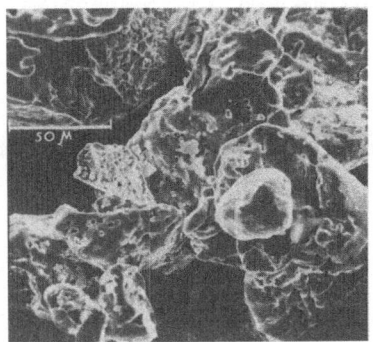

Figure 11A.   Surface.

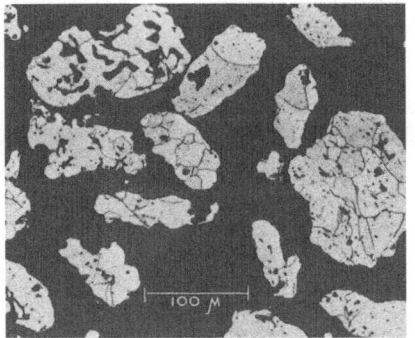

Figure 12.   Cross section of a sample
of RP5 iron powder.
2 percent nital etch.

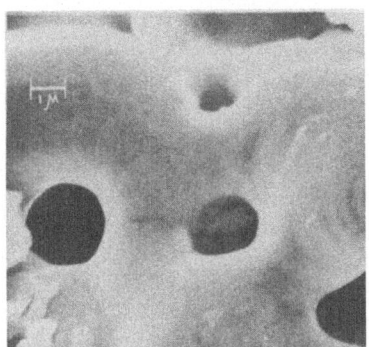

Figure 11B.   Surface.

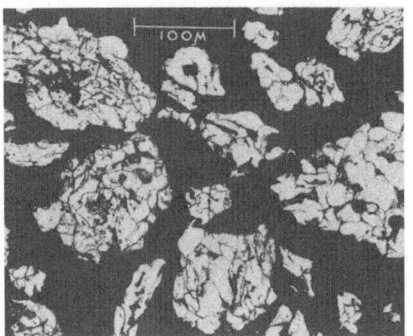

Figure 13.   Cross section of a sample
of RP1 iron powder.
2 percent nital etch.

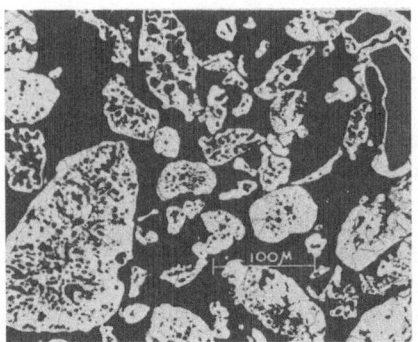

Figure 11C.   Cross section.   2 percent
nital etch.

Figure 11.   Surface and cross section
of a sample of R3 iron
powder.

powder, RP3, and a resulfurized and recarburized powder, RP4, are
shown in cross-sectional view in Figures 14 and 15, respectively.
A typical surface of directly reduced powder that has received an
aftertreatment mechanical densification, RP5, is shown in Figure 16.

   Powder Produced by Electrolytic Deposition. As shown in
Figure 17, a typical commercial molding-grade electrolytical iron
powder (E1) is characterized by solid, elongated irregular shapes.
A close-up view, Figure 17B, shows a very clean surface, with
large and well-crystallized equiaxed grains. Figure 17C, a cross
section of the powder, shows that it is clean and relatively dense.

   Powders Produced by Chemical Reduction. The iron powder shown
in Figure 18 (C1 and C2) is produced by first dissolving either
machine turnings or iron ore in hydrochloric acid to produce iron
chloride. The iron chloride, as either a finely divided solid or
a vapor, is then reduced by hydrogen. The iron powder in Figure 19
(C3) is produced by the well-known carbonyl process.

                              SUMMARY

   The results of the present evaluation of the properties and
metallographic characteristics of commercially available iron
powders may be summarized as follows:

1. The particles of all the reduced iron powders, though
    produced by various processes, were generally irregular,
    smooth-cornered, and spongy, and numerous pipe-like, randomly
    dispersed small voids were observed (at an average spacing of
    5 microns on the surface). Large particles of unreduced
    inclusions, mingled with numerous, irregular-shaped, small
    internal voids were also observed.

2. The particles of the atomized iron powders decarburized
    with iron oxides had nearly the same surface appearance as
    that of the reduced iron powders (because of similar reaction
    kinetics), except that the atomized iron powders generally
    had an agglomerated appearance and only a few large internal
    voids.

3. The particles of iron powders water-atomized from molten
    steel or low-carbon iron were clean and dense with irregular
    surfaces. Such powders give moderate green strength but
    high density.

4. The particles of the gas-atomized iron and alloy powders were
    spherical, with dendritic structures clearly obvious on the
    surface.

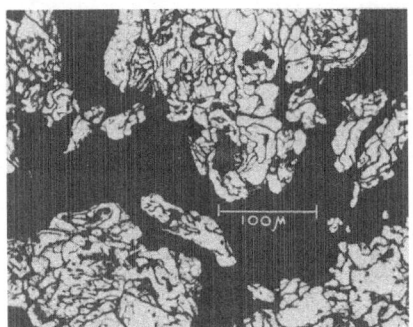

Figure 14. Cross section of a sample
of RP3 iron powder.
2 percent nital etch.

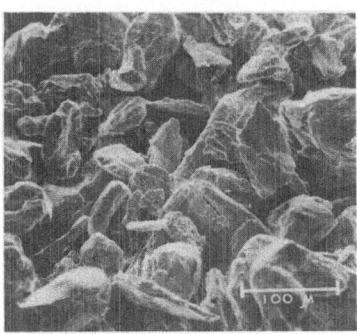

Figure 17A. Surface.

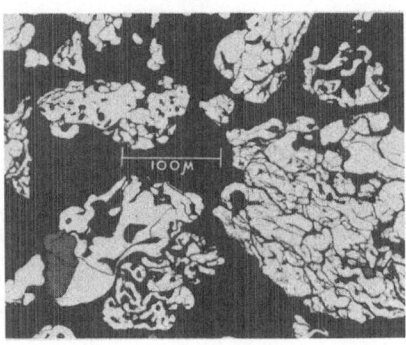

Figure 15. Cross section of a sample
of RP4 iron powder.
2 percent nital etch.

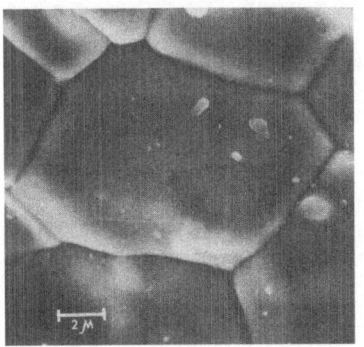

Figure 17B. Surface.

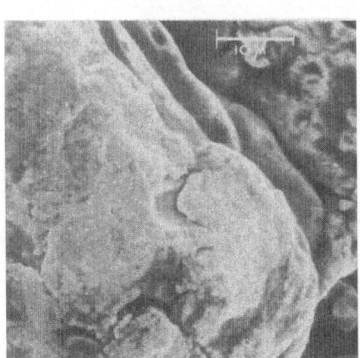

Figure 16. Surface of a sample of
RP5 iron powder.

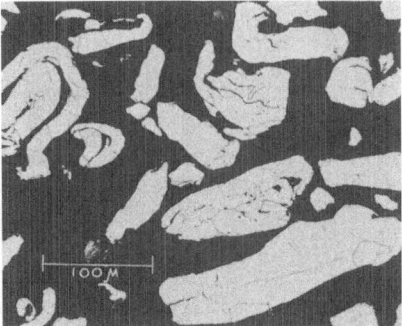

Figure 17C. Cross section. 2 percent
nital etch.

Figure 17. Surface and cross section
of a sample of E1 iron
powder.

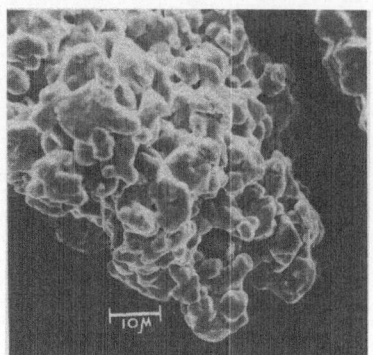

Figure 18A.   Surface of a sample
              of C1 iron powder.

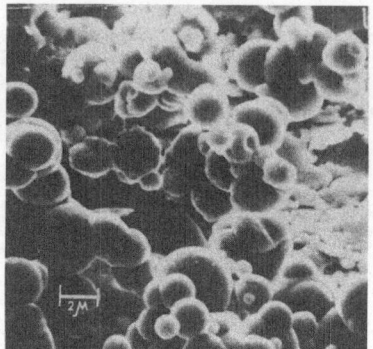

Figure 19A.   Surface.

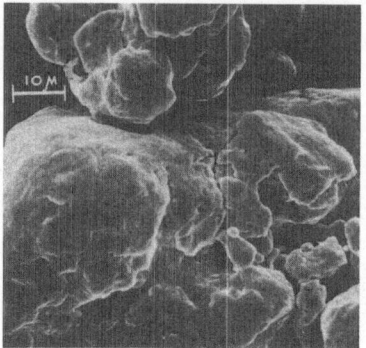

Figure 18B.   Surface of a sample
              of C2 iron powder.

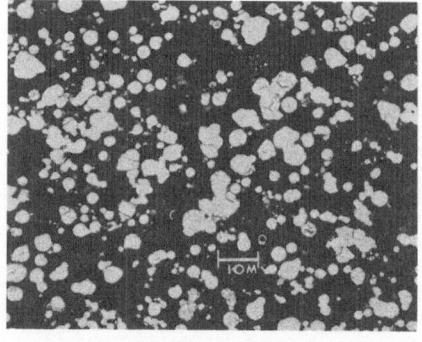

Figure 19B.   Cross section.   2 percent
              nital etch.

Figure 19.   Surface and cross section
              of a sample of C3 iron
              powder.

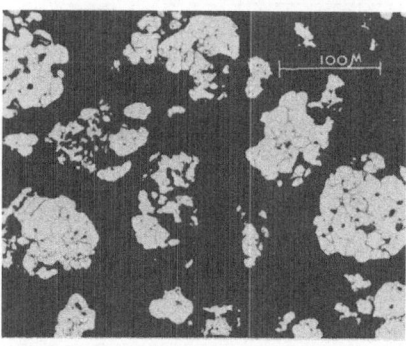

Figure 18C.   Cross section of a sample
              of C2 iron powder.
              2 percent nital etch.

Figure 18.   C1 and C2 iron powders.

5. The particles of the electrolytic iron powder studied were moderately irregular and elongated, and had large grains. No voids or inclusions were observed.

6. The particles of the chemically reduced iron powders had the appearance of agglomerates of numerous small particles with very clean surfaces.

7. Generally, mechanically worked surfaces could be observed on directly reduced or chemically reduced iron powders that were mechanically densified in an aftertreatment. Annealing after mechanical densification resulted in recrystallization and grain growth, and consequently, internal voids were minimized or eliminated.

8. The chemical compositions of the powders evaluated depended on the manufacturing process. The reduced iron powders usually had higher metalloid contents than the atomized iron powders. The oxygen contents and hydrogen losses of the reduced iron powders were particularly high. Oxygen contents of 1/2 to 1-1/2 percent were observed in the reduced iron powders, whereas the atomized iron powders annealed in reducing atmospheres generally contained less than 0.45 percent oxygen.

9. The atomized iron powders usually had better flow rates than the reduced iron powders, and had slightly higher apparent and tap densities and better green densities than powders made by other methods.

10. Considerable variation in green strength was observed. In general, the atomized iron powders had lower green strength than the reduced iron powders. Resulfurization improved the green properties, but recarburization impaired these properties.

GAMMA RAY DENSITY GAUGING

John E. Hoel and Thomas W. Novitsky

IBM Systems Manufacturing Division

Endicott, New York

ABSTRACT

The density of pressed powder metal parts can be rapidly and non-destructively measured by means of gamma ray transmission techniques.

General considerations of gamma-ray density gauging are discussed from an applications viewpoint.

Specific applications and results which are discussed include the density of green, sintered, and repressed iron parts and of bronze and other alloys.

CONTENTS

I.   DESIGNING A GAMMA RAY TRANSMISSION GAUGE

    A.   Basic Design Calculations

       There are three characteristics which are basic to
       every gamma ray transmission application. They are:

       First: The mass-thickness of the sample to be measured,
       Second: The principal elemental constituent in the
                   sample, and
       Third: The required precision of the measurement.

       The variable of interest in sintered metals technology
       is, of course, density. Thus the following discussion
       will center on this parameter. Thickness measurements,
       however, are also amenable to this type of design
       rationale.

      1.   Beer's Law for the transmission of radiation through
          matter can be expressed as in equation (1).

$$\alpha = I/I_o = e^{-\mu\rho x} \tag{1}$$

Where:
I   is the transmitted intensity,
I_o is the incident intensity on the sample,
e   is the base of the natural logarithms,
μ is the mass absorption coefficient for the material
      of which the sample is composed,
ρ is the bulk density of the material,
x   is the thickness of the sample along the direction
      of transmission.

       I/I_o is known as the transmission ratio, α.

      2.   Calculation for optimum energy

          The rate of change of the transmission ratio with
          respect to a change in the density, ρ, is known as
          the sensitivity, s, of the measurement and is ex-
          pressed in equation (2).

          A plot of equation (2) with μ as the independent
          variable is given in Figure I. It is seen that
          there is an optimum value for μ such that the measure-
          ment is most sensitive. Since μ is the only energy
          dependent term in equation (1), the magnitude of
          the gamma ray energy which corresponds to the op-
          timum value of μ yields the best measurement.

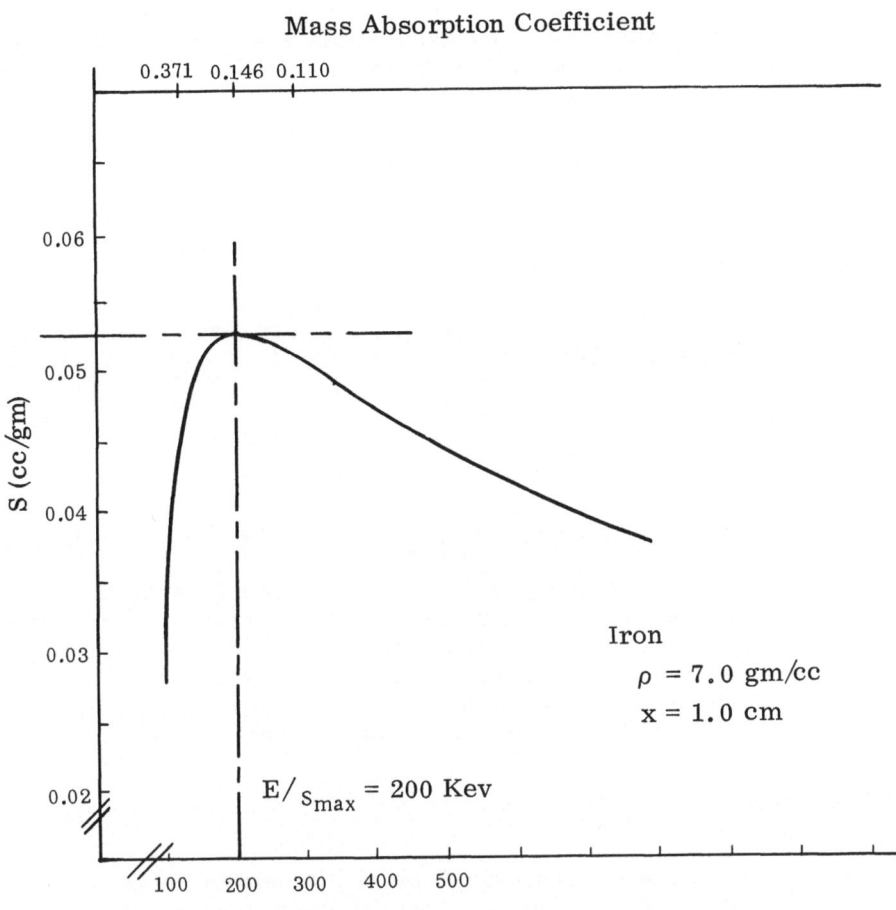

Mass Absorption Coefficient

FIGURE I

MEASUREMENT SENSITIVITY, S, VERSUS GAMMA ENERGY, E,
AND MASS ABSORPTION COEFFICIENT (μ)

$$S \equiv \frac{\partial(I/I_0)}{\partial\rho} = -\mu x e^{-\mu\rho x} \tag{2}$$

3.   Calculation for significant counting level.

The precision required in the application will yield
information on how many gamma rays must be detected
to yield a significant measurement.

Equation (1) can be expressed as follows in
equation (3).

$$\rho = \frac{1}{\mu x} \ln I_0 - \frac{1}{\mu x} \ln I \tag{3}$$

This algebraic relationship for ρ, between independent observable variables, x, I, and I_0 has a net
standard deviation which is approximated in
equation (4):

$$\sigma\rho = \left\{ \left(\frac{\partial\rho}{\partial I_0}\sigma_{I_0}\right)^2 + \left(\frac{\partial\rho}{\partial I}\sigma_I\right)^2 + \left(\frac{\partial\rho}{\partial I}\sigma_x\right)^2 \right\}^{1/2} \tag{4}$$

Evaluating the partial derivatives,

$$\sigma\rho = \left\{ \left(\frac{1}{\mu x}\frac{1}{I_0}\sigma_{I_0}\right)^2 + \left(\frac{1}{\mu x}\frac{1}{I}\sigma_I\right)^2 + \left(\frac{\rho}{x}\sigma_x\right)^2 \right\}^{1/2} \tag{5}$$

Since the decay of a radio nuclide is a random
event, large decay rates have a distribution which
conforms to the normal or Gaussian distribution.
It is a characteristic of Gaussian distributions
that the standard deviation, σ_n, equals the square
root of the mean, $\bar{N}$, or as expressed in equation
(6).

$$\sigma_n = (\bar{N})^{1/2} \tag{6}$$

Thus, approximating $\bar{I}$ by the observed value I and
$\bar{I_0}$ by I_0, σ_I and σ_{I_0} can be expressed as in
equation (7):

$$\sigma_I = I^{1/2} \quad ; \quad \sigma_{I_0} = I_0^{1/2} \tag{7}$$

Thus equation (5) reduces to equation (8):

$$\sigma\rho = \left\{ \left(\frac{1}{\mu x}\right)^2 \left(\frac{1}{I} + \frac{1}{I_0}\right)^2 + \left(\frac{\rho}{x}\sigma_x\right)^2 \right\}^{1/2} \tag{8}$$

$$\sigma\rho = \left\{ \left(\frac{1}{\mu x}\right)^2 \left(\frac{1+\alpha}{\alpha I_0}\right)^2 + \left(\frac{\rho}{x}\sigma_x\right)^2 \right\}^{1/2} \tag{9}$$

Solving for I_o yields equation (10):

$$I_o = \left\{ \frac{\left[\frac{1}{\mu x} \left(\frac{1+\alpha}{\alpha} \right) \right]^2}{\sigma\rho^2 - \left(\frac{\rho}{x} \sigma_x \right)^2} \right\}^{1/2} \qquad (10)$$

Actual values for σx correspond to the precision generally yielded by the means to be used for measuring the thickness of the sample. Eg., a micrometer measurement would yield a σ_x of 0.0002". Evaluating equation (1) for α and substituting in values for $\sigma\rho$, σ_x, α, ρ, x and μ yields the magnitude of I_o necessary to secure a density measurement to the desired precision, $\sigma\rho$.

B.   Additional Considerations

In addition to the three basic application characteristics previously discussed, there are two additional consid- erations which will round out the design criteria for a particular application. These are the effective area of the sample over which the measurement is to be taken and the duration of the measurement.

1.   Activity of the source

Since the radiation from a radio nuclide is directed into all space, it is obvious that some means for its confinement and collimation will be required. Figure II depicts a typical array of collimators.

Collimators must be thick enough in the unperforated region to reduce the transmitted radiation by 10^{-3}. This is necessary so that variations in the overall size of the sample will not influence the measurement.

Aperture diameters can be identical but they must be no larger than the smallest effective cross sectioned area to be measured.

The fraction of all radiation emitted, which is directed through the system of apertures, is cal- culated as follows:

The solid angle, Ω, into which the useful radiation is directed is expressed in equation (11) as:

$$\Omega = \frac{1/4\ \pi a^2}{d^2} \qquad (11)$$

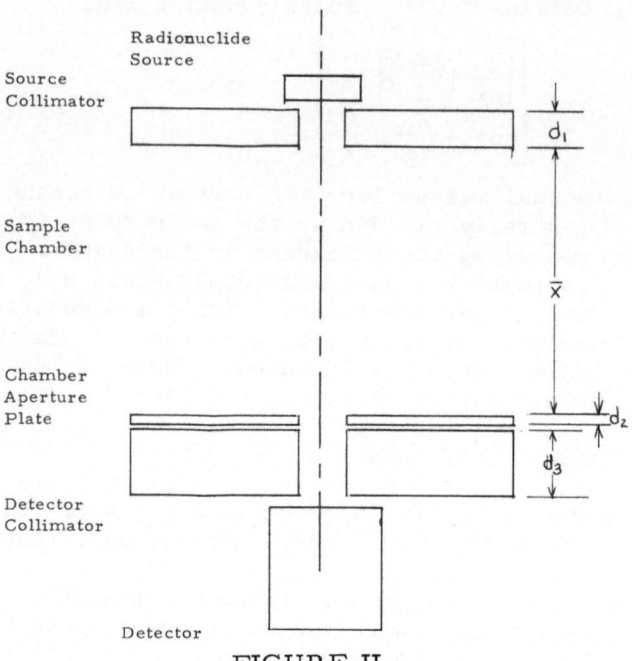

FIGURE II

GAUGE CONFIGURATION

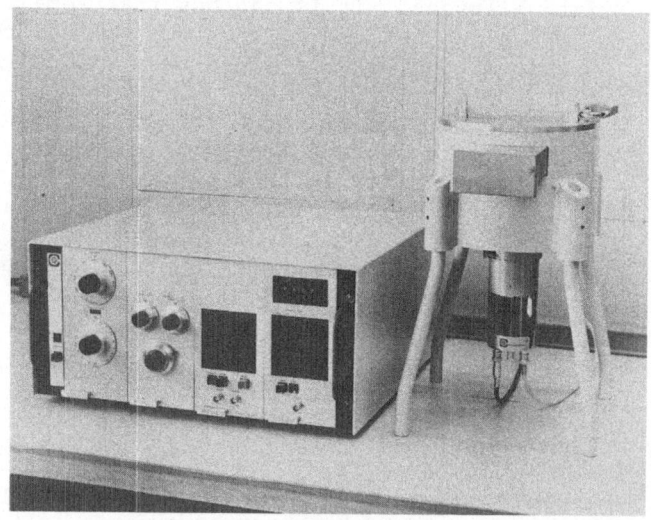

FIGURE III

GAMMA RAY DENSITY GAUGE

The fraction of all space which is represented by Ω is expressed in equation (12) as:

$$f = \frac{\Omega}{4\pi} = \frac{1}{16}\frac{a^2}{d^2} \tag{12}$$

The duration of the measurement will be symbolized by t.

The values of f, I_o, and t are used to calculate the required activity of the radio nuclide as follows:

The unit of activity is the Curie and is defined as the quantity of a radio nuclide yielding 3.7×10^{10} disintegrations per second.

The required activity of a radio nuclide which emits one gamma ray-of-interest per disintegration is expressed in equation (13) as:

$$A = I_o/(3.7 \times 10^{10} \times t \times f) \quad \text{(Curies)} \tag{13}$$

If the fraction, g, of gamma rays-of-interest emitted per disintegration is something other than unity, the activity is expressed in equation (14) as:

$$A = I_o/(3.7 \times 10^{10} \times t \times f \times g) \tag{14}$$

2.   Detector Type

Referring again to Figure II, the detector is shown located in its conventional position, coaxially with the collimator apertures.

There are basically two types of detector which can be employed in gamma ray transmission; gas proportioned and scintillation.

Because of its relatively long recombination (dead) time, the gas proportional detector can be employed only in those applications where there will be less than 10,000 counts per minute. Therefore, unless cost is of a paramount importance, it is technically advantageous to use a scintillation detector.

The diameter of the scintillation detector should be as small as is practicable so as to reduce the probability for detection of photons penetrating the unperforated region of the collimator. This

condition can lead to an undesirable sensitivity
to the size or position of the sample.

The efficiency, ε, for detection for a NaI scintillation crystal is readily evaluated from Beer's
Law, using tabulated values for the thickness,
density, and mass absorption coefficient.

If the detection efficiency, ε, is significantly
less than unity, then the required activity of the
radio nuclide to be used in the application is given
in the equation (15):

$$A = I_o/(3.7 \times 10^{10} \times t \times f \times g \times \varepsilon) \tag{15}$$

II.   An Application - Sintered Iron Parts

An application wherein the gamma ray transmission gauge is
presently employed is the density measurement of sintered
iron parts. Such parts do not exceed 0.25 inches in thickness and must be nondestructively measured to ±1% for a
nominal 7.0 gm/cc density.

Figure III shows the gamma ray transmission gauge built to
fulfill this application. It consists of an Americium-241
gamma ray source having a 45 millicurie activity. The gamma
rays are directed through a 1/16" diameter collimation system
to the detector. Part of the collimated path consists of the
sample chamber where parts may be inserted for measurement.
The transmitted gammas are detected with a 1/8" diameter NaI
crystal coupled to a photomultiplier tube.

The magnitude of the count rate with no sample present is
150,000/min and for a three minute count at this level the
gauge yields a density precision of ±0.07 gm/cc.

POINT DENSITY MEASUREMENT AND

FLAW DETECTION IN P/M GREEN COMPACTS

Ivor Hawkes, C. W. Spehrley, Jr.

Creare Incorporated

Hanover, New Hampshire 03755

The term "non-destructive testing" when considered in relation to P/M green compacts, implies tests
for local density and crack detection. P/M compacts
are unique among metal preforms in that a distinction
can be made between density variations and cracks and
this is the basic reason that conventional non-destructive testing techniques have had such little
success in the P/M industry. In one sense we can think
of a P/M green compact as a mass of pores joined by
particles. The object of testing is to determine when
the density distribution of the pores is greater or
less than some predetermined standard or when the pores
are continuous along one particular plane and form
cracks.

The testing problems are quite different from
those normally encountered in non-destructive metal
testing where a flaw can be much more precisely defined
as a metal discontinuity. For this reason it is essential that, in addition to seeking to modify the more
conventional techniques of non-destructive testing, a
completely new approach be taken.

In this paper some novel techniques are described
for determining comparative point densities and for
detecting cracks in green compacts having interconnected porosity. The techniques are based on the

concepts that in green compacts point density will be
related to point gas permeability and that internal
pore pressures can be built up to rupture the compact
and thus enable tensile strength proof tests to be made
in crack suspected regions. It is not claimed that the
techniques are a panacea for all types of P/M compacts
but they do show great promise as the basis of simple
and reliable tests for a wide range of compacts where
the porosity is largely interconnected.

Porosity and Cracks in P/M Green Compacts

Before describing the actual testing techniques,
it is necessary to briefly discuss the nature of the
porosity and the development of cracks in green compacts as some understanding of these factors is necessary for the application of the techniques.

Density variations in green compacts arise mainly
as a result of the frictional forces generated during
the compaction process and the resulting inability of
the metal powder to flow easily under pressure.

In relation to any individual part, variations in
density have two main sources: friction between
individual grains and between grains and the compaction
surfaces during what might be termed normal compaction;
and, powder displacements during later stages of compaction due to incorrect press motions (Taylor, 1960).

Figure 1 shows the density distribution for a
single and compacted nickel specimen, arising as a
result of frictional forces between the individual
grains and the die and punch surfaces. The wide
density variation illustrates the difficulties of
defining density, even in very simple compacts. The
problems become much greater in stepped compacts where
in addition to the density variations caused by
friction, the actual compression ratio may vary widely
in the different steps. The use of graded powders and
appropriate lubricants may reduce the variations to
some extent, but the basic problem of correlating
density values obtained by different techniques still
remains.

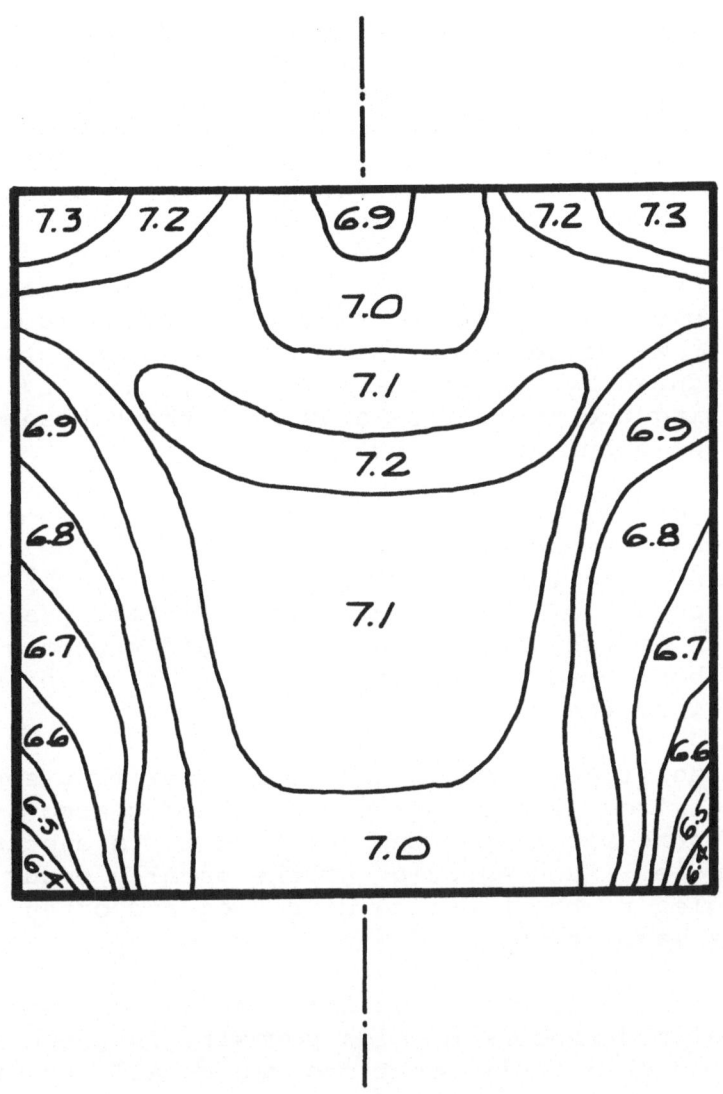

Figure 1.   Actual Green Density (gm/cc) Distribution
for Pure Nickel Compacted at 46 tsi in
Single Action Press.   L/D 0.87
(Reproduced from Hirschhorn, 1969)

The density variations produced by incorrect press motions in multistep compacts are more serious as they can involve complete disruption of the bonds between the grains to produce either shear planes (cracks) or local regions of very high porosity.

Figure 2(a) illustrates such a localized region of very high porosity occurring in this case at the junction of the hub section and tooth flange of a gear compact. Figure 2(b) shows a close up of the high porosity region and illustrates how the porosity is caused by the disruption of the bonds between individual particles along planes parallel to the axis of pressing. We would expect this type of high porosity to be produced by overpressing of the flange regions, i.e. pushing the upper face away from the hub section. It can also occur as an ejection crack.

Figure 3(a) shows an example where high compressive forces have actually produced a shear type crack penetrating into the flange region. Figure 3(b) is a photomicrograph of such a part after sintering showing the nature of the shear plane. We would expect such a crack to be formed by overpressing the hub section to force it into the flange.

Another type of crack that can occur in green compacts arises as a result of the elastic recover of residual stresses in the compact when it is ejected from the die. Such ejection cracks usually occur at right angles to the pressing direction and often appear as laminations.

One important characteristic feature of cracks or very localized regions of high porosity in green compacts is that their locations can usually be predicted in any given part.

Let us now try to determine what the parts manufacturer would like to know about his parts in relation to density distribution and flaws.

At the start of a production run the manufacturer has specified his powder mix and additives. He has established the press cycle and compacting pressures that will produce a green compact which when put

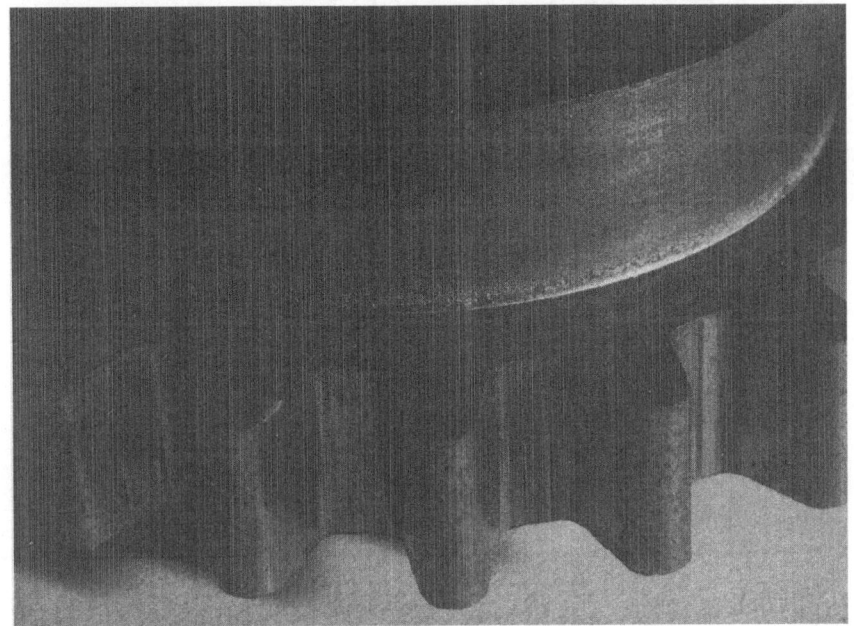

(a)   P/M Green Compact

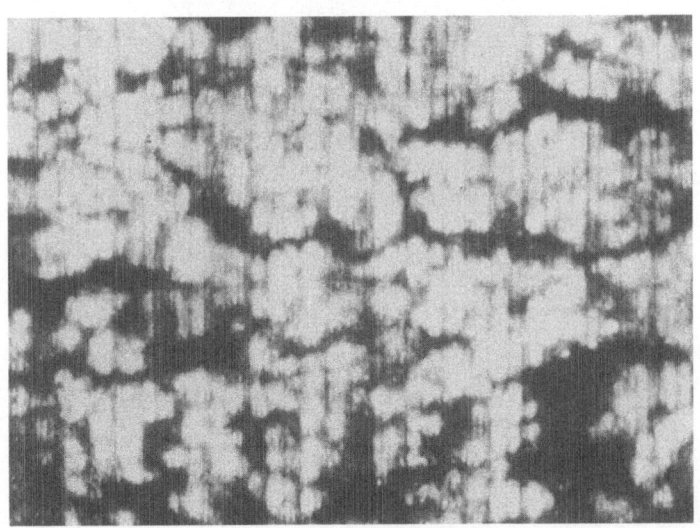

(b) High Porosity Region (X80)

Figure 2.   High Local Porosity at the Junction of a
            Hub Section and Tooth Flange in a P/M
            Green Compact Gear

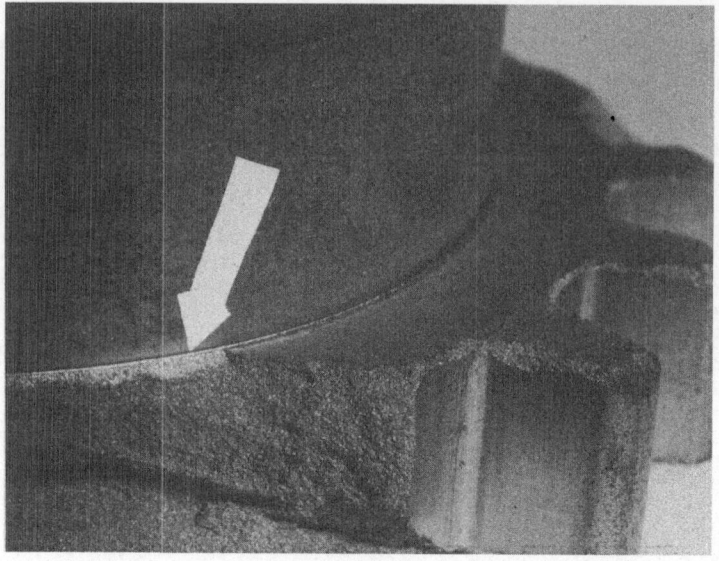

(a)   Crack at Junction of Hub Section and Gear Flange

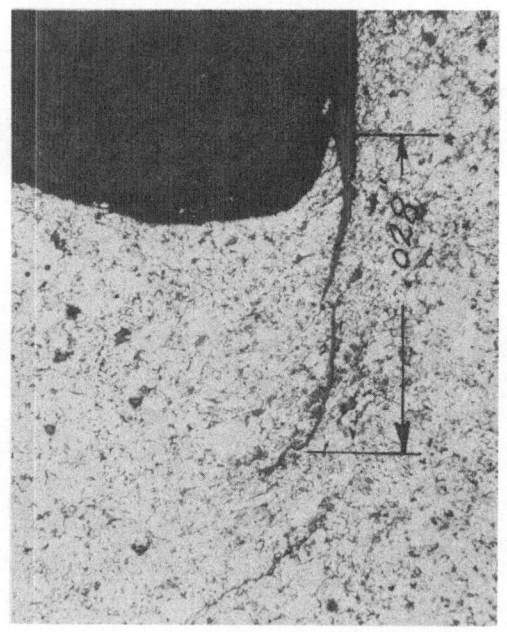

(b) Photomicrograph of a Crack in a P/M Sintered Part

Figure 3.   Cracks in P/M Green Compacts

through the predetermined sintering cycle will produce
a satisfactory part. By satisfactory we mean a part
which fulfills its assigned function to the satisfac-
tion of the user and meets any specifications laid down
by him. It is now the main aim of the manufacturer to
reproduce the compacts as precisely as he is able
within the bounds set by an economic operation.

To do this in addition to controlling the weight
and dimensions of the green compacts the manufacturer
would like to be certain that the average density distribution in certain regions falls within predetermined
tolerance limits and that there are no cracks. At present the only generally accepted way that this can be
done is to cut up selected green compacts into small
pieces for density checks by weighing techniques, and
to carry out relatively crude visual inspections for
cracks. Penetrant crack detection tests are, of course,
possible on sintered parts.

These are direct methods, i.e. the density by
definition is what is measured in the weighing test and
the cracks are what can be seen and measured by eye.
The basis of non-destructive testing is, of course, to
measure some parameter which is sensitive to changes
in porosity and the presence of cracks. The pneumatic
techniques developed for testing P/M green compacts
use the flow of gas and internal pore pressures as the
parameters to be measured, with the additional requirement that the measurements be made in predetermined
regions most susceptible to flaws or density variations.

Point Density Measurements by Local Permeability

The concept behind the pneumatic point density
measurement technique for P/M compacts is that low
density regions have a relatively high permeability and
therefore density differences between similar compacts
can be detected by measuring the gas flow rate into
that region under fixed conditions.

The system is shown schematically in Figure 4.
Air or gas is applied under pressure to the surface of
the compact at the point where a test is required and
then allowed to flow into the compact and out to
atmosphere following any preferred paths. Any change

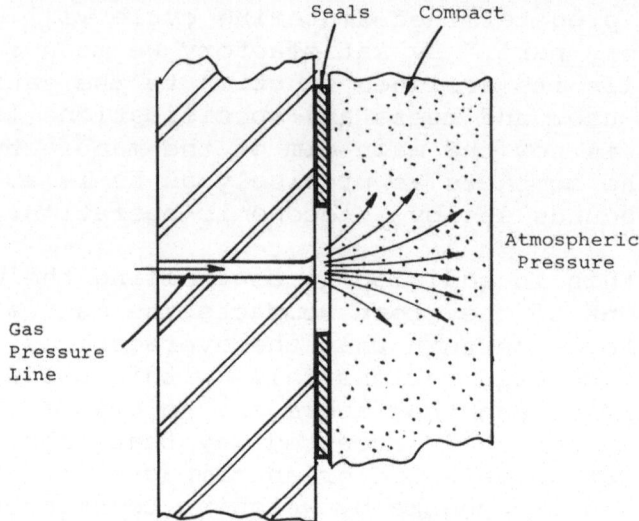

Figure 4.   Point Gas Permeability Test For Local
            Density Measurements.

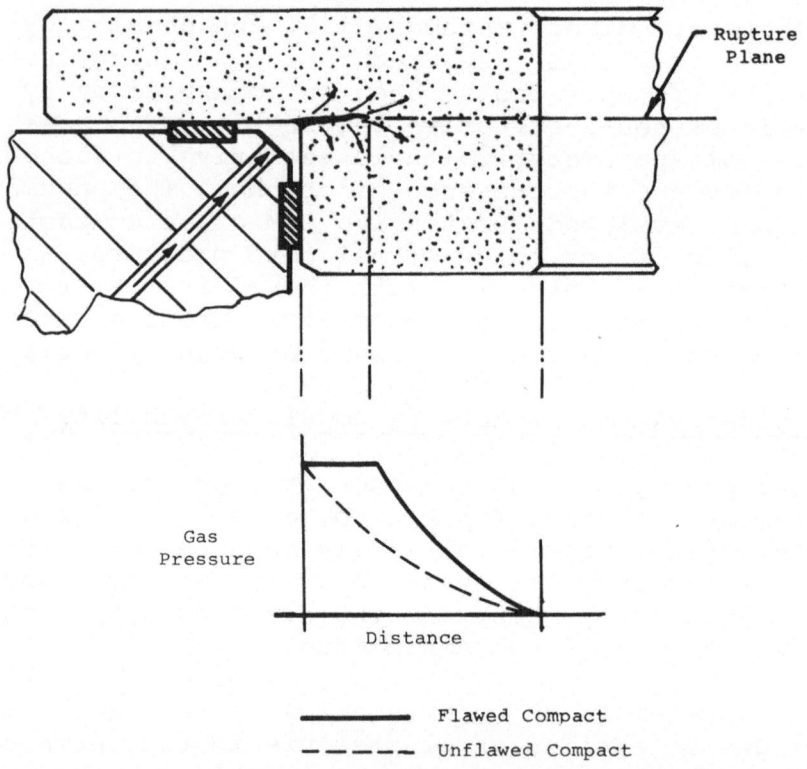

Figure 5.   Pore Pressure Rupture Test For Crack   Detection

in the flow rate between individual compacts is interpreted as representing a change in density in that region.

The actual flow rate of a given gas into a green compact, from a limited region on the surface, is a function of the average porosity in that region, the applied pressure, the area from which the gas is flowing, the geometry of the compact in the test region, the anisotrophy of pore distribution, the average size and shape of the pores and the ratio of interconnected to non-interconnected pore volume.

In practice a number of these influencing factors can be lumped together into one parameter called the permeability constant, K, which is defined by Darcy's Law for laminar flow, i.e.

$$q = \frac{KA}{\mu} \frac{\Delta P}{L}$$

where

    q is the flow rate of an incompressible fluid
    A is the area through which the fluid is flowing
    μ is the viscosity of the fluid
    ΔP is the pressure drop over length L
    K is the permeability constant

The formula can be modified for gas flow, but the flow rate is still directly proportional to the permeability constant, K.

The relationship between the permeability constant and porosity has never been satisfactorily established due to the difficulties of isolating all the contributing factors including pore shape, roughness and surface area.

If the porous body can be represented as an assembly of parallel capillaries of various cross sections then the permeability constant K can be expressed in terms of the permeability by:

$$K = \frac{\phi^3}{cs^2}$$

where

    φ is the porosity
    S is the specific surface area
    C is the Kozeny constant which ranges from 2 to 3
       depending upon the average pore cross-sectional
       shape

The use of this equation in relation to porous
sintered parts has been briefly discussed by Weger and
Greenberg (1959).

The whole problem of predicting the flow through
porous media is very complicated and certainly beyond
the scope of this paper. What is being shown here is
that theoretically the flow rate of a gas into a
permeable body is a very strong function of the
interconnected porosity all other things being equal.

Crack Detection by Pore Rupture Pressure

The concept behind the pore rupture pressure
crack detection technique is that a crack or very high
localized and directional porosity (flaw) in a green
compact will reduce its tensile strength normal to the
general plane of the flaw, as compared to an unflawed
part. If we can introduce gas under pressure into this
flawed region so that the flow paths are roughly normal
to the plane of the flaws then the pore or internal
pressure that can be generated prior to rupture of the
compact will be a measure of the magnitude of the flaw.

The technique is shown schematically in Figure 5.
Gas under pressure is applied as before to specific
regions of the compact where localized high porosity
or cracks are known to occur from experience, as
discussed earlier. Regions of high porosity or cracks,
form relatively low resistance channels for the air flow
and there is relatively little pressure drop as shown
schematically in Figure 5. The total force acting
across any plane is the mean effective pressure and is
determined by the applied pressure and the pressure
profile across the plane being considered. Where there
is a high porosity channel or a crack the mean effective
pressure and therefore the force is obviously much
higher than for an uncracked or constant density part
(Figure 5). In a flawed part, the number of grain to

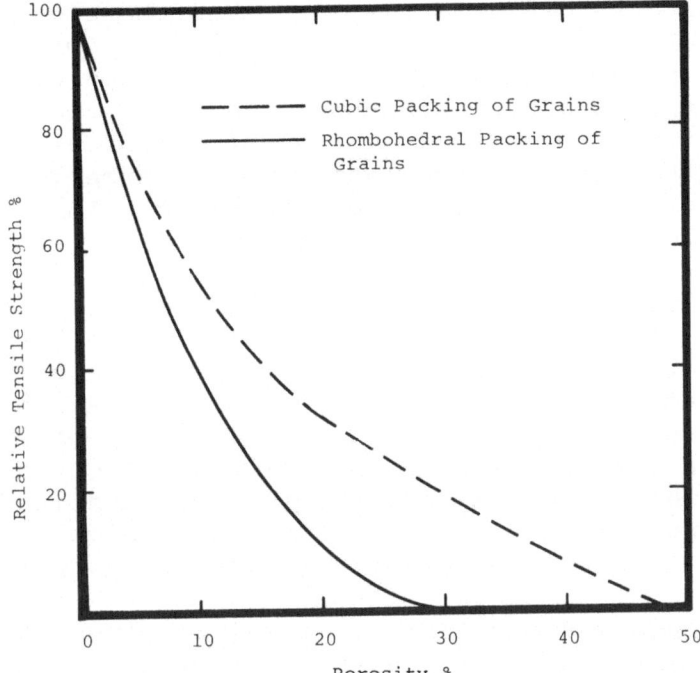

Figure 6.   Relation Between Relative Strength and
Porosity of Cubic and Rhombohedral
Packing of Coalescing Spheres. (After
Knudsen, 1959)

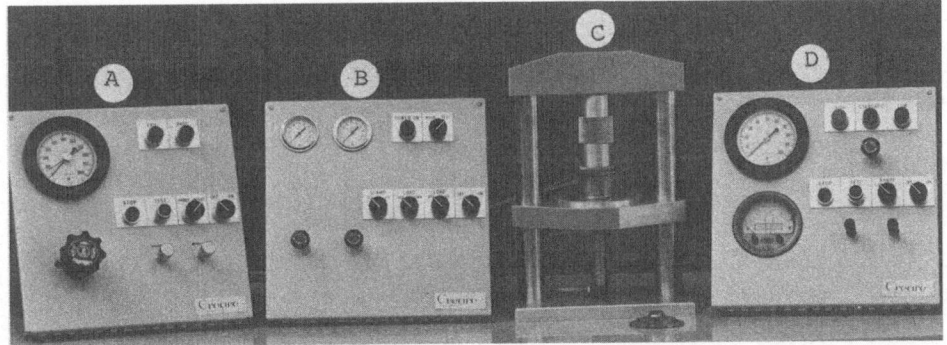

Figure 7.   Point Density and Crack Detection Equipment

    A.   Pore Rupture Pressure Control Console
    B.   Pressure Control Console
    C.   Press
    D.   Point Density Measurement Control Console

grain contact points is also less and therefore the actual metal stress is higher. In this respect the technique is doubly sensitive, i.e. in a flawed compact not
only is the total force acting to rupture the part
greater for a given applied pressure, than with an
unflawed compact but even assuming the same total
applied force the metal stress is greater in the flawed
compact because of the fewer grain to grain contacts.

The actual relationship between tensile strength,
porosity and grain to grain contact area is not simple.
Knudsen (1959) has carried out some studies on this
problem and has shown that if we assume that the inte-
granular contact areas are weaker than the grains, it
is possible to develop an approximate concept of this
relationship by defining a theoretical idealized specimen having very orderly structure, and calculating the
relative increase in contact area and strength with
decreasing specimen porosity. Knudsen made such an
analysis and assumed that his ideal specimen was made
up of a systematic arrangement of equal sized spheres.
The change in porosity was produced by each sphere
flattening at its point of contact with its neighbor
so that at zero porosity each sphere becomes a
polyhedron. Figure 6 is a replot on a linear scale of
Knudsen's results. Over the range 0 to 15% porosity,
the relationship between the tensile strength and
porosity can be expressed as

$$S = S_o e^{-b\phi}$$

where

  S is the tensile strength at porosity ϕ,
  S_o is the tensile strength at zero porosity,
  b is a constant ranging from 6 for the "most
    porous" cubic arrangement to 9 for the "least
    porous" rhombohedral arrangement
  e is the base of Napierian logarithms

Actually test data obtained by Knudsen confirmed
this form of the relationship for polycrystalline
specimens.

This work is very pertinent to an understanding of the pore rupture test as used on green compacts as it shows that quite small changes in porosity will have very appreciable effects on the tensile strength (Figure 6). This, coupled with the fact that the total tensile forces across the test area are greater for a flawed than unflawed specimen, pressurized at the same level, leads to the conclusion that the pore rupture test will be extremely sensitive to porosity changes and the presence of cracks. This has indeed been shown by practical tests.

Application of the Techniques

Figure 7 shows the latest form of the equipment as developed for tests on P/M green compacts in manufacturing plants. There are five components; a press, tooling, a press control console, a point density measurement console and a pore rupture pressure measurement console.

The press is designed to hold the compact being tested in the tooling and allow for the application of clamping forces across different sections of the tooling.

The tooling is basically a chamber designed to hold the compact being tested with provisions for seals and gas ports so that different sections of the compact can be subjected to gas pressure. Figure 8 shows a very simple form of tooling for carrying out tests in two regions of a turbine hub, i.e. in the corner region between the hub and flange and also in the flange. In this particular case simple O-rings prove adequate as seals. In other instances soft rubber surface seals have been used and currently, inflatable seals are being tested. Tooling is currently being built that will allow for sequential testing in five regions of a single green compact.

The press control console is simply equipment for actuating the various clamping modes of the press and controlling the opening and closing times.

The point density control console is designed to precisely measure flow rates down to less than 5 cc/min

Figure 8.   Simple Tooling for a Flange Test

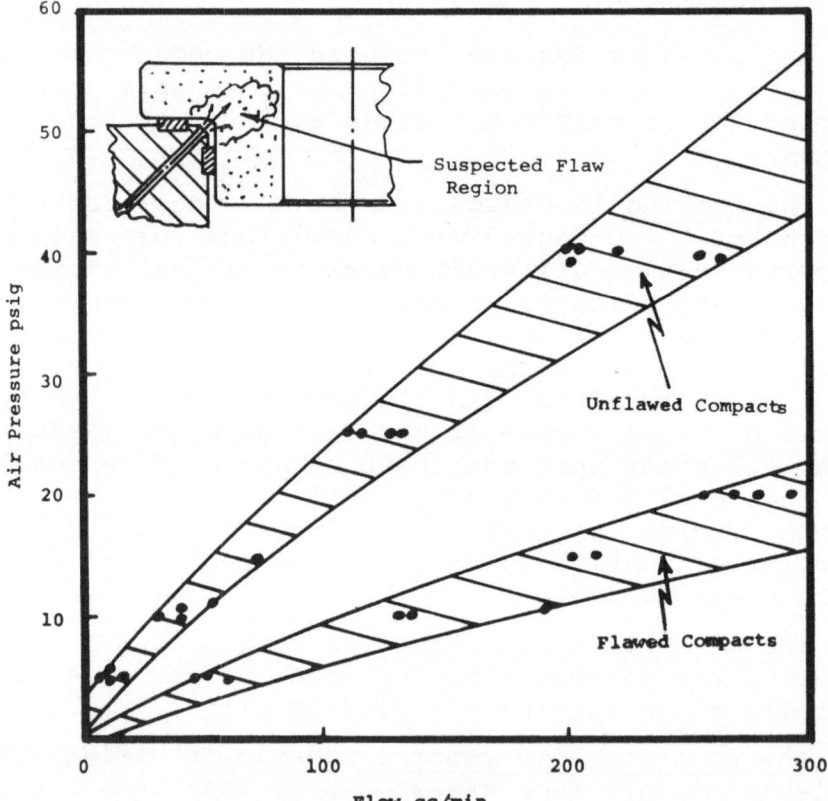

Figure 9.   Detection of Flawed Compacts Using Gas
            Permeability Technique.

and give rapid comparative testing rates (less than
5 seconds). Once a satisfactory green compact has been
produced, the flow rate under a given pressure can be
established for different regions of the compact. These
then become the standard flow rates for that type of
compact. If a compact subsequently tested with flow
rates which vary outside preset limits from the standard
rate for that region, then attention can be drawn to
this fact by indicator lights. If necessary the
numerical value of the flow rate can then be determined.

The relationship between the flow rate and point
density for a given type of compact can only be determined by independent checks of density. This is very
difficult to do in corner regions because of the very
rapid density gradients, as previously discussed. It is
preferable to gain experience on individual types of
compacts and then make a decision as to whether or not
compacts should be scrapped and the press motions and
powder composition checked, based on abnormal flow
readings. Figure 9 shows how this can be done. The
graph shows the results of tests made in the corner
region between the hub and gear flange of two batches
of gear compacts. In one batch high porosity had been
deliberately induced by a slight modification to the
compacting press motion. It will be noted from Figure 9
that the two types of compacts could be very easily
separated. In a production run any compacts produced
outside the scatter bond of the "satisfactory compacts"
would warrant attention.

The pore pressure control console is designed to
either apply a proof pressure to a specific region of
the compact where flaws are suspected or alternatively
to raise the surface pressure in a controlled manner
until the compact ruptures under internal pore pressure.
The rupture of green compacts can be quite noiseless and
so a light indicator system has been built into the unit
to indicate when the compact has ruptured. A pressure
gauge gives the actual surface pressure at rupture. As
before, a standard rupture pressure can be established
for the acceptable compacts and rupture pressures outside the tolerance limits of this standard pressure
indicate changes in powder composition or press action.
The actual rupture pressure varies, with individual
types of compacts and pressures up to 500 psi and higher

have been obtained. The tensile failure of a green compact by internal pore pressure is almost instantaneous
due to the fact that even the smallest local failure
further weakens the compact. For this reason the test
has no deleterious effects on the compacts. Tests have
been made to substantiate this by holding pressures
just short of the rupture pressure for several hours.

As discussed earlier the pore rupture test is very
sensitive to the presence of flaws and porosity in the
predetermined rupture plane. For one set of tests, 30
turbine hubs were compacted in batches of 10 using
three different press cycles. The weights were all
within 1% and the dimensions within .004 inches. Tests
were made to determine the applied pressure necessary
to blow the hub off the main flange. The compacts in
the first group all failed at 345 $\pm$ 15 psi, in the
second 270 $\pm$ 15 psi and in the third at 220 $\pm$ 10 psi.
At the time these tests were made the sensitive flow
rate techniques described earlier were not available
and using less sensitive rotameter type flowmeters
much lower percentage variations in flow rate were
obtained between the compacts.

Conclusions

The pneumatic testing techniques briefly described
herein show great promise as a simple and reliable
tool for monitoring point density and detecting cracks
in P/M compacts. The technique, in its present state of
development, is in essence a comparative test. It
should find widespread use for quality assurance,
either as a spot test or for 100% proof testing of
critical parts.

The technique should also prove very useful during
initial press set up as the relative effects of
changing a given motion can immediately be determined
without the need to cut up and weigh every compact.

Ultimately, it is envisioned that the technique
could be used for real time feedback control of press
movements, using perturbation or other control systems.

References

Hirschhorn, J. S.; INTRODUCTION TO POWDER METALLURGY;
American Powder Metallurgy Institute, 1969.

Knudsen, F. P.; DEPENDENCE OF MECHANICAL STRENGTH OF
BRITTLE POLYCRYSTALLINE SPECIMENS ON POROSITY AND GRAIN
SIZE; J. of the Amer. Ceramic Soc., Vol. 42, No. 8,
pp 376-387, August 1959.

Taylor, H. G.; THE INFLUENCE OF TOOLING METHODS ON THE
DENSITY DESTRIBUTION IN COMPLEX METAL POWDER PARTS;
Powder Metallurgy, No. 6, pp 87-123, 1960.

Weger, E. and Greenberg, D. B.; AN INVESTIGATION OF THE
VISCOUS AND INERTIAL COEFFICIENTS FOR THE FLOW OF GASES
THROUGH POROUS SINTERED METALS WITH HIGH PRESSURE
GRADIENTS; Interim Technical Report No. 1 to Dept. of
Army-Ordnance Corps, ASTIA Document No. 213897,
March 1959.

TEST PROCEDURE FOR THE DETERMINATION OF THE CORROSION RESISTANCE OF SINTERED IRON AND BRONZE PARTS

Alois E. Kindler

Ringsdorff-Werke, Bonn-Bad Godesberg, Germany

The object of the work represented in this paper was to investigate the corrosion behaviour of sintered metals. Powder metallurgy parts are especially exposed to corrosion because, due to the manufacturing process, the surface of such parts is larger than that of massive metals and this surface is activated.

At first we checked whether the already existing corrosion tests for massive metals could be transferred to sintered metals without modification. From these standardized methods were selected the salt spray test, the boiling test and the sulphur dioxide test according to Kesternich, the latter simulating corrosion in an industrial atmosphere.

The corrosion resistance of plated powder metal parts was not yet tested systematically, in spite of the production of nickel, zinc, cadmium and tin plated parts. When summarizing the results obtained to date in this field it is seen that the principal difficulty in getting reproducible data is the electrodeposition of pore-free coatings on the base material.

We were particularly interested in the influence of corrosion on the mechanical properties of the sintered metals. In order to obtain comparable results the tests were carried through with uniform test specimens which allowed to determine the change in weight due to corrosion as well as the thereby caused change in their mechanical properties. The MPIF test-bar was found to

be most suitable for these investigations. Usually the
determination of corrosion resistance is carried through
with flat test specimens of either circular or square
shape. These specimens have a geometrically simple surface which facilitates surface measurements, but test
pieces of this shape are not suitable for the determination of the mechanical properties.

These were characterized by the determination of
the Brinell hardness, tensile strength, elongation, flexural strength and density, measured before and after
the corrosion test. The test pieces were manufactured
in different densities in order to examine also the influence of the density on their corrosion behaviour.
Three different iron powders were used. Test bars were
also milled from stainless steel, grey cast iron, steel,
brass and bronze and exposed to the same tests as the
sintered specimens to obtain a comparison to the corrosion behaviour of massive metals.

The salt spray test was carried through in an aero-
sole-corrosion-test chamber, the sulphur dioxide test
in a standardized apparatus – both sets being commercially available· – and the boiling test in a 5oo ml
flask with reflux condenser.

The composition and density of the sintered materials meet the German standards laid down in the specifications for powder metals Sint B oo, C oo, D oo, B 2o
for sintering iron and Sint B 5o, C 5o for sintered
bronze. The mechanical properties of the specimens before the corrosion tests are characterized by the data
listed in table 1. These data are mean values of at
least five test pieces.

The salt spray test was carried through at a temperature of $25^{\circ}C$, with a 0,05 % aqueous sodium chloride
solution as corroding agent. Spraying was done in 3o minutes intervals. Before weighing the samples were taken
out of the apparatus and dried in an electric oven for
one hour at $105^{\circ}C$.

All sintered test pieces showed an increase in
weight that depended mainly on the density of the samples. The corrosion products formed on the surface
could not be removed mechanically or by rinsing and
washing. Even keeping the specimens in boiling water
for several hours did not reduce the increase in weight
significantly. Apparently the corrosion products were

retained in the pores and voids of the test specimens.

The massive metal test pieces showed an increase
in weight upon corrosion, however, by thorough rinsing
with water and alcohol the corrosion products could be
removed enough that a loss in weight could be observed.

Considering the percentage change in weight as a
measure of the corrosion process this change was plotted
versus the corrosion time. These diagrams gave more or
less straight lines, see Fig. I.

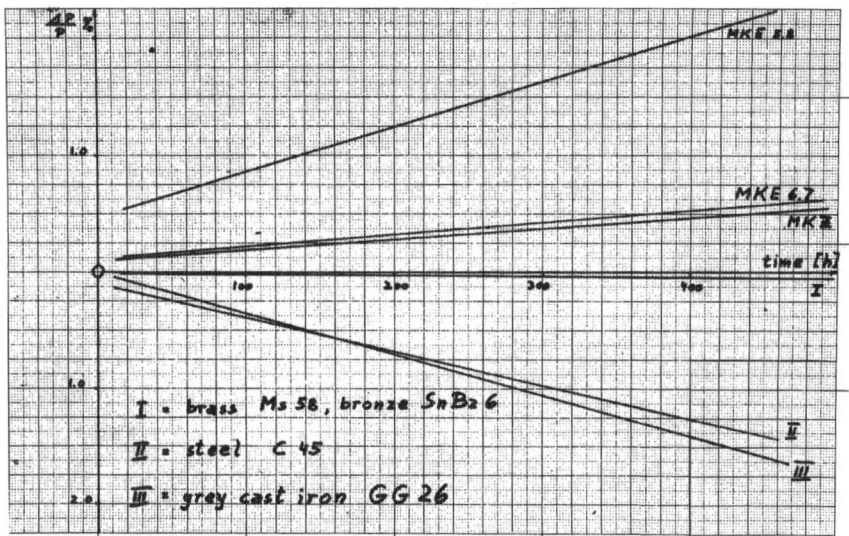

Fig.I. Diagram of percentage change in weight versus
       time.

The slope of these straight lines is positive for the
sintered test pieces and negative for the massive metals,
after the removal of the corrosion products. The absolute amount of this coefficient which can be considered
as a measure of the corrosion rate, was in the same range
for the steel and grey cast iron samples and the sintered iron of low density 5,8. On the other hand the
plot of MKE 6,7 corresponded with that of the sintered
bronze. An influence of the different densities on the
corrosion behaviour was hardly to be observed in these
bronze specimens, whereas the difference in the corrosion rate of sintered iron was obvious. The corrosion
rate of massive brass and bronze was extremly low and

nil for stainless steel.

    The protective influence of steam-treatment on sin-
tered iron MKE 6,7 is illustrated in Fig.2. From the
slope of the curves a marked increase in corrosion
resistance could be observed. Best results for corrosion
resistance were obtained by paraffination of the test
pieces as well as of structural parts for actual use.
Less dense parts had a higher up-take of paraffin due
to their higher porosity. Thus MKE 5,8 in comparision
to MKE 6,7 showed a higher degree of protection, see
Fig.2.

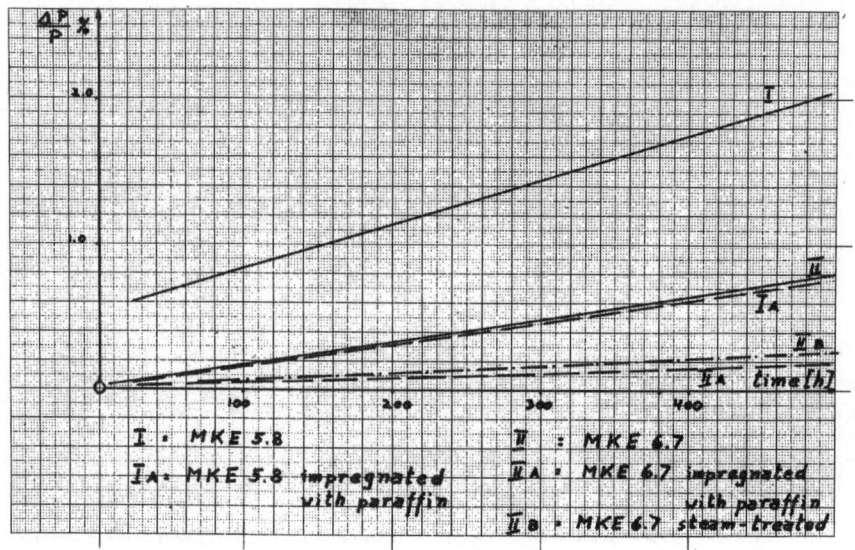

Fig.2. Influence of steam treatment and paraffin impreg-
       nation on the corrosion of sintered iron.

As sintered iron of low density 5,8 is used for slide
bearings we were interested in the influence of an oil
impregnation on the corrosion behaviour. Four different
oils were tested in specimens MKE 5,8, MKZ 6,5 and
MKX 5,8 (sintered iron grade with copper). Machine oil
and naphtha base oil provided an inferior corrosion
protection for the sintered iron samples than silicone
oil or synthetic oil.

    Test pieces were manufactured from a highly compac-
tible iron powder L 144 in order to investigate further

the influence of different densities on the corrosion
resistance of the same material (see Fig.3).

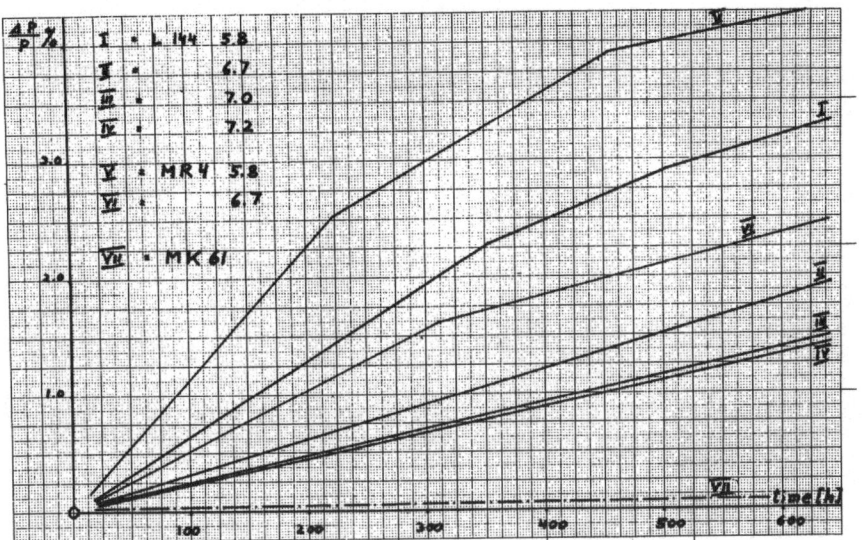

Fig.3. Influence of density on the corrosion of sinte-
       red iron.

The increase in density from 5,8 to 6,7 showed a marked
influence on the corrosion rate. An increase from 6,7
to 7,0 still improved the corrosion resistance to a certain degree whereas by a final increase to 7,2 the corrosion rate could not be reduced further.

For further investigations in samples of a sintered
stainless steel MK 61 (similar to American steel 316 L)
and of a third iron powder grade MR 4 with densities of
5,8 and 6,7, see Fig.3. The specimens of MR 4 showed a
similar behaviour to those of the grades MKE and L 144. The
sintered stainless steel was stable against corrosion.
By means of specimens MR 4 the reproducibility of the
salt test was checked and found to be satisfactory,
(Fig.4).

The boiling test according to DIN 50 906 was carried through with MKE 5,8 and normal tap water containing 128 mg/liter chloride, i.e. one half of the chloride concentration used at the salt spray test. In spite
of the smaller chloride concentration the corrosion rate

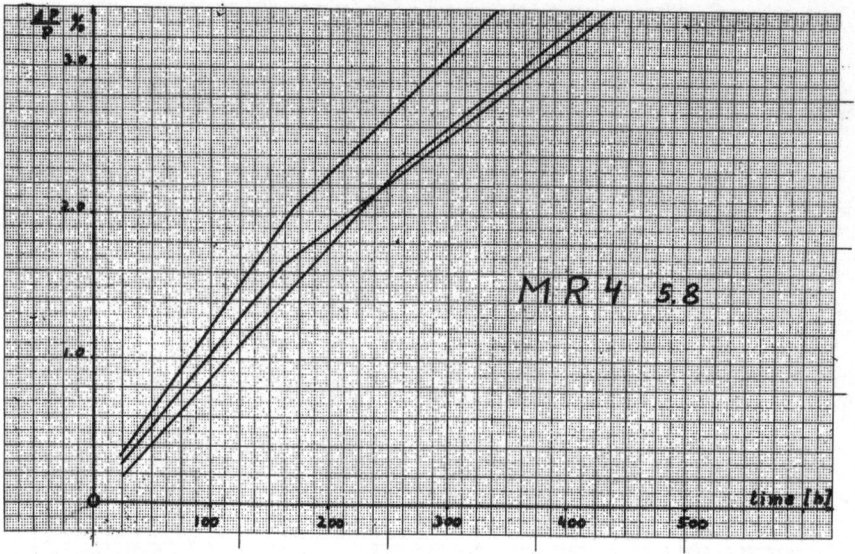

Fig.4. Reproducibility of the salt spray test checked
      by three series.

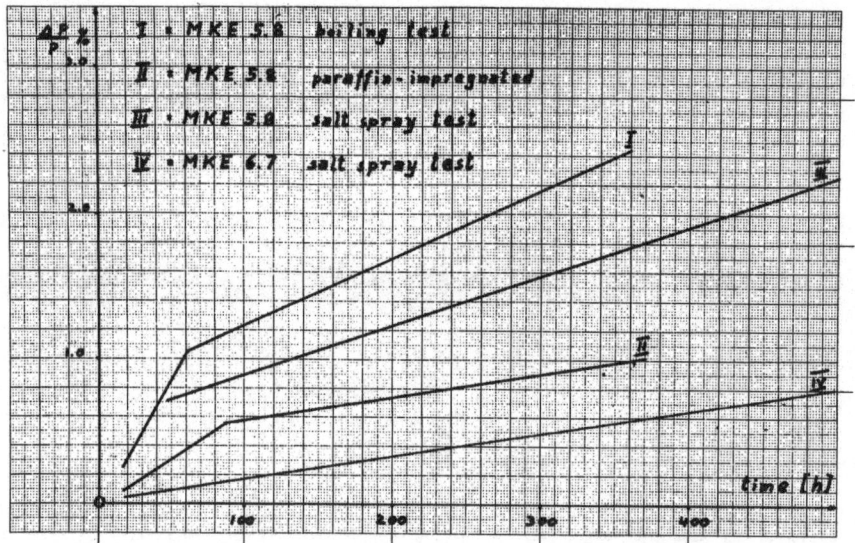

Fig.5. Boiling test in comparison to salt spray test.

was much higher than that observed during the salt spray
test. Specimens of the same grade were also tested after
an impregnation with paraffin as a corrosion inhibitor.
The influence of this impregnation on the corrosion resistance can clearly be recognized (Fig.5).

In Table I are listed the mechanical properties of
the specimens as they were characterized by the determination of density, tensile, strength, flexural strength
Brinell hardness and elongation.

The Brinell hardness of the sintered iron samples
increased similar to the increase in density. The paraffin - impregnated specimens showed the same behaviour
but to a less marked degree. The sintered bronze test
pieces showed an increase in density, too, but a decrease in hardness. The tensile strength and flexural
strength of sintered bronze decreased upon corrosion
whereas the data obtained for sintered iron were at
first increasing before decreasing to the initial
values and beyond.

Parallel to these investigations a series of long
term outdoor corrosion tests were carried through. Since
we did not have a facility to conduct these tests in
a maritime climate, the samples were exposed to industrial atmosphere with a chloride concentration exceed-

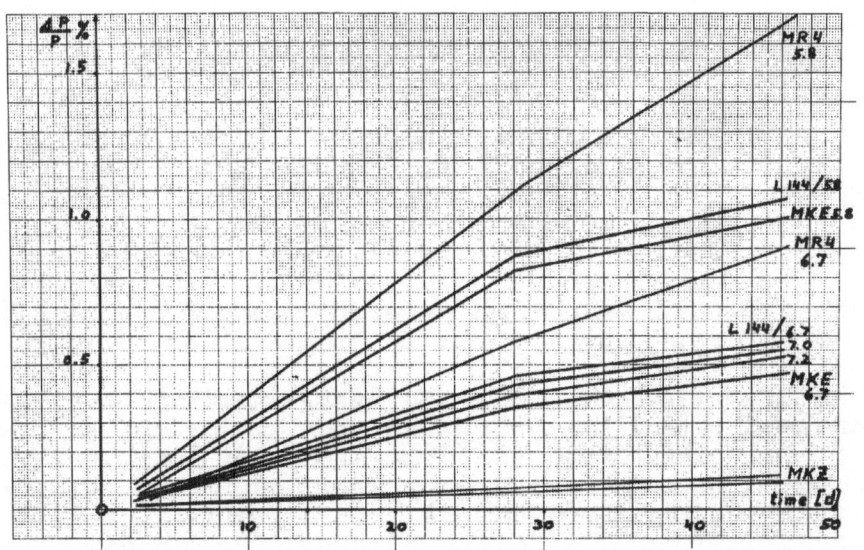

Fig.6. Results of atmospheric corrosion.

ing the normal level due to the immediate neighbourhood
of a garbage combustion plant. Comparing the results of
this atmospheric corrosion (Fig.6) with those obtained
by the salt spray test it was observed that the data of
the outdoor corrosion were scattered over a larger
range, as was to be expected since there were no defined test conditions given.

Nevertheless one could notice that the samples of
sintered bronze MKZ were fairly stable against corrosion
and that the scattering range of the values determined
is less than that of sintered iron specimens.

The salt spray test gave poor results for the low
density iron 5,8 in so far as it was impossible to find
a reasonable correlation between this test and the out-
doors corrosion test. Apparently the corrosion rate under the conditions of the salt spray test is too high
for this material. The same observation was made for a
correlation between boiling test and outdoor corrosion.

For the density range about 6,7 as it is used for
sintered structural parts, a better correlation could
be observed. This would mean that from the range between 200 and 600 hours of short time corrosion test conclusions may be drawn to the actual performance of the
material in a corrosive environment.

The results of the salt spray test for sintered
bronze were not significant enough for a prediction of
the corrosion resistance of this material.

The Kesternich test according to DIN 5o o18 was
carried through at a sulphur dioxide concentration of
0,2 liters at a temperature of $40^{\circ}C$. When lowering the
temperature to $25^{\circ}C$ a marked decrease of the corrosion
rate was observed, but the changes of the mechanical
properties were too insignificant.

The plots of weight gain versus time gave straight
lines as before, (Fig.7).

The corrosion rate of the massive metal samples
of brass and bronze was very low so that in the diagram
the plot coincides with the abscissa. The data obtained
for sintered bronze were higher than those for the massive metals, but regardless of the specimen's density
the corrosion rate was low. Even under the conditions
of the sulphur dioxide test at $40^{\circ}C$ the corrosion rate

of sintered bronze just reached the values obtained for
sintered iron MKE 6,7 upon outdoor   corrosion.

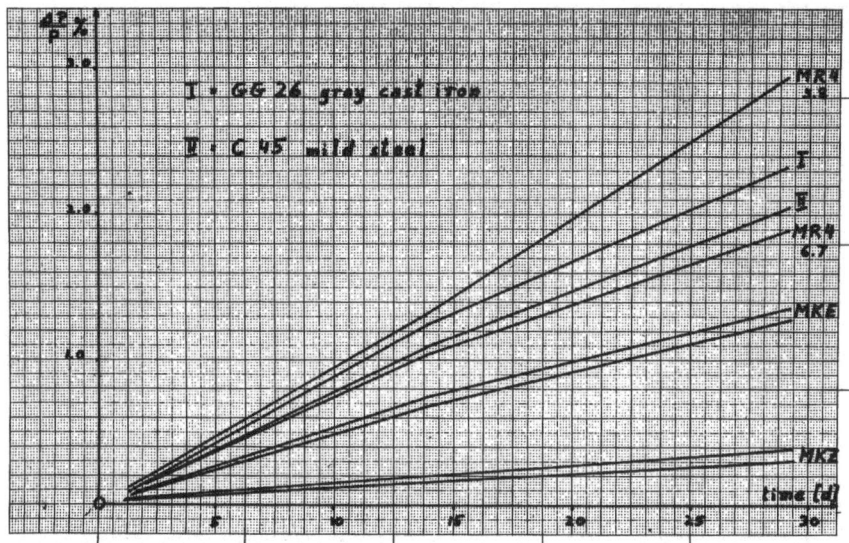

Fig.7. Results of the sulphur dioxide test at 40°C.

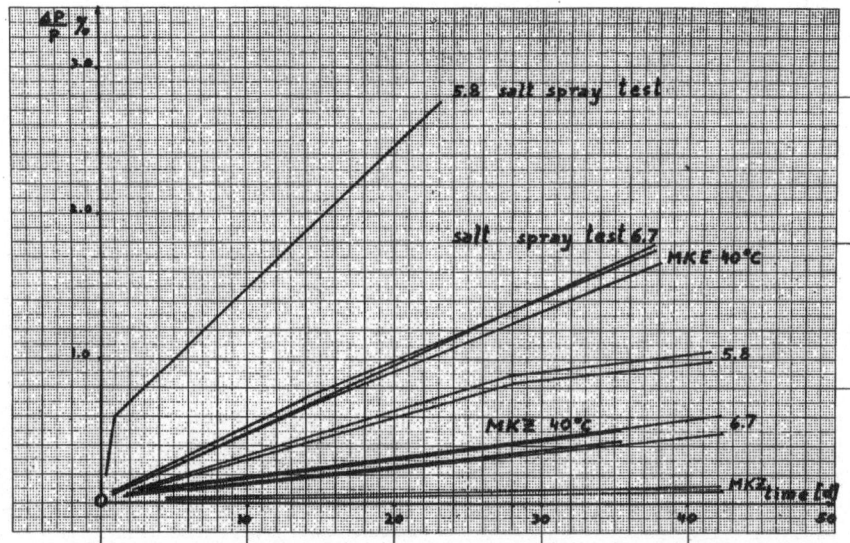

Fig.8. Results of outdoor   corrosion in comparison to
       sulphur dioxide test and salt spray test.

Parallel investigations of specimens under outdoor conditions in normal industrial atmosphere without enriched
chloride concentration yielded nearly the same results
as before, Fig.8. A reasonable correlation between short
time test and long term atmospheric corrosion could only
be found for sintered iron of the density about 6,7, the
material commonly used for sintered structural parts.

Table I.

Grade	Time h	Density g/cm³	Tensile strength kp/mm²	Flexural strength kp/mm²	Hardness HB 5/2,5 kp/mm²
Bronze	0	6,52	9,6	19,5	38,1
	440	6,77	6,0	13,8	32,1
MKZ	0	7,41	20,1	25,4	57,5
	440	7,74	10,8	23,9	56,6
Iron	0	5,88	10,4	21,2	45,4
	170	6,12	11,4	27,1	50,6
	600	6,05	9,8	21,3	76,4
MKE	0	6,76	19,1	36,1	67,3
	170	6,86	21,4	56,1	91,2
	600	6,72	18,3	38,0	91,1
MR 4	0	6,77	2o,6	42,4	80,1
	200	6,88	19,7	39,5	88,3
	400	6,90	18,1	42,0	92,3
	600	6,82	17,5	39,4	110,0
L 144	0	7,03	24,2	49,4	80,4
	200	7,o7	20,3	40,3	95,7
	400	7,11	20,8	44,4	98,5
	600	7,10	21,4	50,8	113,0
	0	7,21	27,7	57,4	87,9
	200	7,15	26,7	48,2	1o4,0
	400	7,20	25,2	40,0	101,4
	600	7,20	23,8	36,7	118,0
MK 61	0	6,80	39,6	74,5	94,5
	600	6,86	37,0	75,0	108,0

Table I. Mechanical properties of specimens before and
after the salt spray test.

THE EFFECT OF MATERIAL CHARACTERISTICS ON THE COMPACTION BEHAVIOUR OF METAL POWDERS[+]

John L. Brackpool

The B.S.A. Group Research Centre, Birmingham, England

INTRODUCTION

The process of compaction whereby a loose powder is pressed into a coherent mass is of fundamental importance to many industries. The widespread use of the process has promoted a considerable amount of experimental study; but a literature survey has revealed only isolated attempts at studying the underlying principles relating the density of the compact with the applied pressure. Moreover, several workers were concerned merely with the formulation of empirical relationships of compressibility with pressure having only limited applicability. The few attempts made to correlate material properties with compaction behaviour have had only limited success.

The main objective of this work was to study the factors governing the densification of metal powders, particularly at high levels of density. In the initial stages of compaction, the amount of porosity is controlled largely by the external characteristics of the particles; such as their size, shape and surface conditions. However, as compaction proceeds, the inherent properties of the material emerge as the dominant factors. For metals this implies the ability of the particles to deform, the yield stress required for deformation, and the influence of strain-hardening on the stress requirements for continued deformation. A range of powders was studied, therefore, having a wide variation in material properties. Isostatic compaction was selected initially in order to eliminate die-friction effects and uneven density distributions, but several powders were also die-compacted for comparison.

[+]This paper is based on a M.Tech thesis submitted to the University of Technology, Loughborough.

The measurements of material properties of metal powders are extremely difficult. The number of material properties that can be determined on a particulate solid is clearly limited, and although many more properties can be determined on the bulk material the results are not necessarily related to the original powder properties. Many workers have taken published values of bulk yield strength and related these to experimentally determined compaction behaviour but it can be shown that values can be selected to fit almost any set of data.

A material property which can be measured, even on a small particle, is hardness (micro-hardness) which can be defined as the resistance of a material to indentation by another body under the action of an applied force. For metals, hardnesses are simply measures of their resistance to permanent or plastic deformations and are given by forces per unit area and defined actually as pressures. Therefore, a relationship would be expected to exist between compaction behaviour, which can be regarded as mutual particle indentation, and hardness.

The correlation between powder properties and compaction behaviour requires accurate pressure and density measurements particularly at the high levels of density where large pressure increases produce relatively small density changes. Therefore, particular attention has been paid to accurate measurements of density and pressure.

POWDERS

The eight powders given in Table I were selected for the investigation covering a broad spectrum of material and particulate properties. The iron powders were selected to cover two commercial powders having different levels of impurities and different particle characteristics. The copper and nickel powders provided examples for the deformation behaviour of face-centred cubic metals and the austenitic and ferritic stainless steels extended the range of hardness and yield strength whilst having different comparative work-hardening rates. Silver and lead, because of their low hardness and high compressibilities can be compacted to higher densities than the other powders.

It was noted that considerable errors could occur in the measurement of porosity without a precise knowledge of the theoretical density at zero porosity. Values cannot be taken from the literature, since metal powders contain some non-metallic inclusions which usually lower the maximum attainable density. The solid densities were measured on accurate symmetrical forms machined from compacts of the respective powders which had been thoroughly melted under a protective atmosphere, cold-worked to eliminate any internal porosity, and finally annealed.

TABLE 1

General Description of Powders

Powder Number	Material	Type	Grade or Batch	Source	Sieve Analysis (U.S. Std. Mesh)					
					-70 +100	-100 +140	-140 +200	-200 +270	-270 +400	-400
1	Iron	Sponge	MH100 24	Höganäs	–	19.1	30.8	23.3	15.1	11.7
2	Iron	Electrolytic	-100 mesh	Cohens	–	23.4	23.2	21.4	18.8	14.2
3	Copper	Atomised	H5076	B.S.A.	–	14.6	23.3	26.7	21.2	14.2
4	Nickel	Hydro-metallurgical	S. Grade	Sherritt Gordon	–	59.4	40.6	–	–	–
5	310 Stainless Steel	Atomised	N1621	B.S.A.	0.8	3.9	3.9	21.3	37.2	32.9
6	410L Stainless Steel	Atomised	N6937	B.S.A.	1.8	12.7	14.5	17.3	20.1	33.6
7	Lead	Atomised	-100 mesh	Hopkins & Williams	–	1.5	2.8	7.3	20.0	68.4
8	Silver	Electrolytic	Type EC	Johnson Matthey	–	12.5	87.5	–	–	–

TABLE II

Particle Properties Determined on Experimental Powders

Powder Number	Solid Density (g/cm³)	Particle Density (g/cm³)	Apparent Density (g/cm³) (ρ_a)	Tap Density (g/cm³) (ρ_t)	$\dfrac{\rho_t}{\rho_a}$	Shape Factors [1]			Specific Surface (M²/g)
						Elongation X	Bulkiness Y	Surface Z	
1	7.75_4	7.32	2.42	3.25	1.34	1.46	0.55	4.15	0.0902
2	7.84_3	7.81	3.06	3.91	1.28	2.59	0.68	2.06	0.0301
3	8.83_9	8.60	2.98	3.66	1.23	1.37	0.62	2.85	0.0675
4	8.91_4	8.83	4.10	4.79	1.17	1.19	0.70	1.21	-
5	7.76_7	7.74	2.46	3.37	1.37	1.49	0.58	2.54	0.0850
6	7.74_9	7.71	2.90	3.73	1.29	1.58	0.64	1.77	0.0805
7	11.28_0	11.20	4.95	7.04	1.42	1.34	0.75	1.24	0.1025
8	10.49_7	10.45	2.10	2.58	1.23	1.54	0.58	3.90	0.0425

The particle densities were determined by helium/air pycnometry, and the apparent and tap densities using a Hall Flowmeter according to MPIF standards. The results are given in Table II together with a quantitative measure of particle shape which was made by using the method described by Hausner [1].

HARDNESS

Reproducible comparative micro-hardness measurements can readily be made using accurate equipment. A Reichert micro-hardness tester was used for the measurement of particle hardness having a diamond indentor in the form of a square pyramid with an apex angle of 136°, and a load which could be varied up to 100 g.

The following expression has been given relating load and indentation diagonal in hardness testing.

$$F = ad^{\gamma} \qquad\qquad (1)$$

where
F = load
d = indentation diagonal
a and γ are constants

Macro-hardness should give an index of $\gamma = 2.0$ showing that hardness is independent of load. However, in the micro-hardness range γ is generally less than 2 and the hardness is therefore load-dependent. A minimum of six indentations were made at six loads ranging from 1.6 g to 32 g depending on the sizes of the particles and their hardnesses. The values of micro-hardness determined for an indentation diagonal of 10μm and accurately computed index values are given in Table III.

TABLE III

Micro-hardness of Powders

Powder Number	Micro-hardness (kg/mm^2) 10μm diagonal	Index (γ)
1	168	1.70
2	143	1.65
3	115	1.85
4	148	1.76
5	222	1.78
6	276	1.60
7	11	1.88
8	58	1.96

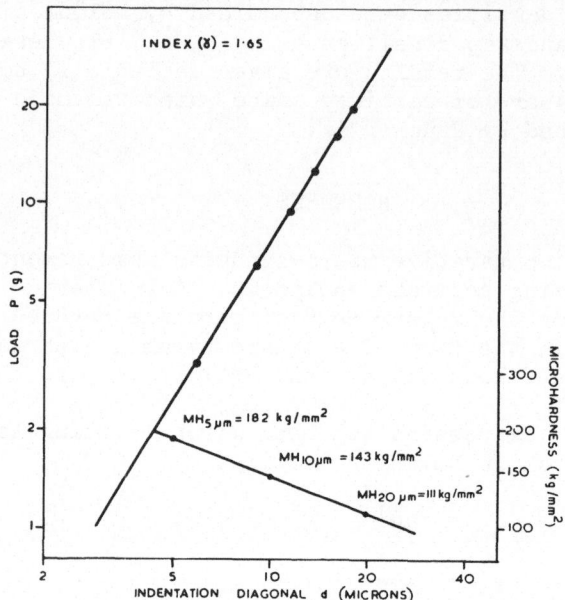

FIG. I   MICRO – HARDNESS   ANALYSIS OF POWDER 2

Buckle [2] has stated that in general the index may be fortuitous
and variable.   Bergsman [3] disputes the validity of the Meyer method
of analysis for micro-hardness.   Several workers [4,5] have shown
that the dependence of hardness with load is affected by test
conditions such as external vibrations, surface finish and even
humidity.   However, the analysis has shown that under comparative
conditions an accurate index value can be obtained.   Fig. 1 shows
a typical micro-hardness analysis of powder 2.

COMPACTION

The powders were isostatically compacted at pressures up to
100 kg/mm^2 in cylindrical rubber membranes having initial volumes
of about 2.5 cm^3.   The hydraulic pressing medium, castor oil, which
is known to be sufficiently fluid at the pressures which were
employed, was pressurized directly from a hydraulic press acting on
a plunger in a closed vessel.   The method of internal pressure
measurement using a Manganin coil and the plunger seal and terminal
seal design had been developed by the National Engineering Laboratory [6].

The compact densities were measured using an accurate mercury
displacement technique employing a contact indication system to
ensure a constant sinker volume immersed in the mercury bath.
Fractional density in relation to compacting pressure is summarised
in Table IV and Fig. 2.

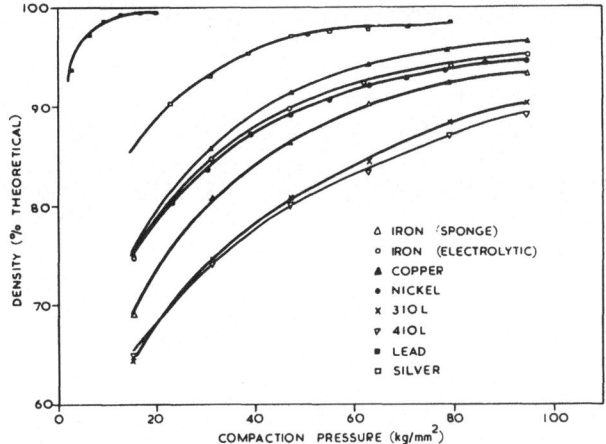

FIG. 2   PRESSURE – DENSITY RELATIONSHIP FOR THE
ISOSTATIC COMPACTION OF METAL POWDERS

TABLE IV

Fractional Density of Isostatically Compacted Powders

Compacting Pressure (kg/mm^2)	Fractional Density (D)							
	1	2	3	4	5	6	7	8
3.3							0.937	
6.6							0.974	
9.8							0.986	
13.3							0.991	
15.8	0.690	0.745	0.750		0.642	0.648		
16.4							0.993	
20.0							0.995	
23.6				0.803				0.902
30.5				0.839				
31.5	0.808	0.848	0.857		0.774	0.740		0.931
39.4				0.871				0.956
47.3	0.864	0.899	0.914	0.894	0.808	0.800		0.971
55.1				0.907				0.973
63.0	0.902	0.921	0.941	0.922	0.843	0.831		0.977
70.9				0.929				0.980
78.8	0.923	0.937	0.957	0.938	0.882	0.870		0.984
86.6				0.945				
94.5	0.934	0.950	0.966	0.949	0.902	0.890		

PRESSURE-DENSITY RELATIONSHIPS

In order to relate powder properties quantitatively to compaction behaviour it is necessary to express pressure-density data in mathematical terms. The compaction results were tested by numerous existing formulae with but limited success; due mainly to the narrow density range over which they apply. However, the formula proposed by Kawakita [7] was found to apply widely over the pressure range. Kawakita's equation is

$$C = \frac{V_o - V}{V_o} = \frac{abP}{1 + bP} \qquad (2)$$

where C = relative reduction in volume
           V_o = initial volume
           V = volume at pressure P
           a is a constant related to initial porosity
           b is a second constant.

Because of the difficulties which arise from determining accurately the initial volume in isostatic compaction, due to the irregular shape and the variation in packing, the volume reduction was calculated using the initial porosities derived from apparent density measurements. The analysis of the data following Kawakita is given in Table V where it can be seen that the constant 'a' is nearly equal to the initial porosity: $(1-D_o)$.

TABLE V

Analysis of Data Following Kawakita

Powder Number	Initial Porosity $(1-D_o)$	Calculated Constants from experimental data	
		a	b
1	0.688	0.697	0.235
2	0.610	0.618	0.220
3	0.663	0.676	0.284
4	0.540	0.554	0.147
5	0.683	0.689	0.164
6	0.626	0.628	0.122
7	0.561	0.564	5.463
8	0.800	0.804	1.364

There is a tendency for the constant "b" to increase as the hardness decreases, but variations in this inverse relationship appear to follow variations in initial porosity. However, the author found that if the two constants are combined according to equation 3 a better inverse relationship exists between micro-hardness and β.

$$\beta = \frac{b(1 - a)}{a} \qquad (3)$$

Moreover, if particle hardening is taken into account, the hardness then being inversely proportional to $\beta\gamma$, the relationship is further improved as shown in Table VI.

$$\text{As} \quad MH_{(10\mu m)} \alpha \frac{1}{b} \text{ or } \frac{1}{\beta} \text{ or } \frac{1}{\beta\gamma} \qquad (4)$$

then from Table VI it can be seen that the best relationship is taking

$$MH_{(10\mu m)} \alpha \frac{1}{\beta\gamma} \qquad (5)$$

$$\text{where} \quad \beta\gamma \, MH_{(10\mu m)} = Q_3 \qquad (6)$$

TABLE VI

Products (Q) Correlating
Kawakita's Constants and Material Properties

Powder Number	Products Q		
	Q_1 $bMH_{10\mu m}$	Q_2 $\beta MH_{10\mu m}$	Q_3 $\beta\gamma MH_{10\mu m}$
1	39.4	17.2	29.2
2	31.4	19.4	32.1
3	32.7	15.6	29.0
4	17.3	13.9	30.8
5	36.4	16.4	29.3
6	33.7	19.9	31.8
7	61.1	45.8	67.2*
8	78.3	19.3	37.7*

* Lead and silver powders, soft, and more subject to error.

Kawakita's equation would appear valid as an empirical equation of state but its usefulness in describing compaction behaviour for studying the effect of powder variables is clearly limited.

The importance of plastic deformation and the influence of strain-hardening suggests the application of a stress-strain formula to the compaction of metal powders. The stress distribution around individual pores is complex, but essentially pore closure will be influenced by the stress required for particle deformation which is controlled by the compacting pressure. The following general expression for stress-strain has been adapted to describe the compaction of metal powders.

$$P = K\epsilon^n \tag{7}$$

where P = pressure of compaction
          ϵ = volume strain of compaction
          K and n are material constants.

The most suitable expression for compaction-strain was derived from the instantaneous change in volume from the tap density to the final density. Namely,

$$\epsilon = \mathrm{Ln}\left[\frac{V_T-V}{V_T-V_\infty}\right] = \mathrm{Ln}\left[\frac{D-D_T}{D(1-D_T)}\right] \tag{8}$$

where Ln = natural logarithm
          V_T = tap volume
          V = volume at pressure P
          V_∞ = volume at theoretical density
          D = fractional density at pressure P
          D_T = fractional "tap-density"

The tap density was chosen as being close to the density where particle deformation commences.

The application of equation 7 to the isostatic compaction of the eight powders is shown in Fig. 3 where all the powders can be represented by straight lines over the pressure range covered. The intercept "K" is related to particle hardness and the slope "n" to the index (γ) by the following values when the pressure is expressed in kg/mm^2.

$$K = 0.054 \, MH_{10\mu m} \tag{9}$$

$$n = (\gamma - 2.57) \tag{10}$$

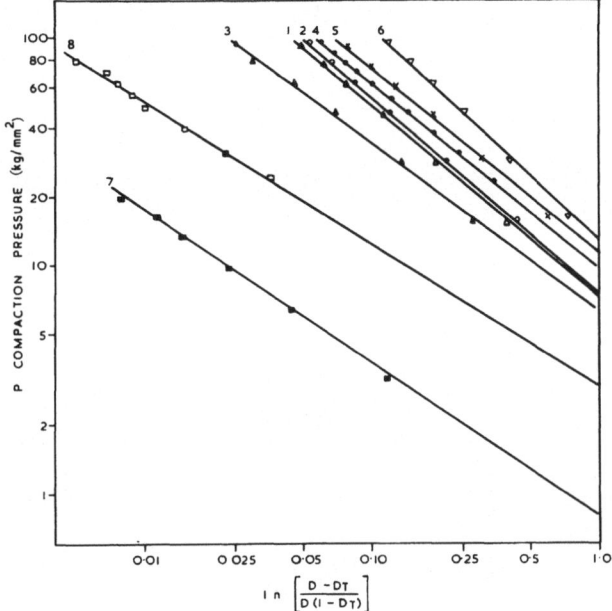

FIG. 3   COMPACTION STRAIN FORMULA APPLIED TO
THE ISOSTATIC COMPACTION OF METAL POWDERS

The values of n and K are given in Table VII together with the
values calculated using equations 9 and 10.

TABLE VII

Analysis of Data from Compaction Strain Formula

Powder Number	Measured		Calculated	
	K	n	K	n
1	0.87	7.2	0.87	9.0
2	0.88	7.4	0.92	7.7
3	0.73	6.6	0.72	6.2
4	0.83	9.8	0.81	8.0
5	0.81	12.5	0.79	12.0
6	0.93	13.4	0.97	14.9
7	0.68	0.6	0.69	0.6
8	0.62	2.9	0.61	3.1

DISCUSSION

The experimental results suggest that for the materials used
and the pressure range studied the isostatic compaction behaviour
can be described as a two-stage process. The first stage is

TABLE VIII

Factors Affecting Compaction
of Metal Powders

Compaction Stage	Controlling Factors	Measured Properties
1. Initial Packing	Particle size distribution Mean size Particle shape Porosity Specific surface Surface conditions	Tap Density
2. Particle Deformation	Structure Purity Grain-size Plasticity Lattice Defects Gas content Alloy additions	Particle Hardness and Index

concerned with the initial packing of the particles and the second stage with particle deformation. The factors affecting the two stages are summarised in Table VIII.

There will of course be a certain amount of overlapping between the factors affecting the two stages. For instance, the particle factors will determine the pore-size distribution and therefore the amount of deformation required for densification.

Stage 1

The extent of particle rearrangement processes which have been used by many authors to explain deviations at low pressures is clearly very limited. This confirms recent investigations on the extent of particle movement by Shapiro [8] and Bockstiegel [9]. Pelzel [10] has shown that the particle contact area at high porosity levels is exceedingly small and therefore deformation of particle asperities probably occurs at very low applied pressures. This means that the surface topography could influence the extent of Stage 1; the tap density being correct for smooth spherical particles but giving a low value in the case of rough, spiky, irregular particles. Kostelnik, Kludt and Beddow [11] have shown that at very low pressures (less than 0.001 kg/mm^2) fragmentation of flake copper powders occurred.

Stage 2

The second stage is controlled by the plastic deformation of particles, which has been quantitatively measured in terms of particle hardness taken as a measure of yield strength and the index (γ) which is taken to be indicative of work-hardening. The particle hardness indicates the pressure requirement for the commencement of Stage 2 and index (γ) indicates the rate at which densification proceeds. Quantitative relationships have been produced relating hardness and index (γ) to densification but it must be emphasised that the hardness refers to an indentation diagonal of 10 microns. Furthermore, the index (γ) has a value which is unique to the narrow range of loads used for the hardness analysis and cannot be compared with Meyer exponents determined by macro-hardness measurements.

SUMMARY

A compaction-strain formula has been proposed which successfully describes the compaction behaviour of a wide range of metal powders above the tap density. The constants have been related to results obtained from micro-hardness analyses on individual particles.

ACKNOWLEDGEMENTS

The author wishes to thank Mr. R. L. Sands for help and advice during the course of the investigation and to Mr. D.A. Oliver C.B.E., Director of Research, for his encouragement.

REFERENCES

1.   Hausner H.M. Planseeberichte Für Pulvermet, 14, 75-84, (1966).
2.   Bückle H. Met. Reviews, Inst. Met., 4(13), 49, (1959).
3.   Bergsman E.B. A.S.T.M. Bull. 176, 37, (Sept. 1951).
4.   Yoshino T. Bull. of J.S.M.E. 8 (31), 291, (1965).
5.   Shpunt A.A. and Nabutovskaya. Soviet Physics - Solid State, 9 (2), 387, (1967).
6.   Pugh H. Ll D. N.E.L. Report No. 142, (March 1964).
7.   Kawakita K. J. Japan Soc. Powder Met. 10 (6), 236-46, (Dec. 1963).
8.   Shapiro I. Tech. Doc. Rept. No. ASD-TDR-63-147, U.S. Dept. of Commerce, (1963).
9.   Bockstiegel G. Modern Developments in Powder Met. Vol. 1, Ed H.M. Hausner, p. 155-187 (1966).
10.   Pelzel E. Z.f. Metallk, p.813, (1955).
11.   Kostelnik M.C. Kludt F.H. and Beddow J.K. Int. J. of Powder Met. 4 (4), (1968).

THE EFFECT OF PRESINTERING ON THE
DIMENSIONAL CHANGE OF IRON POWDER COMPACTS

T. Krantz

Domtar Limited, Research

Centre, Senneville, Quebec, Canada

Introduction

In conventional ferrous powder metallurgy, admixed organic lubricants are used to facilitate compaction and ejection. These lubricants are subsequently removed in the early stages of sintering by decomposition and/or volatilization. To assure a gradual decomposition of the lubricant and evolution of the gases, the heating cycle may be interrupted at an intermediate temperature. This type of operation is called presintering. Though presintering is widely used, there is little information in the literature concerning its effect on product properties.

The nature of the physico-chemical processes occurring during the heating of compacts is equally vague. To obtain a better understanding of these, several mensuration techniques seldom applied in powder metallurgy were employed: high temperature dilatometry, thermogravimetric and differential thermal analyses, and electrical resistivity measurement. These in turn aided interpretation of the relationships between presintering and the dimensional stability and strength of iron compacts.

Materials and Procedures

The study has been based on two commercial grades or iron powder: a "sponge" iron and an "atomized" powder.

437

Mixtures were made with a conventional lubricant, zinc
stearate, and also with both lubricant and graphite.
Properties of raw materials are shown in Table 1.

Physical properties were measured by MPIF standards
except where noted.

Components were mixed in a twin-shell blender for 30
minutes. The straight powders and mixtures were compacted to transverse rupture bars, 1.25" x 0.50" x 0.25",
at 6.65 + 0.02 g./cc. green density, in a double acting
die set. Presintering and sintering of the compacts
were carried out in a 1" diameter vertical tube furnace.
Retention times at elevated temperatures were chosen to
approximate those used in conventional practice.
Specimens were treated one at a time, using a wire
basket to raise the specimen into the hot zone and then
to lower it into the cool section of the tube. The presintering temperature was 650 + 5°C and the specimen
heating rate was 160°C/min. At the end of the presintering period, the specimen was cooled to room temperature at the rate of 100°C/min. The furnace was
supplied with 0.5 l./min. of simulated endogas atmosphere - 40% H_2, 40% N_2 and 20% CO with a dew point of
-22°C. Sintering was carried out in the same furnace
at 1120 $\pm$ 5°C using the same atmosphere. The heating
and cooling rates in this case were 320°C/min. and
110°C/min., respectively.

Densities were determined from the weight and physical
dimensions of the compacts. Dimensional change measurement was carried out on sintered and cooled compacts
with the expansion or contraction computed from the
length of the test bar on a die-size basis. Electrical
resistances were measured perpendicular to the pressing
direction on the central one inch length of the transverse rupture bar. The instrument used was a Pye
Double Kelvin Bridge.

High temperature dilatometry was carried out by the
Powder Metallurgy Section of the Department of Mines,
Ottawa, on a Leitz Universal Dilatometer, Model UBB,
utilizing a light beam optical level photographic
recorder. The dilatometric specimens were prepared by
slicing and grinding green transverse rupture bars into
1" long, 0.15" diameter cylinders. The specimens were
heated to 1130°C at a rate of 4.2°C/min. flow of endogas and their change in length was recorded continuously.

TABLE 1

CHARACTERISTICS OF POWDERS EMPLOYED

	Iron Powder A	Iron Powder B	Graphite	Zinc Stearate
Manufacturer	Metal Powders Div. Domtar Ltd.	A.O. Smith Corp. Powder Met. Div.	J. Dixon Co.	Fisher Scient. Co.
Grade	MP-32	500 MM	200-09	U.S.P. Z-65
Chemical comp.				
Fe %	98.4	98.9	-	-
Mn %	0.16	0.18	-	-
Si %	0.13	0.03	-	-
C %	0.006	0.005	99.05*	-
S %	0.018	0.016	-	-
P %	0.014	0.007	-	-
Weight loss in H_2, %	0.33	0.22	0.20* (volat.)	1* (moist.)
Acid insolubles, %	0.26	0.15	0.75* (ash)	13.6* (ash)
Mesh size, fraction, %				
+80	-	4.8	-	-
-80 +100	1.2	9.1	-	-
-100 +150	17.7	15.1	-	-
-150 +200	24.8	20.4	-	-
-200 +230	15.3	12.7	-	-
-230 +325	18.3	16.7	-	-
-325	22.7	21.2	3-8**	99*
Flow time, sec.	29.2	23.8	-	-
App. density, g./cc	2.59	3.15	-	-
BET surface area, cm.^{2}/g.	630	310	-	-

 * manufacturer's data.

** microns, min. - max.

Thermal analyses were carried out by the Centre de
Thermoanalyse of the Ecole Polytechnique, Montreal, on
a Mettler instrument. The samples used were ground,
green transverse rupture bars. DTA and TGA data were
obtained simultaneously while the specimen was heated
to 1200°C at a rate of 10°C/min. in an endogas flow of
160 cc./min.

All data points represent averages of at least two
samples except for the dilatometry where single experiments were carried out.

Discussion of Results

The total dimensional change of a compact is the result
of a number of successive and/or simultaneous changes
occurring in the processing of a metal powder compact.
There is a "spring-back" or growth when the compact
leaves the die due to the relaxation of elastic forces.
Further growth might occur when the lubricant decomposes on heating and finally the compact shrinks as
pores are eliminated during high temperature treatment.
In addition, any chemical or alloying reaction between
components of the compact, or between the compact and
the atmosphere, can result in either shrinkage or
growth depending on the type of reaction.

Thermo-analytcial methods were used here to study the
reactions which occur on heating a compact. A detailed
description of the theoretical and practical aspects
of these techniques have been given by Wesley(1).

Differential Thermal Analysis

Differential thermal analysis, DTA, measures the
difference between the actual sample temperature and
that of an inert reference material both of which are
heated at a constant rate. The thermal effect of
physical and chemical changes occurring at temperature
in the specimen result in peaks of magnitude proportional to the event.

The DTA curve of a sample containing both graphite and
zinc stearate is shown in Figure 1. It is corrected
for the shift of base line resulting from the change
in specific heat of iron on heating. The peaks observed
correspond to the following reactions and transfor-
mations.

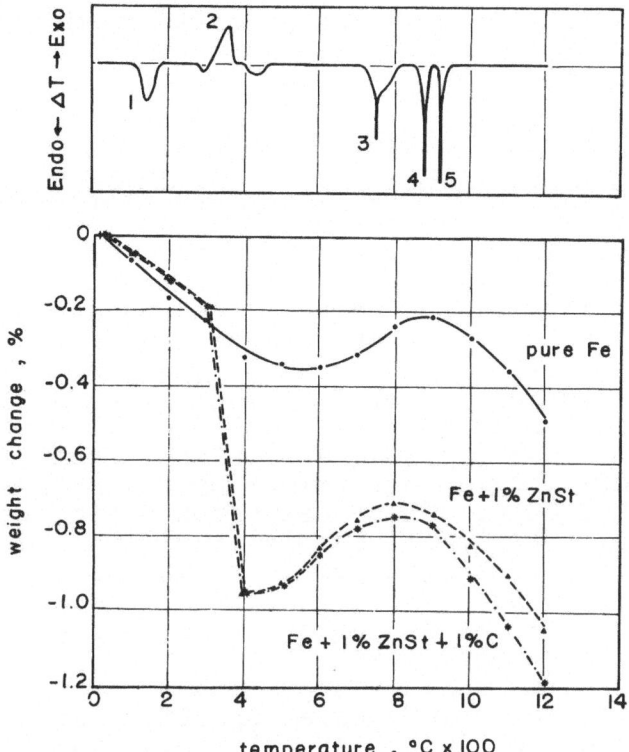

Figure 1. Heat content and weight changes
upon heating of iron compacts.

1. desorption of adsorbed moisture, 100-180°C.

2. decomposition and volatilization of the lubricant,
    280-380°C.

3. demagnetization of iron, 750-770°C.

4. $\alpha \rightarrow \gamma$ allotropic transformation, 880-890°C.

5. graphite-iron interaction, 910-940°C.

For samples free of lubricant and graphite, reactions
2 and 5 respectively are absent. These transitions
are of major interest however.

Our measurements show that zinc stearate decomposes at
around 350°C. Meyer et al(2) using similar measuring
techniques, found complete decomposition of zinc
stearate in hydrogen at 440°C; the manufacturer's
figure is 300°C.

The reaction between iron and graphite seems to be limited to a rather narrow temperature range just above the transformation temperature.

Thermogravimetric Analysis

A thermogravimetric sample is continuously weighed while being heated at a constant rate. The resulting weight change versus temperature curve, as shown in the lower part of Figure 1, is an indication of the thermal stability of the sample. According to this curve, there is a continuous weight loss which starts near room temperature and reaches a maximum value at 500-550°C. The rate of weight change for an un-lubricated powder is uniform until 500°C. The value of the weight loss at 550°C approximates that of the "hydrogen loss" figure determined by the MPIF standard procedure at 1050°C. This unexpectedly high value is probably due to the small sample size (1 g.) and to the low heating rate employed in these tests. Under these conditions the desorption of chemiadsorbed oxygen-hydrogen layers[3] might proceed at a fast rate even at relatively low temperatures.

The weight loss of a lubricated specimen is similar to that of iron up to 300°C. At this point the lubricant starts to volatilize. At 400°C the total weight loss is 0.95%, of which perhaps 0.7-0.8% is due to the lubricant, the balance being surface oxide. According to Meyer et al[2], the solid residue (ZnO) after a burn-off at 750°C is approximately 18% of the initial weight of the lubricant, a figure similar to ours if allowance is made for the higher temperature.

Upon further heating, there is a weight gain which reaches a maximum at about 800-850°C. The gain is apparently, the result of carburization from the atmosphere. Although the maximum solubility of carbon in α-iron is of the order of 0.02 w. %[4], this does not exclude an interaction between CO and iron resulting in the formation of Fe_3C. Above 900°C the carbon potential of the endogas atmosphere changes to decarburizing and the carbon reaction reverses. It is interesting to observe that the presence of graphite in the lubricated sample has only a minor effect on the basic behaviour of the sample; the rate of gaseous carburization slightly decreases while that of decarbuization, above 800°C, moderately increases.

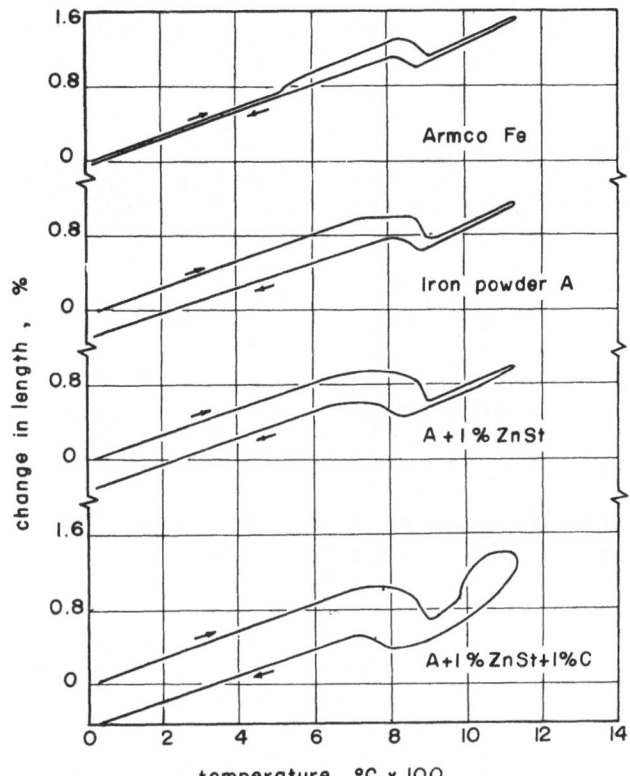

Figure 2.   Dilatometric change versus
            temperature.

Powders "A" and "B" behaved similarly in DTA and TGA
trials.   Differences in chemistry (except for H_2 loss),
surface area, and/or particle size were not of significant magnitude to affect the results.

Dilatometry

The linear dimensional change measured continuously as
green compacts were heated and cooled at a constant
rate is given in Figure 2.   The expansion curves of
the three basic powder compositions are compared with
that of dense pure iron.   Up to about 500°C the linear
expansions are the same in all cases.   Above 600°C,
the decreasing slopes of the powder sample curves
indicate the initiation of shrinkage due to sintering.
The presence of lubricant and graphite does not seem
to have any significant effect on the shape of the

curves in the α-iron region. Above the α → γ transformation, the expansion is linear again with no
visible shrinkage up to the maximum temperature
employed. The slope of this part of the curve is
about the same for all specimens except for the one
containing graphite. This shows a significant expansion followed by an equally rapid shrinkage due
to the reaction between the iron and the graphite, at
approximately the same temperature, 930°C, indicated
by the DTA results. The cooling curves, at least
below the γ → α transformation, are also similar.
The transformation temperature anomaly indicated
between the heating and the cooling parts of the curves
is false. The temperature was recorded during the
heating cycle only, and this scale does not coincide
with that of cooling.

It is important to note that the decomposition of the
lubricant at about 350°C had no apparent effect on the

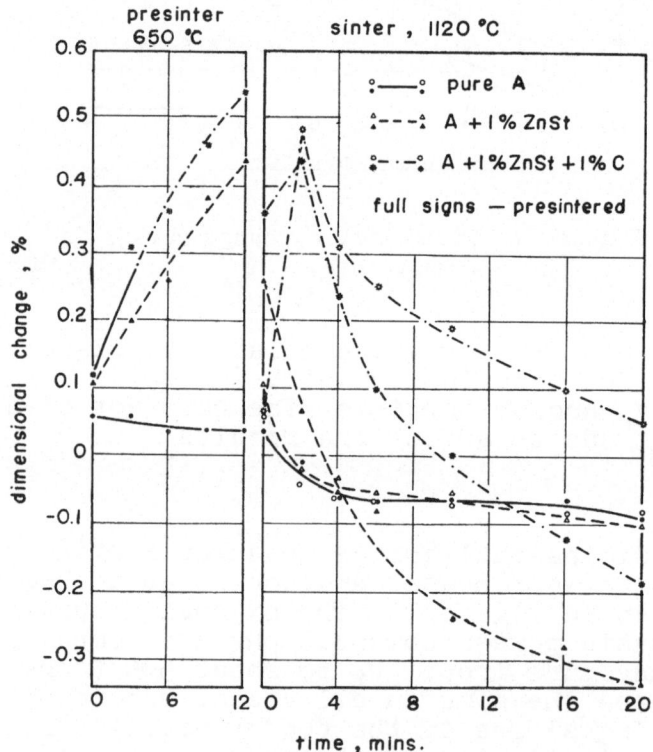

Figure 3.   Linear dimensional change versus
            presintering and sintering time.
            (Powder A)

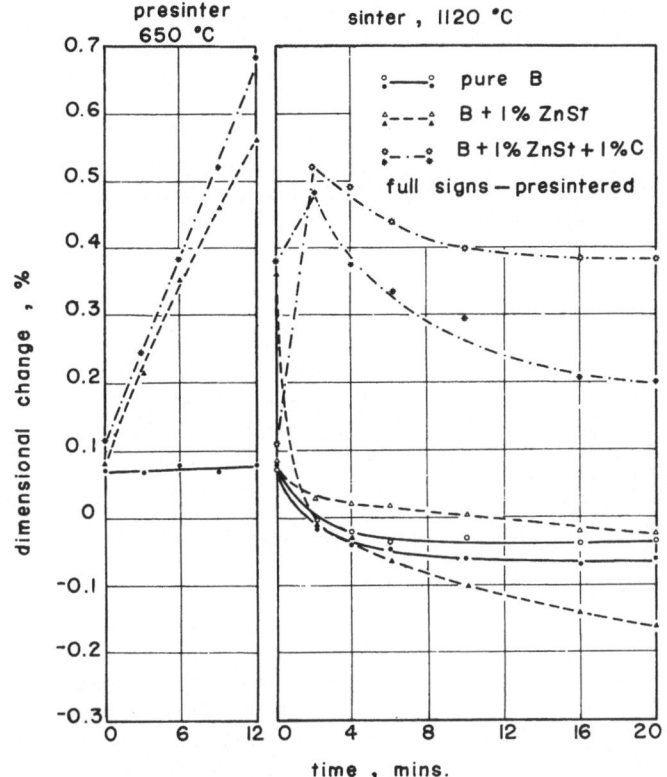

Figure 4. Linear dimensional change versus
presintering and sintering time.
(Powder B)

rate of expansion at such a slow heating rate. There
were no significant differences between the two powders
studied.

Dimensional Change in Presintering-Sintering

The linear dimensional change measurements carried out
on sintered and cooled compacts are considerably
different than those obtained by continuous high
temperature dilatometry. Two types of curves are
shown in Figure 3. Those at the left indicate the
dimensional change occurring during presintering.
The size changes resulting from subsequent sintering
are shown on the right. Two types of specimens were
used in all sintering experiments: as compacted, or
green, and as presintered for six minutes.

As Figure 3 indicates, the dimensions of compacts made

of nonlubricated powder "A" are not affected by presintering. There is no change in length after presintering only and the sintering curves of presintered
and nonpresintered compacts are also identical, if
experimental scatter is disregarded.

When zinc stearate is present in the compact, the
springback is higher, and in addition, there is a
significant growth during presintering. In the
presence of graphite, the growth is even more pronounced. When specimens presintered for 6 minutes
are sintered, the shrinkage observed is considerably
greater than that of nonpresintered specimens. The
graphite-containing samples show large expansions in
the first stage of sintering whether presintered or
not, but again, if presintered, the shrinkage rate is
considerably higher.

In the case of powder "B", the same relationships were

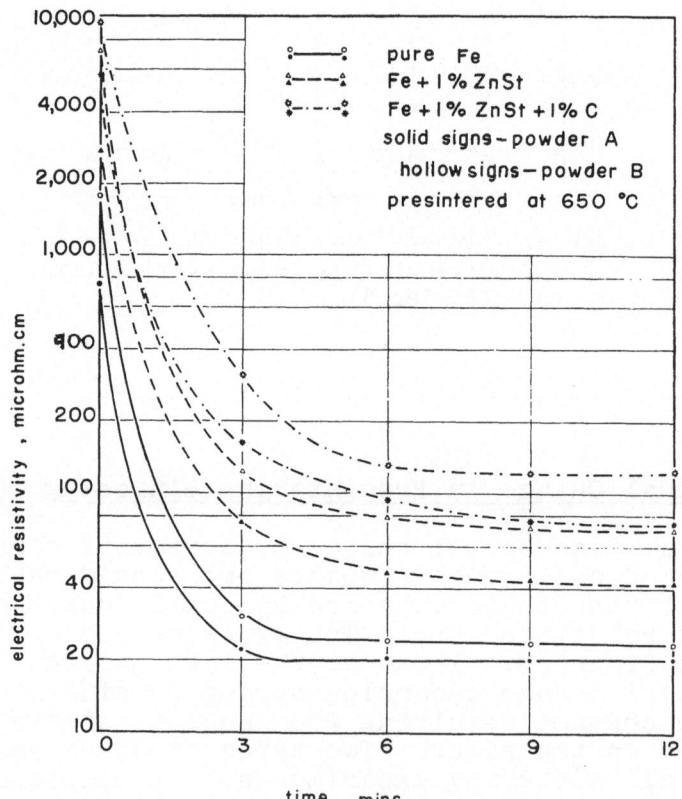

time , mins.

Figure 5. Electrical resistivity versus
           presintering time

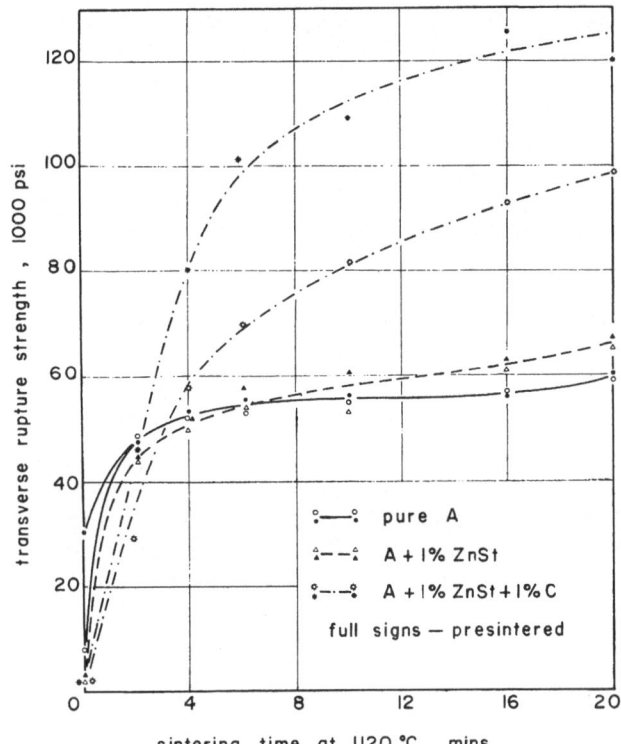

Figure 6.   Strength versus sintering time.
             (Powder A)

found although the magnitudes of the changes were
somewhat different, (Figure 4).   Springback and growth
on presintering of lubricated samples were greater.
Densification during sintering was less pronounced.

These effect might tentatively be tied to decomposition of the lubricant.   The rates of heating in both
presintering and sintering were very high.   Consequently, despite the interconnected nature of the
porosity within the compact, high gas pressures could
develop in the pores for short periods.   This pressure
would destroy some of the particle contacts and expand
the compact.   A more spongy powder would have a greater
number of inter-particle contacts and so greater
resistance to expansion by the trapped gases.   Thus,
the same amount of lubricant would cause less expansion.
At the same time ex-contact surfaces would have a
highly disturbed, stressed structure.   These areas
could then become centres of activation on subsequent
sintering and have a marked accelerating influence on

the process of shrinkage. The magnitude of this
activation would be a function of the size of the active
surface area of the powder. At slow heating rates this
expansion-activation mechanization would be inoperative,
and thus the phenomena complately missing during dilatometric experiments.

Forss(5) claims that the sintering rate of pure iron
compacts is accelerated by sintering in both the alpha
(900°C) and gamma (1120°C) ranges, but his data indicate
this to be the case only for sintering times (30-40
mins.) longer than those used in this work. This could
be the result of significantly lower specimen heating
rates.

Electrical resistivity measurements were also used to
follow the course of presintering (Figure 5). Powder
"A" showed lower resistivity than "B" under all conditions. This is again due to the difference in surface
area between the two powders. Compacts made of irregular particles with internal porosity display less

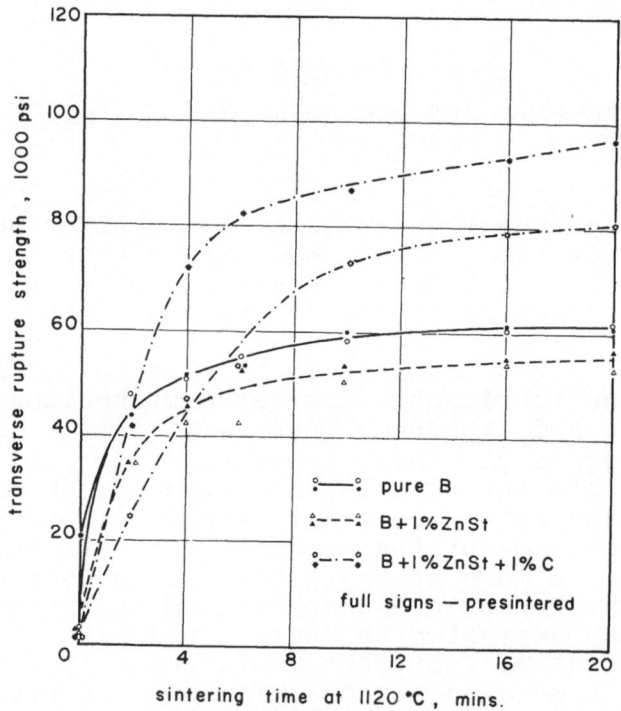

Figure 7.   Strength versus sintering time.
            (Powder B)

resistance to electrical current than those of more regular powders at equal densities(6).

Powder "A" having twice the surface area of "B" when compacted without a lubricant has only half the electrical resistance of "B". The addition of 1% zinc stearate doubles the original difference indicating that the lubricant coating is effectively twice as thick on the regularly shaped powder.

Differences caused by the lubricant do not disappear with volatilization of the zinc stearate even after 12 minutes at the presintering temperature.

Sintered Strength

The effect of presintering on the strength developed after various periods of sintering is shown in Figures 6 and 7. As strength is also dependent on density, this data is shown in Table 2.

The strengths obtained for either powder without graphite, whether lubricated or not, seem to be insensitive to presintering. Forss(5) using electrolytic powder, found an increase in strength for pure iron when presintering proceeded sintering but these were undertaken at a temperature, 900°C, where considerable diffusion could take place in the α-phase. In our tests, however, there are differences in behaviour between powders "A" and "B". In the case of powder "A", the lubricated compacts are stronger; the opposite is true for powder "B". This anomaly might be due to the lubricant residue being twice as much per unit inter-particle contact area for powder "B" as for "A". A residue could hinder the formation of sintered bonds.

TABLE 2

SINTERED DENSITIES

(after 20 mins., in g./cc)

		A	B
not presintered	(straight) iron	6.67	6.69
	iron + 1% ZnSt	6.62	6.63
	iron + 1% ZnSt + 1% C	6.55	6.53
presintered	(straight) iron	6.67	6.70
	iron + 1% ZnSt	6.69	6.66
	iron + 1% ZnSt + 1% C	6.63	6.56

The strength of graphite-containing compacts is strongly presintering-dependent. Both powders, if presintered, show a 20-25% improvement in strength. The reason for this is not clear, especially since the graphite does not react with the iron at the temperature of pre-sintering. There is a possibility that a lubricant residue impedes the formation of interparticle bonds. This condition could arise if the lubricant was being eliminated as the graphite began to react with the iron.

Summary

Based upon the experimental observations made using high heating rates, the following summation can be made:

1. Presintering has no effect on the dimensional change of nonlubricated powder compacts.

2. Compacts containing zinc stearate show increased shrinkage when presintered. The magnitude of this shrinkage is proportional to the surface area of the powder.

3. The presence of graphite in the compact does not change the direction of the dimensional change but does affect its magnitude.

4. Presintering has a considerable effect on the sintered strength of graphite containing mixtures and may improve it by up to 25%.

Though used to a limited extent in these tests, thermo-analytical techniques proved to be a very useful method of obtaining a clearer understanding of sintering and alloying phenomena. They should certainly warrant more attention from powder metallurgists in the design of new research programs.

Acknowledgement

The author would like to thank Mr. R.T. Holcomb for many helpful discussions in the course of this study, and Mr. J.S. Cunnington who carried out part of the experimental work. He also would like to express his gratitude to Mr. N. Spence and Mr. C. Webster of the Department of Mines, Ottawa, for the dilatometric and to Dr. J.C. Sisi of the Ecole Polytechnique, Montreal, for the thermal analyses.

References

1. Wesley, W.W., Thermal Methods of Analysis.
    Interscience Publ. New York (1964).

2. Meyer, R., Pillot, J., and Pastor, H., Powder
    Metallurgy, 12, 298 (1969).

3. Dawihl, W., and Doerre, E., Powder Metallurgy
    International, 1, 11 (1969).

4. Hansen, M., Constitution of Binary Alloys,
    McGraw-Hill Book Co., New York (1958).

5. Forss, L., Iron Powder Metallurgy 3, 211, Plenum
    Press, New York (1968).

6. Klar, E., and Shafer, W.M., International Journal
    of Powder Metallurgy 4, 5 (1969).

SINTERED AND HEAT TREATED 4640 STEEL

F. W. Heck

The International Nickel Company, Incorporated

ABSTRACT

Atomized and mixed elemental powders, prepared to a nominal 4640 steel composition (2.0Ni-0.25Mo-0.40C-Bal. Fe), were pressed and sintered to comparable density levels. Sintered density levels ranged between 6.45 and 7.25 g/cc. Sintering was carried out at 2050 F/30 minutes/dissociated ammonia (representative American practice) and at 2400 F/60 minutes/dissociated ammonia (typical European practice). Test specimens were subsequently heat treated and tempered. The effect of these initial sintering and post sintering heat treatments on mechanical properties and micro-structure of the different starting materials is discussed.

In the as-sintered condition, atomized powder offers strength advantages over mixed elemental powder. After quenching and tempering, better mechanical properties are obtained with mixed elemental powders and appear to be related to the retained heterogeneity of the mixed elemental powders. Final properties and optimum tempering conditions are dependent on specimen density.

INTRODUCTION

In order to improve the mechanical properties of sintered P/M parts, the powder metallurgist has recourse to a number of processing conditions. Among these are:

1. Higher density.
2. Higher sintering temperatures and/or longer sintering times.

TABLE I

RAW POWDERS USED

Alloy or Element	4600	Iron	Nickel	Molybdenum	Graphite
Type	Atomized	Atomized	Carbonyl	Electrolytic	Natural
Apparent Density, g/cc	3.28	2.95	2.17	1.50	-
Flow, Sec.	22.2	25.0	No Flow	No Flow	No Flow
Avg. Particle Size, Microns	-	-	5.1	3.90	0.7
Screen Analysis, %					
-80 + 100	0.1	3.0	-	-	-
-100 + 150	15.6	14.0	0.1	-	-
-150 + 200	25.5	22.0	2.3	-	-
-200 + 250	9.4	10.0	1.5	-	-
-250 + 325	14.9	21.0	5.0	-	-
-325	34.5	30.0	91.1	100.0	100.0
Chemical Analysis, %					
Fe	Bal	99.5	0.003	0.01	-
Ni	1.90	0.03	99.6	0.005	-
Mo	0.25	-	-	99.8	-
Carbon	0.15*	0.01	0.10	0.01	Bal
Oxygen	0.20	0.08	0.125	0.150	-

* Additional graphite added to get 4640 composition.

3. Heat treatment of the finished P/M part.
4. The use of atomized or mixed elemental powders.

The relative merits of atomized powders and mixed elemental powders has been discussed by Goetzel(1), Jones(2), and others; while a heat treating study of sintered nickel steels has been made by Bobo(3). Very little information is available on any powder alloy system that encompasses all the techniques outlined above.

The purpose of the present investigation was to evaluate one particular alloy system under the above conditions. The system selected was a nominal 4640 steel composition (2.0Ni-0.25Mo-0.40C-Bal. Fe). The mechanical properties and structures of both atomized and mixed elemental powders were examined over a range of densities and sintering treatments. Following sintering, the effect of heat treatment on these properties was also studied.

EXPERIMENTAL PROCEDURE

Standard transverse rupture bars and tensile bars (in accordance with MPIF standards 13-62 and 10-63 respectively) were pressed to green density levels of 6.4, 6.8, and 7.1 g/cc. The raw materials used for the investigation are shown in Table I. Zinc stearate (3/4% by weight) was added to the atomized and mixed elemental powders as a lubricant. Sintering was carried out at both 2050 F and 2400 F for periods of both 30 minutes and 60 minutes using either hydrogen or dissociated ammonia atmospheres. Test results on specimens in the "as-sintered" condition showed no evidence that one atmosphere had a particular advantage over the other in regard to mechanical properties. As would be expected, the longer sintering cycles did promote an increase (about 10%) in the strength levels of both the atomized and mixed elemental powders. Based on these preliminary results, two sintering cycles were selected for evaluation in the heat treated condition. These cycles were: 2050 F (1120 C) for 1/2 hour in dissociated ammonia (representing American sintering practice); and 2400 F (1315 C) for 1 hour in dissociated ammonia (simulated European practice). The specimens that were heat treated were held for 1 hour at 1550 F (843 C) in a carbon containing atmosphere, oil quenched and tempered for 1 hour at temperatures of 300 F (149 C), 400 F (204 C), 500 F (260 C) and 600 F (316 C). The carbon potential of the atmosphere used for heat treatment was varied between 0 and 0.40% carbon, depending on the density of the specimen and the subsequent tempering temperature, to prevent decarburization or excessive carbon pick-up. The data resulting are based on duplicate specimens for each test condition and have been reproduced in additional studies.

TABLE II

MECHANICAL PROPERTIES OF SINTERED ATOMIZED AND MIXED ELEMENTAL 4640 POWDER

Powder Origin	Compacting Pressure (TSI)	Sintered Density (g/cc)	Ultimate Tensile Strength (ksi)	Yield Strength (0.2% Offset) (ksi)	Tensile Elongation (%)	Transverse Rupture Strength (ksi)	Hardness (Rb)	Combined Carbon Content (%)
			SINTERED 2050°F/30 MINUTES/DISSOCIATED AMMONIA					
Atomized	26.6	6.51		A		77.8	50.6	0.35
	40.7	6.91	62.6	45.5	2.0	110.9	66.3	0.34
	60.0	7.21		B		138.0	77.4	0.35
Mixed Elemental	20.0	6.48	33.1	19.7	4.0	67.6	32.1	0.34
	30.0	6.87	52.4	31.3	6.0	95.0	46.5	0.34
	42.5	7.15	58.8	33.2	7.0	122.7	60.4	0.35
			SINTERED 2400°F/60 MINUTES/DISSOCIATED AMMONIA					
Atomized	26.6	6.59		A		96.7	55.2	0.34
	40.7	6.97	64.8	54.1	2.0	132.3	70.1	0.34
	60.0	7.28		B		165.9	78.3	0.35
Mixed Elemental	20.0	6.58	44.7	31.0	6.0	91.8	44.6	0.34
	30.0	6.98	54.5	40.4	6.0	119.6	21.8	0.32
	42.5	7.22	70.8	44.5	7.0	149.9	72.5	0.38

A = Green strength of these atomized bars too low to permit adequate handling and sintering.

B = Excessive lamination occurred in the atomized bars at this compacting pressure.

RESULTS AND DISCUSSION

The mechanical properties of the mixed elemental and atomized specimens were examined in the as-sintered, oil quenched and quenched + tempered conditions. The microstructures of these specimens were also examined and are discussed by Fischer et al(4) in greater detail.

As-Sintered Condition

In order to achieve the desired green density levels, it was necessary to compact the atomized powder at higher pressures than required for the mixed elemental powders. This is due to the innately higher resistance to plastic deformation in the fully homogeneous, atomized powder. As a result, tensile bars could not be pressed from the atomized powder at the low density level because of poor green strength, nor at the high density level where excessive lamination occurred. The data reported is therefore based primarily on transverse rupture bars.

After sintering, the atomized powders exhibited higher strength properties than the mixed elemental parts sintered under identical conditions. These results are shown in Table II. Strengths, as would be expected, increased as the sintering cycle (temperature and time) was increased due to greater diffusion bonding between the powder particles. Temperature, however, had the greater effect in increasing strength levels than did time. The mixed elemental powder while exhibiting strengths lower than the atomized prealloy powder, showed greater property improvement with the increase in temperature.

The results also showed that mixed elemental powder sintered to the same density levels exhibits 2 to 3 times the ductility of the atomized powder.

Typical structures of the atomized and mixed elemental specimens at a medium density (6.90-7.00 g/cc) level are shown in Figure 1. An even distribution of phases is present in the homogeneous atomized specimens while the mixed elemental parts reveal a heterogeneous structure. All specimens contain a mixture of pearlite and ferrite with the mixed elemental specimens also containing areas richer in nickel, as verified by microprobe analyses. The microstructures also give further evidence that, while the increased temperature promotes greater homogenization in the mixed elemental parts, the longer time and higher temperature associated with European sintering practice is still insufficient to achieve full homogenization.

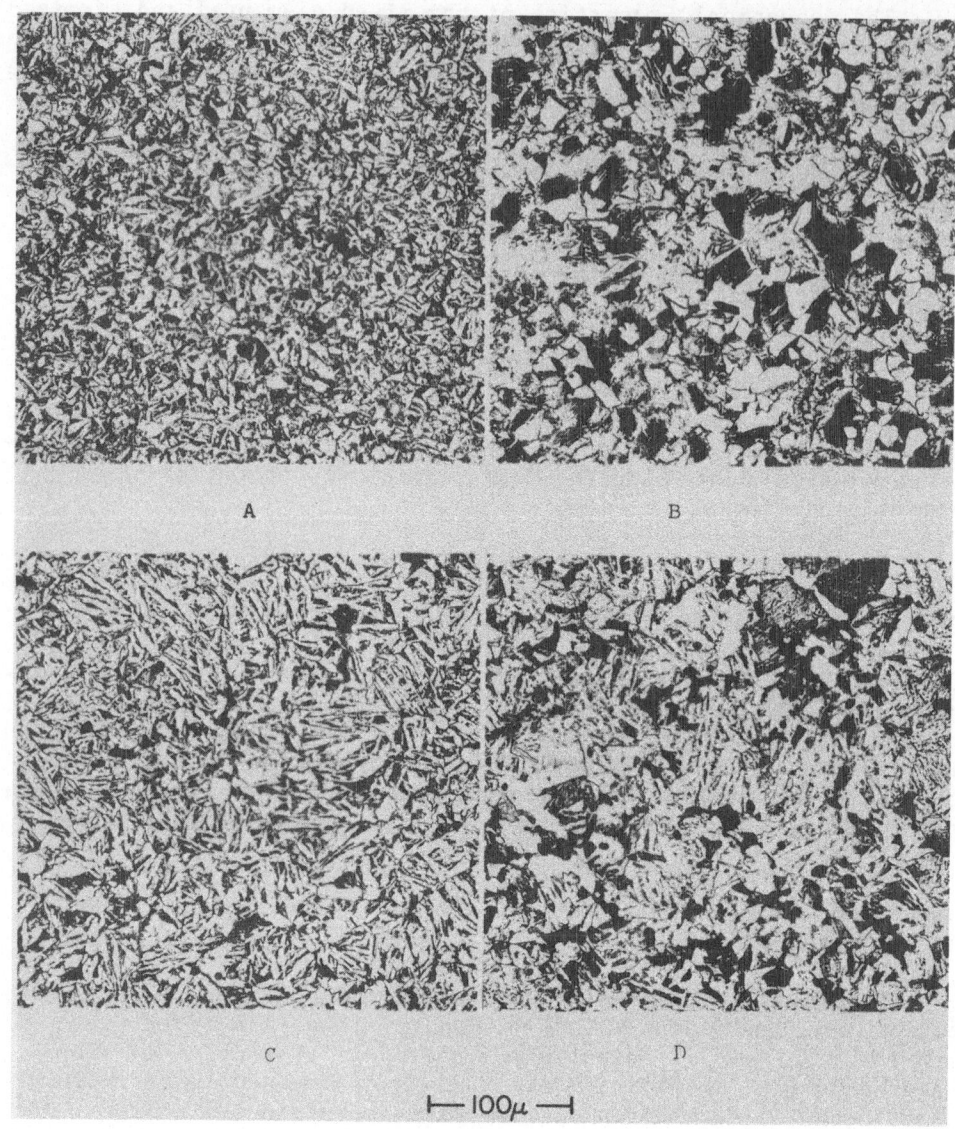

<u>FIGURE 1</u>

ATOMIZED AND MIXED ELEMENTAL 4640 STEEL POWDERS

```
A - Atomized        - Sintered 2050°F/30 min/Dissociated Ammonia - 6.91 g/cc
B - Mixed Elemental - Sintered 2050°F/30 min/Dissociated Ammonia - 6.87 g/cc
C - Atomized        - Sintered 2400°F/60 min/Dissociated Ammonia - 6.97 g/cc
D - Mixed Elemental - Sintered 2400°F/60 min/Dissociated Ammonia - 6.98 g/cc
                    ETCHANT: 2% NITAL
```

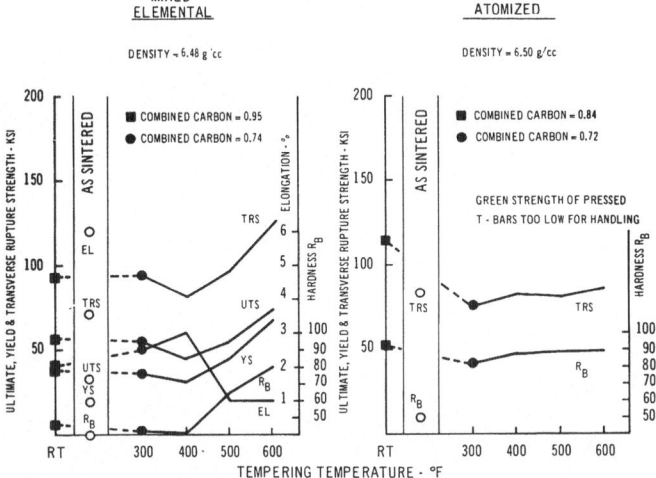

AUSTENITIZED 1500°F/60 MIN./OIL QUENCHED
NOMINAL COMBINED CARBON CONTENT: 0.38% (UNLESS OTHERWISE NOTED)

FIGURE 2 – EFFECT OF HEAT TREATMENT ON MECHANICAL PROPERTIES OF LOW DENSITY
ATOMIZED AND MIXED ELEMENTAL 4640 STEEL POWDERS AFTER SINTERING AT
2050°F FOR 30 MINUTES IN DISSOCIATED AMMONIA (AMERICAN PRACTICE).

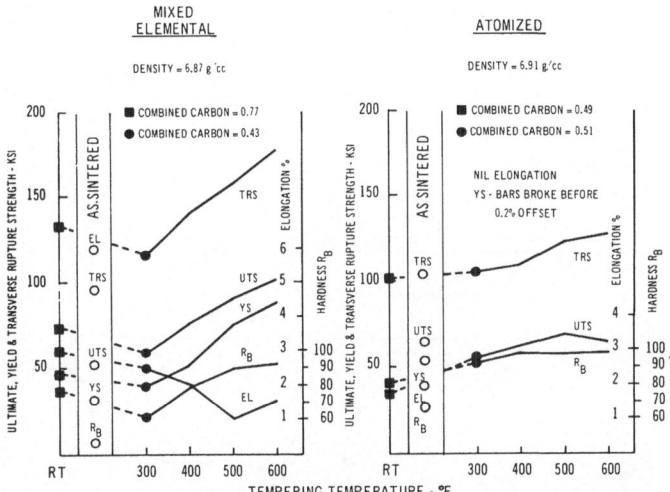

FIGURE 3 – EFFECT OF HEAT TREATMENT ON MECHANICAL PROPERTIES OF MEDIUM DENSITY
ATOMIZED AND MIXED ELEMENTAL 4640 STEEL POWDERS AFTER SINTERING AT
2050°F FOR 30 MINUTES IN DISSOCIATED AMMONIA (AMERICAN PRACTICE).

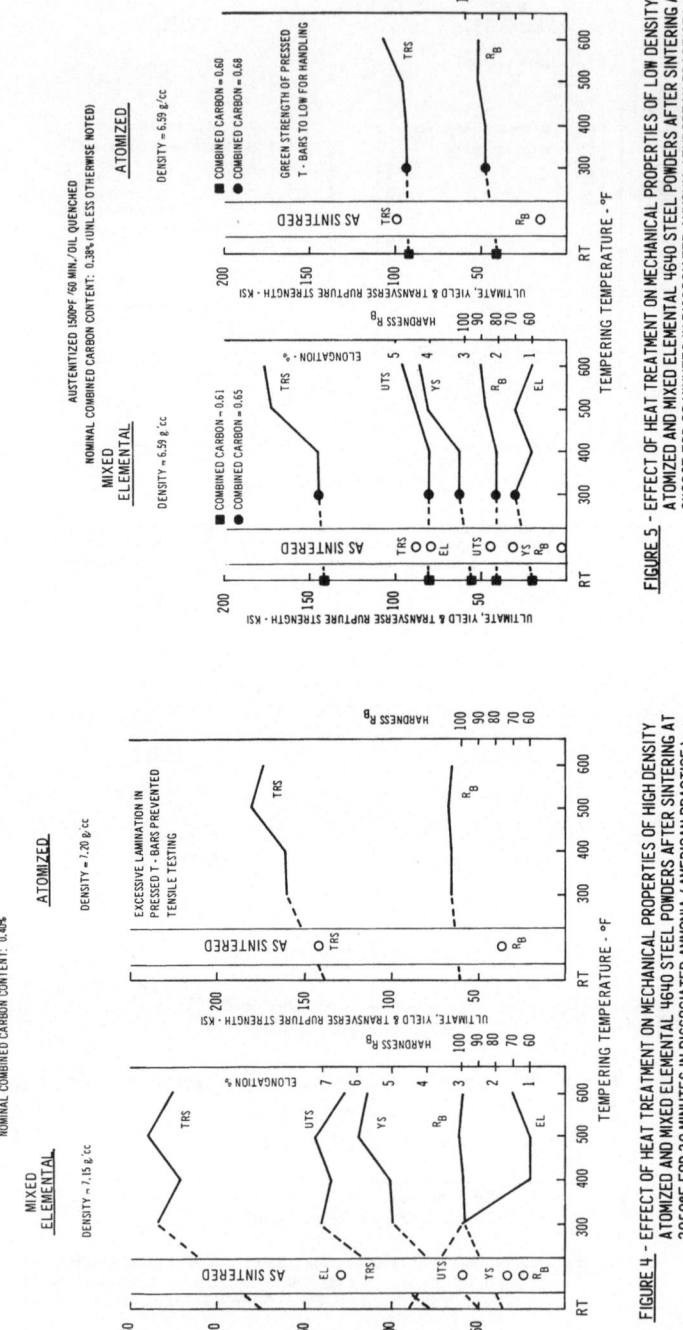

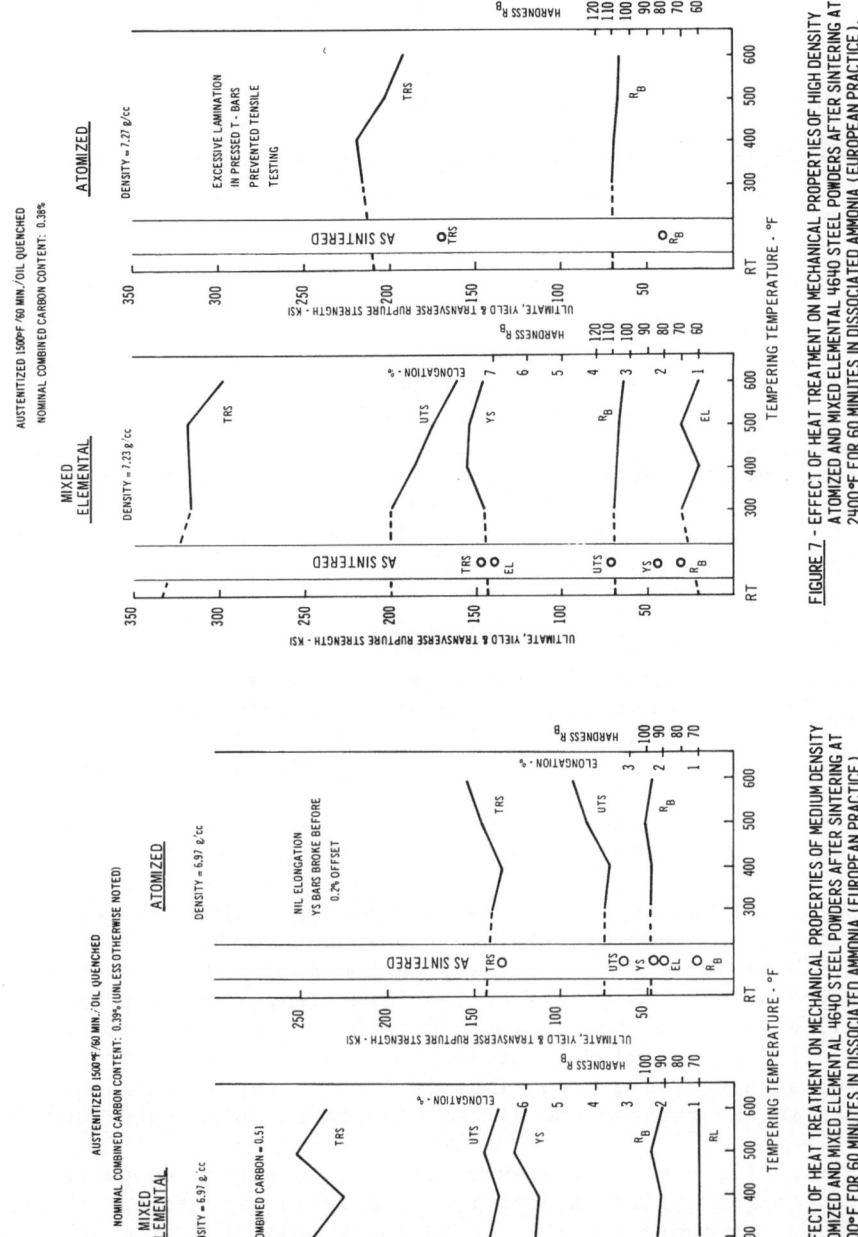

FIGURE 7 - EFFECT OF HEAT TREATMENT ON MECHANICAL PROPERTIES OF HIGH DENSITY ATOMIZED AND MIXED ELEMENTAL 4640 STEEL POWDERS AFTER SINTERING AT 2400°F FOR 60 MINUTES IN DISSOCIATED AMMONIA (EUROPEAN PRACTICE).

FIGURE 6 - EFFECT OF HEAT TREATMENT ON MECHANICAL PROPERTIES OF MEDIUM DENSITY ATOMIZED AND MIXED ELEMENTAL 4640 STEEL POWDERS AFTER SINTERING AT 2400°F FOR 60 MINUTES IN DISSOCIATED AMMONIA (EUROPEAN PRACTICE).

Heat Treated Condition

Heat treatment, after sintering under either representative
American (2050 F/30 min) or typical European (2400 F/60 min)
sintering practices, caused parts prepared from mixed elemental
powders (having a nominal 4640 steel composition) to exhibit mechanical properties superior to those of atomized powders over all the
density ranges evaluated. These results are illustrated graphically
in Figures 2 through 7, and show that the properties of the mixed
elemental powders are 30 to 75% greater than those exhibited by the
atomized powders.

This improvement in mechanical properties of the mixed elemental powders over the atomized powders is also greater at the
higher sintering cycle (typical European practice). There was no
evidence of any internal cracking in the atomized alloy specimens
that may have accounted for their lower properties. Under both
sintering cycles however, the improvement observed in the mixed
elemental powders becomes increasingly less as the density of the
parts is increased. As an example, following American sintering
practice, the maximum strengths observed in the heat treated mixed
elemental parts are approximately 45%, 40% and 35% stronger than
identically heat treated atomized powders at density levels of
6.5 g/cc, 6.9 g/cc, and 7.2 g/cc respectively. A similar trend is
noted in the heat treated parts after initial sintering under
typical European practice. The results indicate that at nearly
full density, there would be little difference in the mechanical
properties of either mixed elemental or atomized alloy powders.

In addition to the obvious trends noted above two other
interesting observations were made. First, as density is decreased,
there was an increasing tendency to absorb the quenching oil into
the specimens. In spite of careful cleaning techniques, all of
this oil was not removed prior to subsequent tempering. As a
result, there was a marked pick-up in the combined carbon contents
in the low density specimens at the lower tempering temperatures.
Increasing tempering temperatures facilitated the burn out of this
excess oil. The results point out the need for special handling
techniques under these conditions (such as lower initial carbon
contents in the pressed and sintered parts to compensate for this).

Secondly, the figures show that the tempering temperature
required to get optimum properties appears to decrease as the specimen density is increased. It would appear that the greater amount
of porosity in the specimen (and therefore the less particle to
particle contact within the specimen) retards the response to the
draw temperatures. The optimum properties obtained in the fully
heat treated condition, for the various density levels evaluated,
are summarized in Tables III and IV.

TABLE III

HEAT TREATED PROPERTIES OF ATOMIZED AND MIXED ELEMENTAL
4640 STEEL POWDERS

Sintered: 2050°F/30 Minutes/Dissociated Ammonia
Heat Treated: 1550°F/1 Hour/Oil Quenched +
Tempered 1 Hour/Air Cooled

ATOMIZED POWDER

Density, g/cc	6.50	6.91	7.20
Tempering Temperature, °F	600	600	500
Ultimate Tensile Strength, ksi	A	64.1	C
Yield Strength (0.2% Offset), ksi	A	B	C
Tensile Elongation, %	A	Nil	C
Transverse Rupture Strength, ksi	85.7	127.5	180.0
Hardness, R_b	88.3	98.3	106.7
Combined Carbon Content, %	0.38	0.37	0.40

MIXED ELEMENTAL POWDER

Density, g/cc	6.48	6.87	7.15
Tempering Temperature, °F	600	600	500
Ultimate Tensile Strength, ksi	73.4	101.1	143.6
Yield Strength (0.2% Offset), ksi	67.6	88.1	119.5
Tensile Elongation, %	1.0	1.5	1.0
Transverse Rupture Strength, ksi	124.1	176.4	239.5
Hardness, R_b	80.2	91.7	101.2
Combined Carbon Content, %	0.38	0.37	0.40

A = Green strength of these atomized bars too low to
permit adequate handling and sintering.

B = Bars broke before 0.2% offset.

C = Excessive lamination occurred in atomized bars.

TABLE IV

HEAT TREATED PROPERTIES OF ATOMIZED AND MIXED ELEMENTAL
4640 STEEL POWDERS

Sintered: 2400°F/60 Minutes/Dissociated Ammonia
Heat Treated: 1550°F/1 Hour/Oil Quenched +
Tempered 1 Hour/Air Cooled

ATOMIZED POWDER

Density, g/cc	6.59	6.97	7.27
Tempering Temperature, °F	600	600	400
Ultimate Tensile Strength, ksi	A	92.0	C
Yield Strength (0.2% Offset), ksi	A	B	C
Tensile Elongation, %	A	Nil	C
Transverse Rupture Strength, ksi	107.2	153.9	219.3
Hardness, R_b	91.7	99.7	109.0
Combined Carbon Content, %	0.38	0.39	0.38

MIXED ELEMENTAL POWDER

Density, g/cc	6.59	6.95	7.23
Tempering Temperature, °F	600	500	400
Ultimate Tensile Strength, ksi	96.1	145.5	186.7
Yield Strength (0.2% Offset), ksi	85.7	127.4	155.6
Tensile Elongation, %	1.0	1.0	1.0
Transverse Rupture Strength, ksi	176.5	252.5	317.1
Hardness, R_b	90.5	100.0	108.3
Combined Carbon Content, %	0.38	0.39	0.38

A = Green strength of these atomized bars too low to
        permit adequate handling and sintering.

B = Bars broke before 0.2% offset.

C = Excessive lamination occurred in atomized bars.

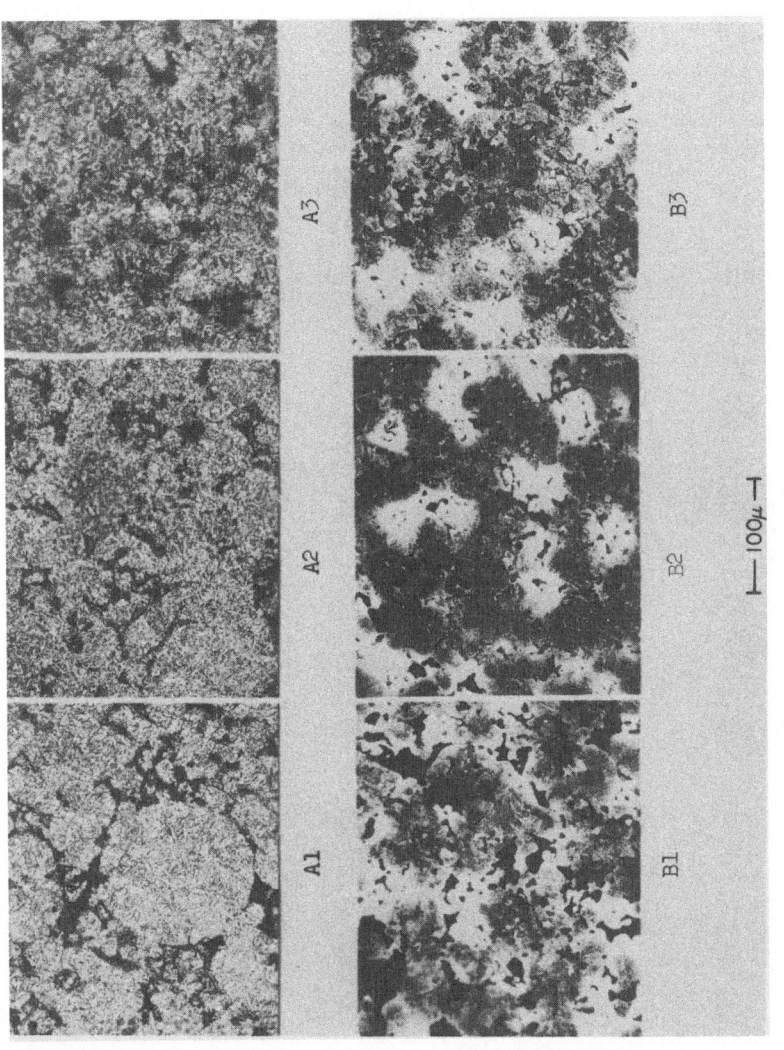

FIGURE 8

ATOMIZED AND MIXED ELEMENTAL 4640; SINTERED 2050°F/30 MIN.,
AUSTENITIZED 1550°F/1 HOUR, OIL QUENCHED AND TEMPERED

Atomized A1 – 6.50 g/cc, Tempered 600°F   Mixed Elemental B1 – 6.48 g/cc, Tempered 600°F
          A2 – 6.91 g/cc, Tempered 600°F   B2 – 6.87 g/cc, Tempered 600°F
          A3 – 7.20 g/cc, Tempered 500°F   B3 – 7.15 g/cc, Tempered 500°F

ETCHANT: 2% NITAL

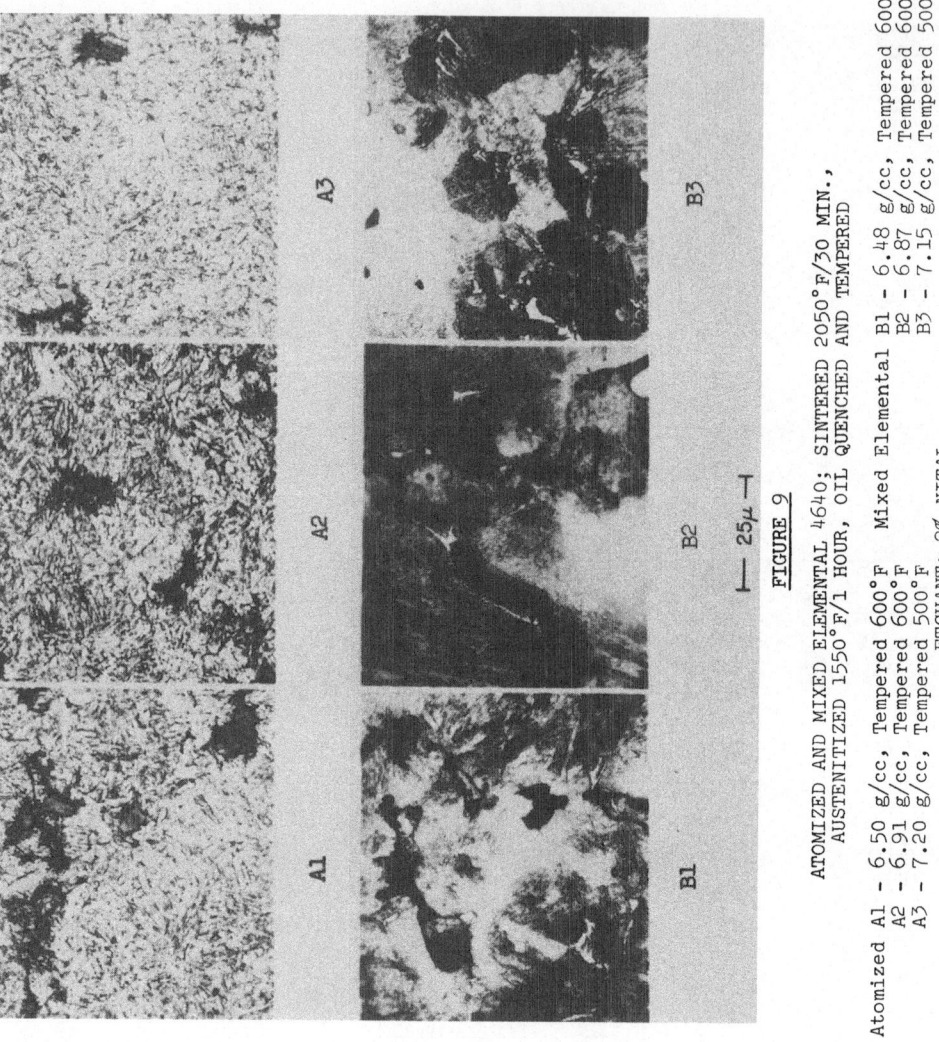

├── 25μ ──┤

FIGURE 9

ATOMIZED AND MIXED ELEMENTAL 4640; SINTERED 2050°F/30 MIN.,
AUSTENITIZED 1550°F/1 HOUR, OIL QUENCHED AND TEMPERED

Atomized A1 – 6.50 g/cc, Tempered 600°F   Mixed Elemental B1 – 6.48 g/cc, Tempered 600°F
          A2 – 6.91 g/cc, Tempered 600°F   B2 – 6.87 g/cc, Tempered 600°F
          A3 – 7.20 g/cc, Tempered 500°F   B3 – 7.15 g/cc, Tempered 500°F
                         ETCHANT: 2% NITAL

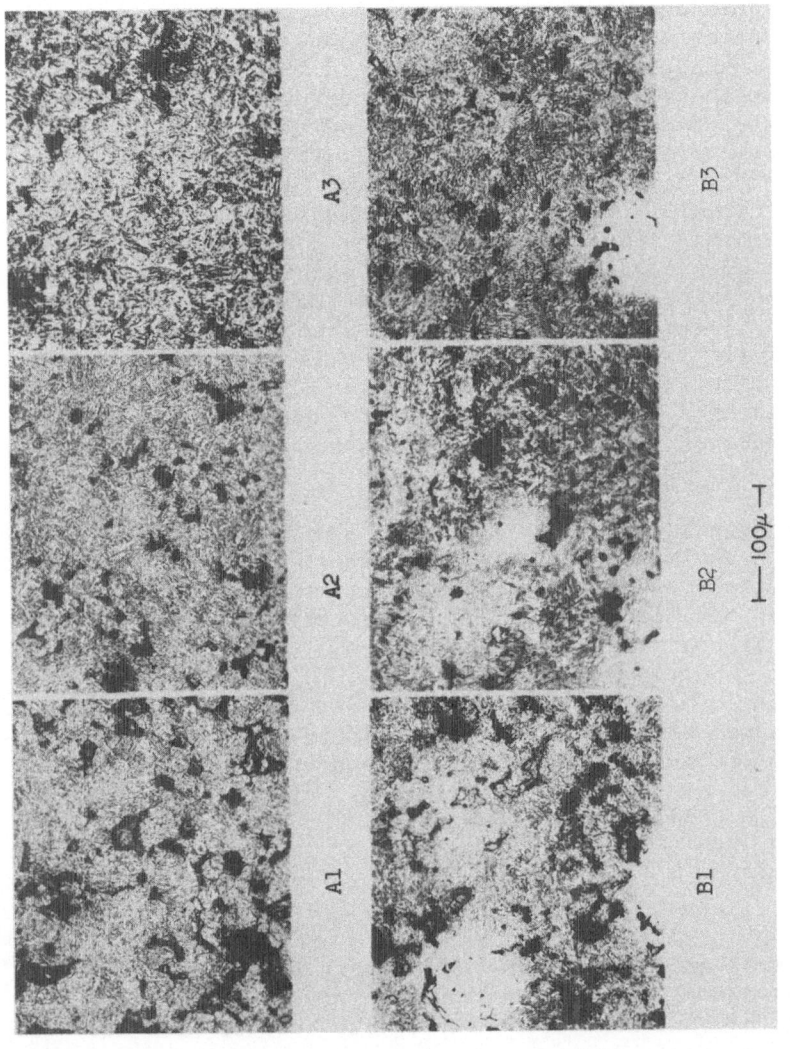

<u>FIGURE 10</u>

ATOMIZED AND MIXED ELEMENTAL 4640; SINTERED 2400°F/60 MIN.,
AUSTENITIZED 1550°F/1 HOUR, OIL QUENCHED AND TEMPERED

Atomized A1 – 6.59 g/cc, Tempered 600°F   Mixed Elemental B1 – 6.59 g/cc, Tempered 600°F
         A2 – 6.97 g/cc, Tempered 600°F   B2 – 6.95 g/cc, Tempered 500°F
         A3 – 7.27 g/cc, Tempered 400°F   B3 – 7.23 g/cc, Tempered 400°F

ETCHANT: 2% NITAL

It is interesting to note that after heat treatment, those mixed elemental parts initially sintered at 2050 F/30 minutes (American practice) are still 10-20% stronger than the atomized parts initially sintered at 2400 F/60 minutes (European practice), in spite of the fact that the mixed elemental parts were at a lower density than the atomized parts.

Metallographic examination of these heat treated parts shows the atomized powders to have a fully homogeneous structure that appears to be tempered bainite or martensite. The mixed elemental powders on the other hand reveal heterogeneous structures. The areas in these structures have been identified by Fischer et al(4) as ferrite, pearlite, martensite and bainite, this latter phase being predominant. While a substantial amount of homogenization appears to have taken place in the mixed elemental parts due to the initially higher sintering cycle (typical European practice) there still remains a large amount of light etching areas. These areas are not austenite, since x-ray diffraction examinations have shown that there is less than 2% retained austenite in all the specimens. Typical microstructures are shown in Figures 8 through 10. It is felt that it is this retained heterogeneity with the wide distribution of phases present, that has contributed significantly to the improved properties exhibited by the mixed elemental parts.

It must be pointed out that it is not the purpose of this paper to propose that mixed elemental powders will always be superior to atomized powders. Both systems have an important purpose in powder metallurgy. Rather, the intent is to show the type of properties that are available.

As a result of the higher properties attainable, the 4600 type composition can be more effectively used in the form of mixed elemental powder rather than atomized. This applies in spite of, and perhaps because of, the very complicated microstructures characteristic of the mixed elemental product.

SUMMARY

Both atomized and mixed elemental powders having a nominal 4640 steel composition were examined over a range of densities, sintering conditions and heat treatments. The results have led to these following conclusions:

1. In the as-sintered condition, under either American or European sintering practices, atomized powders exhibit higher strength properties than mixed elemental powders.
2. After oil quenching and tempering, mixed elemental

powders exhibit better mechanical properties than atomized powders. This occurred at all density levels examined and under both initial sintering practices.

3.   The advantage mixed elemental powders exhibit over atomized powders, in the oil quenched and tempered condition, decreases as specimen density increases. This would indicate that similar properties would be realized at nearly full density regardless of the starting system.

4.   The tempering temperature required to achieve optimum strength properties is dependent on specimen density (and independent of the starting system); decreasing as density increases.

5.   Atomized parts have a uniformly distributed homogeneous structure, while mixed elemental parts have a heterogeneous mixture of phases. In the heat treated condition bainite appears to be the major phase.

6.   The higher sintering cycle increases homogenization in the mixed elemental parts, but their heterogeneous nature is still very evident.

Based on these conclusions, it appears that the retained heterogeneity in the mixed elemental parts, with the distribution of phases present, contributes significantly to the results observed and is therefore beneficial. This type of structure is most readily available to the powder metallurgist, and as such, every advantage of its usefulness should be employed.

ACKNOWLEDGEMENTS

The author wishes to express his thanks to Doctors J. H. Brophy and C. R. Cupp and to Messrs. J. J. Fischer, J. T. Casey, and S. L. Keresztes for the interest and valuable assistance they have taken in this work.

REFERENCES

1.   C. G. Goetzel, "Treatise on Powder Metallurgy, Vol. II", Interscience Publishers, Inc., 1950.

2.   W. D. Jones, "Fundamental Principles of Powder Metallurgy", Edward Arnold (Publishers) Ltd., 1960.

3.   L. L. Bobo, "Expanding the Application of Powder Metal Parts Through Heat Treatment", ASME Design Engineering Conference, 1968.

4.   J. J. Fischer and F. W. Heck, "Characterization of the Microstructure of a Sintered 4640 Steel", International Powder Metallurgy Conference, 1970.

CHARACTERIZATION OF THE MICROSTRUCTURE OF A SINTERED 4640 STEEL

J. J. Fischer and F. W. Heck

The International Nickel Company, Incorporated

ABSTRACT

The various phases present in a sintered 4640 steel prepared from both mixed elemental and prealloyed powders have been identified. The results indicate that only ferrite and pearlite are present in either material in the as-sintered (1/2 hour at 2050 F) condition. Electron microprobe results indicated that the nickel concentration varied from 0-11% and molybdenum varied from 0-0.5% in the premixed specimens. Little or no compositional variation was found in the prealloyed specimens.

In the oil-quenched condition, prealloyed specimens were found to be almost entirely bainitic. In specimens prepared from premixed powders, however, areas containing ferrite, pearlite, bainite and martensite were identified. Microprobe examination indicated that the phase formed in a given area was dependent on the nickel concentration in that area. Therefore, consideration of isothermal transformation diagrams enables a direct correlation to be made between composition and microconstituents.

INTRODUCTION

Sintered steels prepared from mixed elemental powders are characterized by complex heterogeneous microstructures. Recently, Heck(1) has shown that heat treated specimens of 4640 steel prepared from premixed powders exhibit superior strength and ductility compared to similarly prepared specimens from prealloyed powders. The mechanical properties of the premixed powder specimens are undoubtedly strongly influenced by their heterogeneous structure.

471

Although some metallographic studies have been made(2,3), little information is available concerning the nature and extent of the various phases formed in heat treated sintered steels.

The present study was aimed at identifying the products found in a sintered 4640 steel prepared both from prealloyed and mixed elemental (premixed) powders. In addition, the distribution of the principal alloying elements, nickel and molybdenum, was examined.

EXPERIMENTAL PROCEDURE

The specimens examined in this investigation were transverse rupture bars, 1-1/4-inches long by 1/2-inch wide by approximately 1/4-inch thick. The prealloyed specimens were made from a commercially available atomized 4600 steel powder. The premixed specimens were prepared by blending elemental iron, nickel and molybdenum powders in the desired composition (2%Ni, 0.25%Mo, balance Fe). Graphite powder was added to both the prealloyed and premixed powders so that the final carbon content after heat treating was approximately 0.4 percent.

The prealloyed specimens were pressed at 60 tsi (84.3 kg/mm) and the premixed specimens at 42.5 tsi (59.7 kg/mm). The sintering was done in dissociated ammonia for either 1/2 hour at 2050 F (1120 C) representing American sintering practice or 1 hour at 2400 F (1315 C) to simulate typical European sintering practice. The cooling rate of the specimens after sintering approximately that of an air cool (600 F/min.). The resulting sintered densities were nearly equal as shown below:

    Prealloyed (1/2 hr. at 2050 F) = 7.21 g/cc
    Prealloyed (1 hr. at 2400 F)   = 7.28 g/cc
    Premixed (1/2 hr. at 2050 F)   = 7.15 g/cc
    Premixed (1 hr. at 2400 F)   = 7.22 g/cc

The specimens that were heat treated were held for 1 hour at 1550 F (843 C) in a carbonaceous atmosphere, oil quenched and then tempered for 1 hour at various temperatures.

RESULTS AND DISCUSSION

The microstructures of the premixed and prealloyed specimens were examined in the as-sintered, oil quenched and quenched + tempered conditions. The specimens sintered for 1/2 hour at 2050 F are discussed in detail since these structures were found to be considerably more heterogeneous than those sintered for 1 hour at 2400 F.

As-Sintered Condition

The structures of a prealloyed and premixed specimen sintered at 2050 F are shown in Figure 1. Both specimens contain a mixture of pearlite with excess ferrite, although a more uniform distribution of the phases is evident in the prealloyed specimen. Several electron microprobe scans for nickel and molybdenum were made on both the prealloyed and premixed specimens. The results showed an average composition of 1.75%Ni and 0.3%Mo for the prealloyed specimen   The variation in nickel and molybdenum content was less than 0.3%. In the premixed specimen a wider variation in composition was found. The nickel content varied from 0 to 11% and molybdenum from 0 to 0.5%. Attempts to determine the distribution of carbon were also made, but failed to show any variation above the background counts.

The grain size of the prealloyed specimen shown in Figure 1 is cosiderably finer than that of the premixed specimen (6 microns compared to 20 microns). The larger grain size of the premixed specimen may be the result of a faster rate of grain growth during sintering due to the heterogeneous distribution of the elements. The fact that the prealloyed specimens were pressed at a higher pressure would also contribute to a finer recrystallized grain size.

Oil Quenched Condition

Figure 2 shows the microstructure of the prealloyed and premixed specimens after oil quenching from 1550 F.

The microstructure of the premixed specimen is considerably more heterogeneous than that of the prealloyed. The structure is shown at high magnification (1000X) in Figure 3. Microprobe analyses and microhardness measurements were made on three selected areas designated A, B and C in Figure 3.

The pure white area (A) was found to have a nickel concentration in the range 0-1.5% and a hardness of about R_b 74. These results indicate these areas are ferrite.

The dark etching areas designated (B) had a hardness of about R_b 90 or R_c 10. Microprobe results indicated the nickel content ranged from 1 to 3 percent. It was concluded that these areas are mainly composed of a lower pearlite or bainitic product. The presence of fine lamellae in some areas supported this conclusion.

The light etching areas designated (C) were found to contain between 3 and 11 percent nickel. In a given area, the highest nickel concentration was generally found near the center. Micro-hardness measurements showed a range of R_c 30 to R_c 60. The

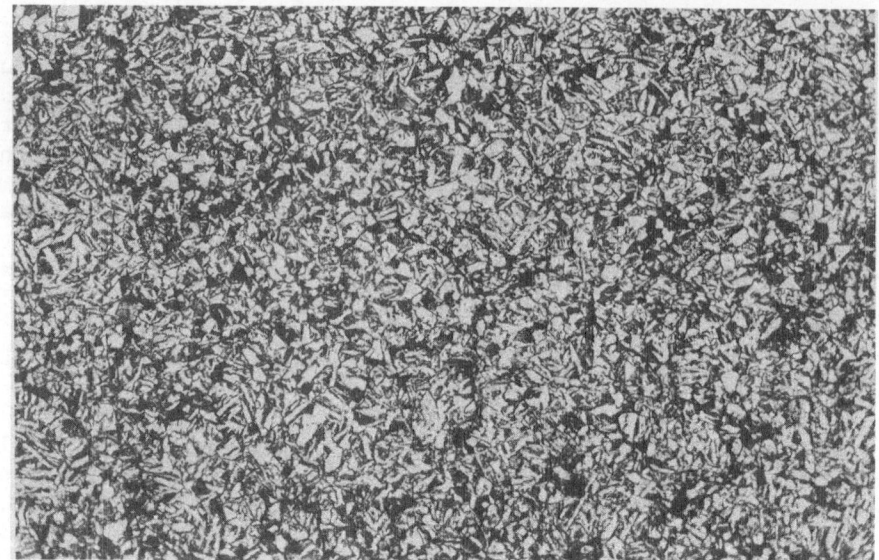

Prealloy

Premix

Figure 1 - Microstructures of prealloyed and premixed
          specimens in as-sintered condition (1/2 hour
          at 2050 F).

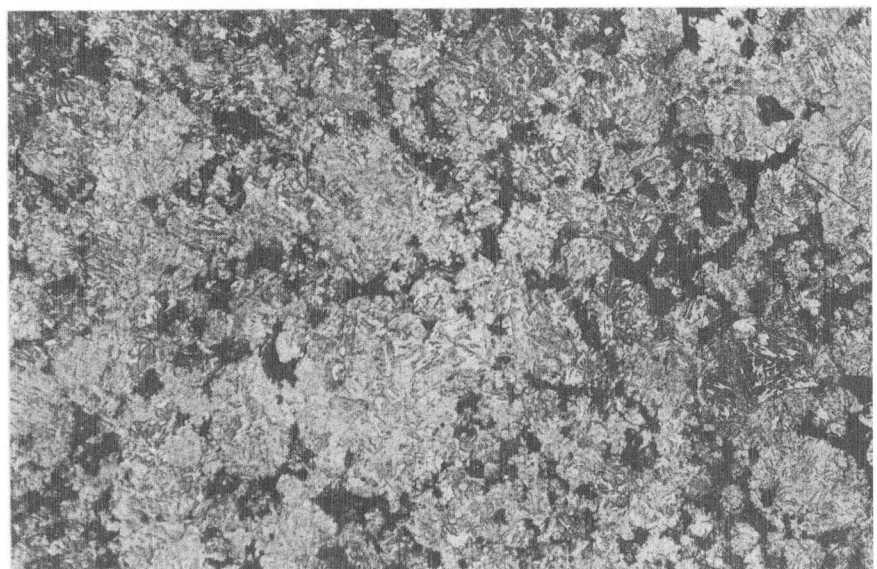

Prealloy

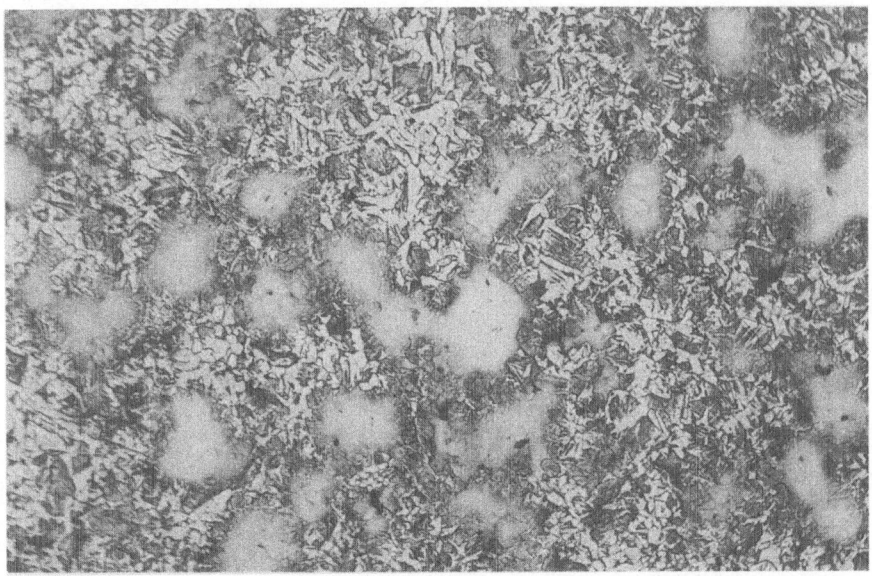

Premix

Figure 2 - Microstructures of prealloyed and premixed
specimens after oil quenching from 1550 F.

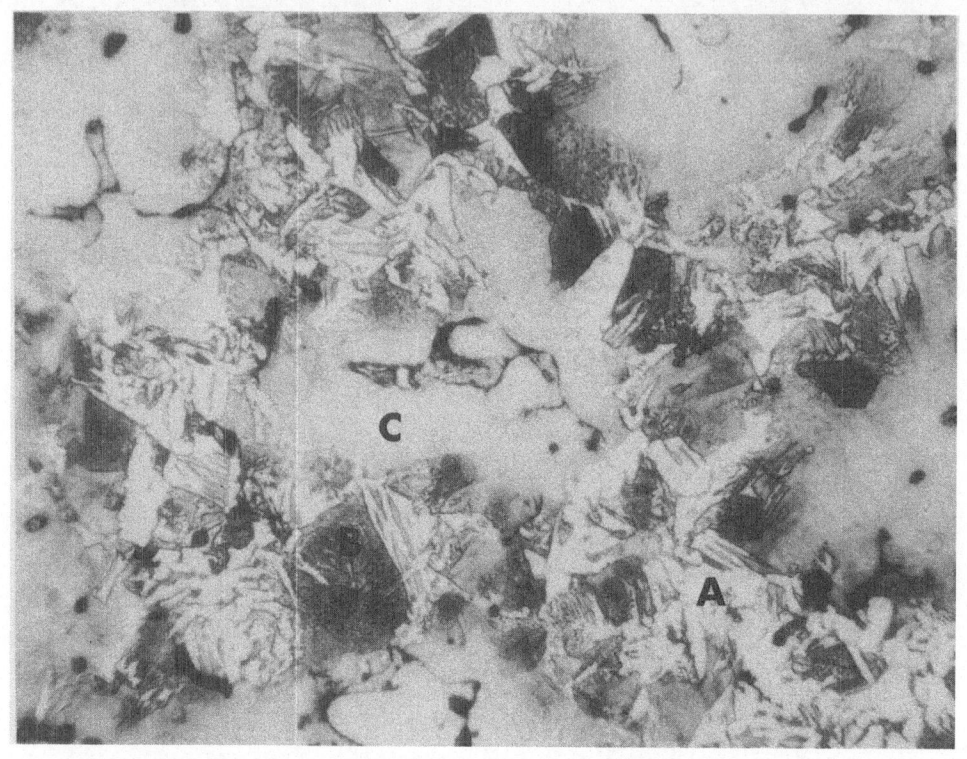

FIGURE 3 - Microstructure of premixed specimen in oil
          quenched condition showing areas A, B and C.

hardest regions occurred near the center of area C and corresponded to the high nickel concentrations measured with the microprobe. These areas are undoubtedly martensitic near the center
with bainite possibly occurring near the outer periphery. X-ray
diffraction results showed that all specimens examined contained
less than 2% austenite.

Although measurements of the molybdenum concentration were
also made on the same selected areas, there was no consistent
change from a nominal level of 0.25 percent. The results, therefore, indicate that the amount of nickel present in a particular
area controls the product that is formed after oil quenching.
Examination of the isothermal transformation and cooling transformation diagrams for a 4640 and 2340 steel support this argu-
ment(4). The cooling curve for a 2340 steel (3.34%Ni) indicates
the product formed after oil quenching will be martensite. This
is in agreement with the microprobe results for area (C) where
the nickel content was 3-11 percent and the product formed had
the hardness of martensite. The cooling curve for a 4640 steel
(1.8%Ni) indicates the product formed after oil quenching will
contain both pearlite and bainite, supporting the characterization
of area (B).

The oil quenched specimens were also examined by x-ray
diffraction. Since the formation of body-centered tetragonal
martensite results in a shift of the diffraction peaks, measurements of the d-spacings of the peaks can be used to estimate the
amount of martensite present. For comparison purposes two speci-
ments of wrought 4640 steel were heat treated to produce fully
martensitic and fully bainitic structures. The martensitic specimen was prepared by water quenching and the bainitic specimen
was made by isothermally transforming above the M_s temperature in
a salt bath at 700 F.

The d-spacings of the (110) diffraction peak for the prealloyed and premixed specimens and wrought standards are shown
in Figure 4 along with the values obtained after tempering (to be
discussed later). It is evident that the amount of martensite
present is relatively small; approximately 14% for the premixed
and 24% for the prealloyed*. The balance of the structure is
bainite in the prealloyed specimens, while the remainder of the
premixed structure is composed of the phases discussed previously.
Measurements were also made on specimens sintered at 2400 F.
(Figure 5) and showed an increase in the amount of martensite
(approximately 20% martensite in the premixed and 33% in the prealloyed specimen).

_____

*   The amount of martensite was estimated by assuming a linear
relationship between the d-spacings for the martensitic and
bainitic wrought specimens.

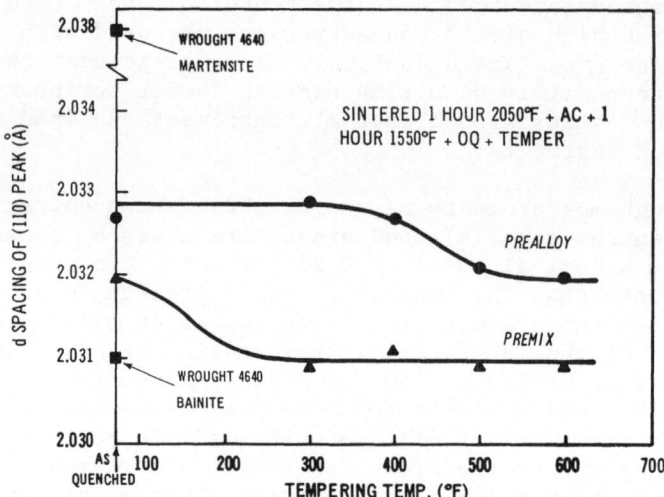

FIGURE 4 - X-ray d spacing vs. tempering temperature for specimens sintered at 2050 F.

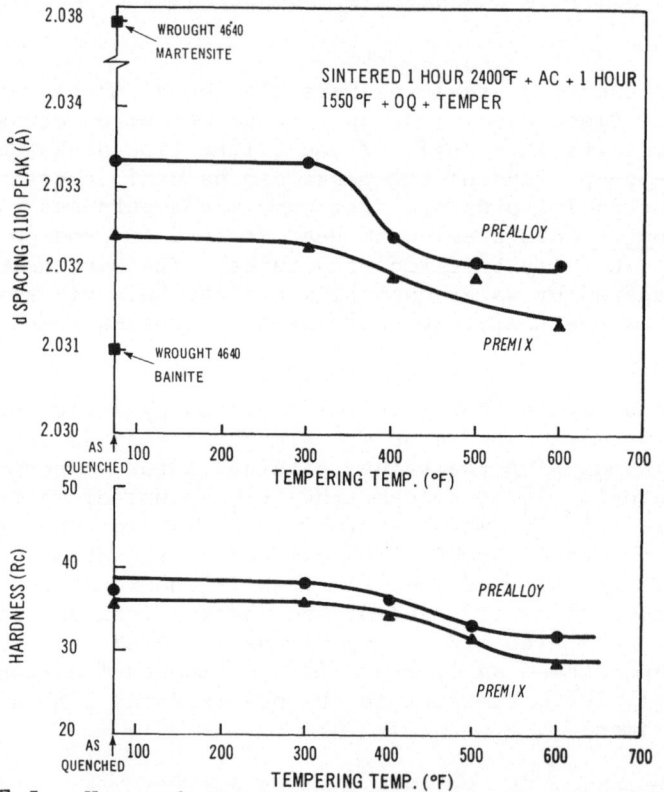

FIGURE 5 - X-ray d spacing and hardness vs. tempering temperature for specimens sintered at 2400 F.

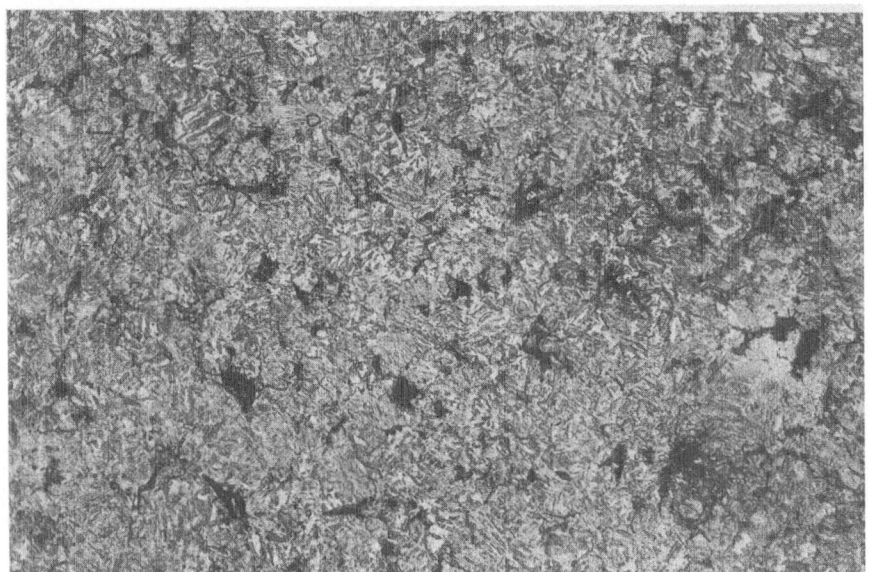

Prealloy

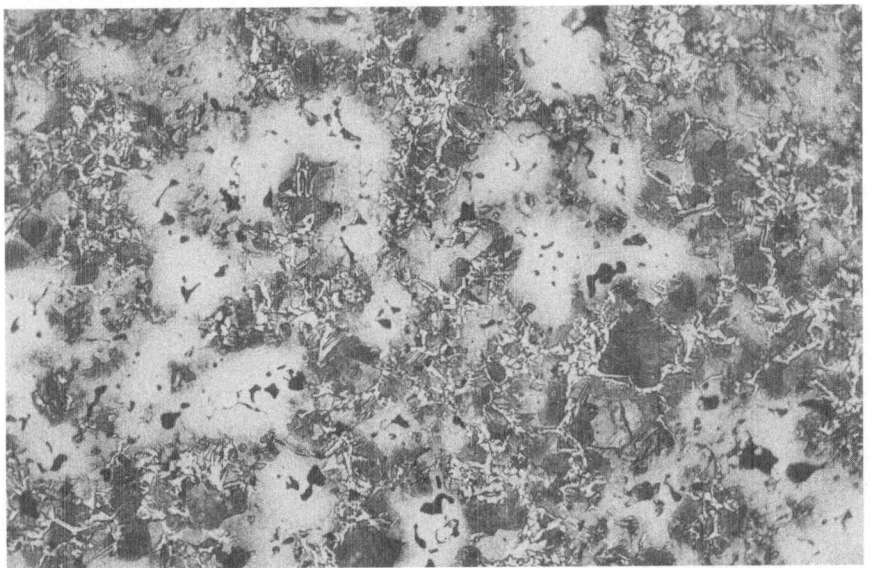

Premix

Figure 6 - Microstructures of prealloyed and premixed
specimens after oil quenching from 1550 F
and tempering at 400 F.

Oil Quenched + Tempered Condition

Typical structures of quenched and tempered (400 F) specimens
are shown in Figure 6. Qualitatively, there is little change in
the structures compared to the as-quenched condition (Figure 2).
The variation in d-spacing of the (110) diffraction peak with
tempering temperature is shown in Figures 4 and 5. The hardness
values of the specimens sintered at 2400 F have also been plotted
in Figure 5. It is evident there is a close relationship between
the curves of hardness and d-spacing vs. tempering temperature.
Both curves indicate the tempering reaction occurs at about 400 F.

SUMMARY

Both the premixed and prealloyed 4640 steel are pearlitic
with an excess of ferrite in the as-sintered condition.

After oil quenching, the major phase present in the prealloyed
specimens was bainite with the balance being martensite. In the
case of the premixed specimen, areas of ferrite, pearlite, bainite
and martensite were identified. The product formed in a given
area was found to depend on the nickel concentration in that area.
Thus ferritic areas contained 0-1.5%Ni, pearlitic and bainitic
areas 1-3%Ni and martensitic areas 3-11%Ni. These conclusions
were supported by available transformation diagrams.

The tempering reactions were followed using the d-spacing of
the (110) diffraction peak and hardness measurements. The curves
for both types of measurements were in close agreement and indicated a minimum temperature of 400 F is necessary to insure
tempering.

REFERENCES

1.   F. W. Heck, "Sintered and Heat Treated AISI 4600 Steel",
International Powder Metallurgy Conference, 1970.

2.   A. Stosuy, "A Metallographic Study of Sintered Steel",
Precision Metal Molding, May 1967, pp. 76-79.

3.   S. W. McGee and E. R. Andreotti, "Modern Developments
in Powder Metallurgy", 3, Plenum Press, New York, 1966, pp. 206-
226.

4.   "Isothermal Transformation Diagrams of Nickel Alloy
Steels", The International Nickel Co., Inc., 1965.

A NEW APPROACH FOR THE PRODUCTION OF MARAGING STEEL P/M PARTS

Joel S. Hirschhorn[*] and David A. Westphal

University of Wisconsin
Madison, Wisconsin

[*]also Director of Research
    Friction Products Co.
    Medina, Ohio

INTRODUCTION

The manufacture of maraging steel P/M parts is of interest for
a number of reasons. Primarily, there are numerous potential applications for this high strength-high toughness type of material involving parts that could be most economically made by P/M techniques.
With the 300,000 to 400,000 psi tensile strength range of certain
wrought maraging steels, it is quite conceivable that a P/M material
could be made with a strength in excess of 200,000 psi and with
substantial ductility, using conventional press and sinter techniques.

In addition to favorable properties that might be achieved with
such materials, that would be difficult to obtain with conventional
carbon alloy steels, there are a number of processing factors that
make a P/M approach advantageous. First, the entire area of maintaining proper carbon control during sintering and subsequent heat
treating would be eliminated because the best maraging steels do
not contain carbon, except as an impurity. The heat treatment of
these steels is markedly simpler than for more conventional ones.
A typical treatment consists of solution annealing at 1500°F for
one hour followed by aging at 900°F for 3 hours; in both cases the
material is air cooled. The elimination of oil quenching, usually
done for heat treated P/M steels, is significant. Additionally,
there is excellent dimensional stability during the treatment.

However, the use of P/M techniques for a heavily alloyed
material such as the maraging steels is difficult for two major
reasons. If totally prealloyed powder is used its compressibility

481

Table 1 - Description of -325 mesh Powders Used in Alloy Blends

Number	Powder Type	Average Size (microns) Fisher	Microscopic
1	Fe; -325 mesh fraction, commer-cial atomized	18.9	20.8
2	Fe; carbonyl; spherical	-	2.0
3	Fe-36% Ni; atomized; irregular	10.8	11.5
4	Co;precipitated; angular equiaxed	4.8	5.1
5	Fe-70% Co; atomized; angular equiaxed	-	4.9
6	Mo; reduced; irregular equiaxed	4.3	3.7
7	Fe-50% Mo; atomized; equiaxed	-	1.7
8	Fe-26% Ti; atomized; angular	10.2	3.2
9	Fe-50% Al; atomized; sharply angular	8.0	12.9

Table 2 - Screen Analysis of As-received Commercial Atomized Iron Powder

Mesh Fraction	Weight %
+100	1.8
-100 +150	18.3
-150 +170	15.1
-170 +250	16.0
-250 +270	13.8
-270 +325	13.7
-325	21.3

Table 3 - Description of Six Types of Alloy Blends

Alloy Designation	Apparent Density(g/cc)	Flow Time(sec)	Source of Component (Table 1) Fe	Ni	Co	Mo	Ti	Al
1	3.15	26.6	AR	3	4	6	8	9
2	2.59	NF	2	3	4	6	8	9
3	2.79	NF	1	3	4	6	8	9
4	3.55	27.9	1	3	CC	6	8	9
5	2.56	NF	1	3	5	6	8	9
6	2.20	NF	AR	3	5	7	8	9

AR = as-recieved atomized Fe (Table 2)
CC = coarse Co; precipitated; -200+325 mesh fraction
NF = no flow behavior in standard test

is poor, and the somewhat high levels of density required cannot be easily obtained. With a mixture of elemental powders it is very difficult to achieve adequate diffusional alloying during commercially attractive types of processing. When this study was initiated only these two approaches had been reported on, with rather poor results[1-4]. An intermediate approach using mixtures of partially prealloyed powders was deemed appropriate and was the main theme of the present study. A similar approach was later described in the literature[5]; however, the details of it were substantially different than this one.

MATERIALS AND EXPERIMENTAL PROCEDURES

Most of the powders used were of the -325 mesh type; they are described in Table 1. In addition to these, most of which were commercial grades, a commercial atomized iron powder was used in the as-received condition; the screen analysis for this is given in Table 2. Various powders were blended together to yield a composition of: Fe-18% Ni-9% Co-5% Mo-0.4% Ti-0.1% Al (in weight %). The component powders were blended sequentially, in order of decreasing size with intermittent blending. The total blending time was 2 hours, and the mixtures appeared quite uniform. A normal "V" blender was used.

Compaction was done at 50 or 60 tsi using a suspension of stearic acid in acetone for die wall lubrication. Slugs with 1/2 in. dia. and 1/4 in. thickness were made with single action tooling. Standard flat tensile specimens were made in a floating die.

Both sintering and heat treatment were done in an atmosphere of 95% argon-5% hydrogen (very high purity). Specimens were furnace cooled from the sintering temperature (except types 5 and 6). The standard heat treatment used was the one noted in the Introduction. There was a slight oxide coating after this treatment which was removed by grinding prior to density measurements and testing. Densities are reported in % theoretical using 8.0 g/cc as the value for the pore-free material.

RESULTS

Because of space limitations the results for the six most important types of alloy blends are presented. These are described in Table 3; they will be referred to in terms of their designation numbers.

Alloys 1, 2 and 3 make use of different types of base iron powders. The dependence of density on sintering time for these is shown in Fig. 1a for the same compaction pressure and sintering

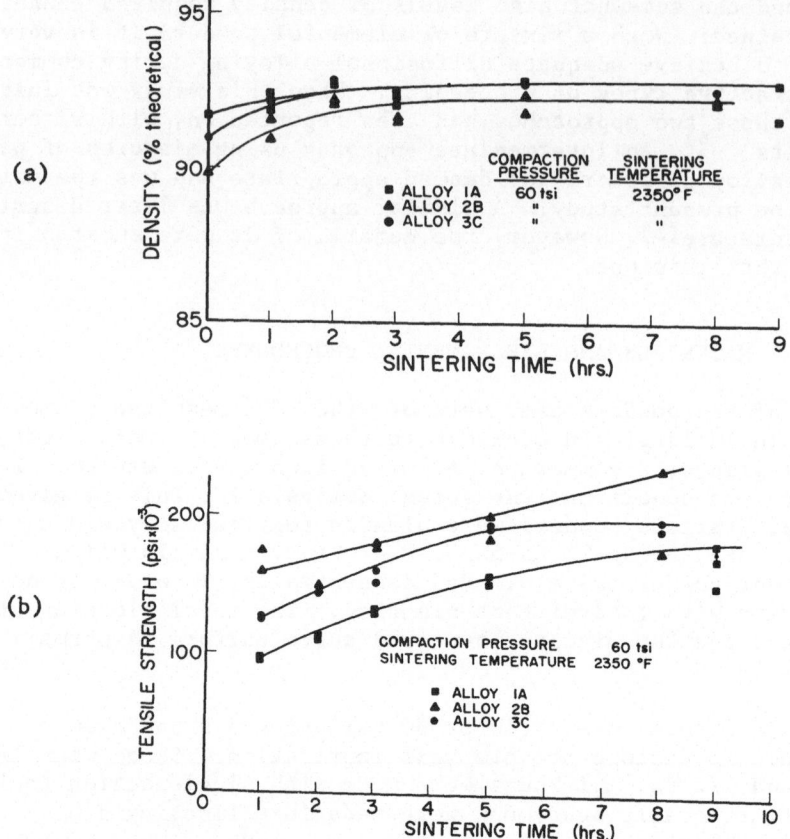

Figure 1.   (a) Dependence of density on sintering time; (b) dependence of tensile strength on sintering time; both for Alloys 1,2 and 3.

temperature.   The letters after the alloy number refer to tensile
bars as the source of the data rather than slugs.   Densification
and shrinkage is greatest for the two atomized base powders and
least for the carbonyl.   In all cases there is little change after
about 2 hours.   Also, the green densities were all near or above
90% theoretical and increased with increasing size of the base iron
powder.   The dependence of tensile strength on sintering time for
these alloys is shown in Fig. 1b.   The strengths increase with decreasing size of the base iron powder.   Alloy 2 with the carbonyl
iron yields strengths near or over 200,000 psi for times of 5 or
more hours.   The continued increases in strength beyond the times
corresponding to density stabilization signify the continuation of
diffusional alloying and homogenization, particularly for Alloy 2.

All the tensile specimens tested exhibited only 1 to 2% elongation.   A distinct mottled appearance was evident just prior to

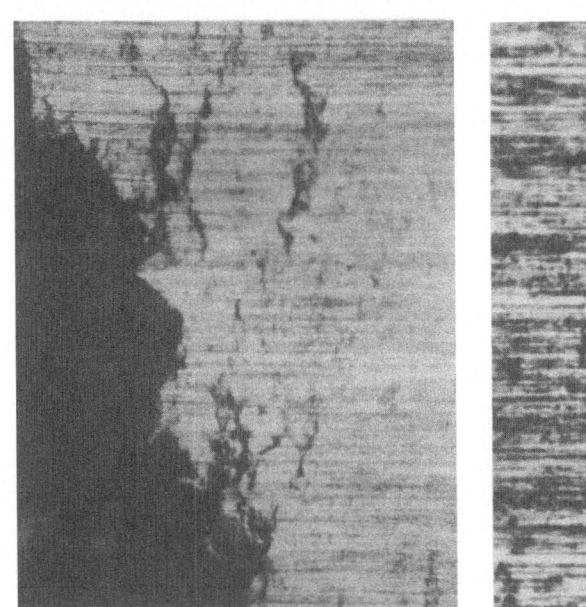

Figure 2. Micrographs of typical surface cracking on fractured tensile specimens: (left) surface near fracture (50X); (right) surface 1/4 to 1/2 in. from fracture (100X).

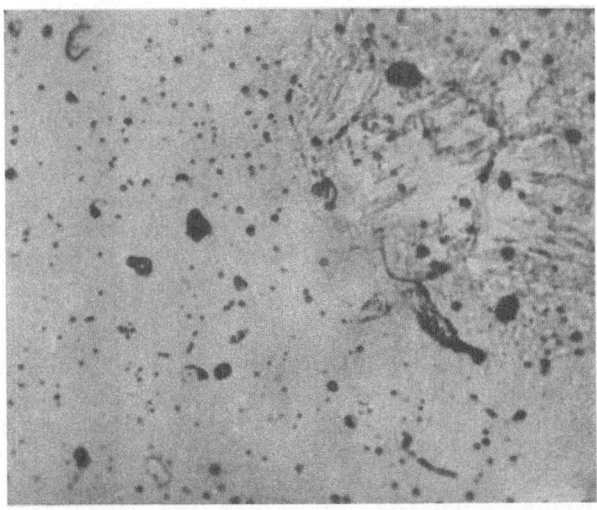

Figure 3. Micrograph of Alloy 2B illustrating retained austenite (upper right) and typical isolated and well rounded porosity, after heat treatment; etchant: 60 vol.% nitric acid–40 vol.% acetic acid (400X).

fracture on all specimens during tensile testing. Examples of this
are given in Figure 2; it can be seen that fine surface cracking
is the cause of the mottled appearance. This cracking is believed
to be caused by an embrittling precipitation, and explains the limited elongations observed and the brittle nature of the fracture.

Normal metallographic examination of heat treated materials
did not reveal precipitates. There was some evidence of unalloyed
powder particles in some cases; in most there was some retained
austenite as shown in Fig. 3. Also evident is the isolated and
well rounded residual porosity present in all materials. Such
porosity is normally conducive to significant ductility and supports
the above conclusion on the presence of an embrittling precipitate.

Alloys 1, 2 and 3 were also studied for a compaction pressure
of 50 tsi and sintering temperatures of 2100 to 2300°F. The densities were low (86 to 91%) and the tensile strengths moderate
(100,000 to 150,000 psi).

Alloys 3, 4 and 5 represent a group in which the same base
iron powder is used (-325 mesh atomized) but the cobalt is introduced in three different ways. Also both 50 and 60 tsi compaction
pressures and several sintering temperatures are evaluated. The
dependence of density on sintering time is given in Fig. 4a. The
use of the lower compaction pressure and sintering temperature, and
the coarse cobalt powder yields very low densities. Comparing
Alloys 3C and 5A, both processed in the same manner, shows that the
use of the fine pure cobalt rather than the fine iron-cobalt powder
promotes densification. The reduction in compaction pressure and
sintering temperature for Alloy 3 has only a minor effect on density.
For all cases there is a relatively small dependence of density on
time. The tensile results for these alloys is given in Fig. 4b.
The increases in strength up to 5 hours is indicative of continued
homogenization with little change in density. Comparison of Alloys
3C and 5A shows a slight advantage with the use of the iron-cobalt
powder (even though the densities are lower); both alloys approach
200,000 psi strength after about 3 hours. The reduction in compaction pressure and sintering temperature for Alloy 3 causes a
significant reduction in strength.

Alloy 5 was also sintered at 2500°F, and the dependence of density on time is shown in Fig. 5a. There is an initial decrease
(compact expansion) followed by an increase (shrinkage). This is
believed to be linked to a chemical reaction and/or incipient melting. The residual porosity was larger than for the 2350°F sintering.
The tensile strength data for this alloy are given in Fig. 5b. The
strengths are rather constant and considerably lower than for the
2350°F sintering, even though at 4 hours the density is not much
less.

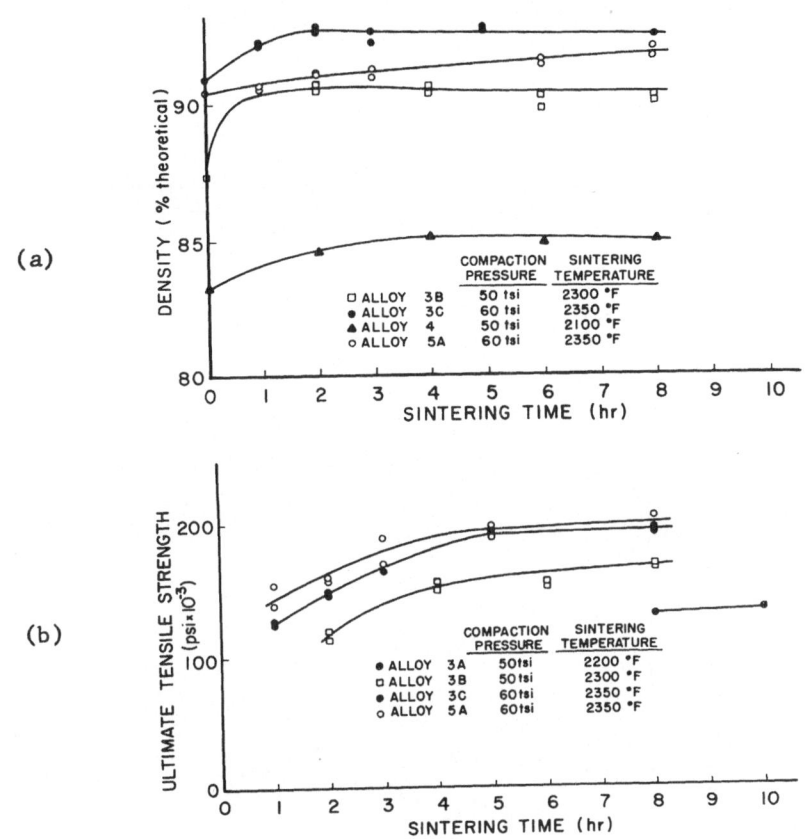

Figure 4.   (a) Dependence of density on sintering time for Alloys 3, 4 and 5; (b) dependence of tensile strength on sintering time for Alloys 3 and 5.

Alloy 6 is similar to Alloy 5 except that an iron-molybdenum powder was used instead of pure molybdenum, and the base iron was as-received instead of the −325 mesh fraction. In Fig. 5a an initial density decrease that does not change is shown for Alloy 6. The tensile strengths shown in Fig. 5b are also constant and at a very low value. This behavior is probably related to some type of chemical reaction and/or melting associated with the Fe-Mo powder.

Alloys 1, 2, 3 and 6 were studied with regard to repressing. After an initial treatment of 60 tsi compaction and sintering at 2350°F for 3 hours, specimens were repressed at 60 tsi. Increases in density ranged from 2 to 5%; the repressing being more effective the lower the original sintered density.

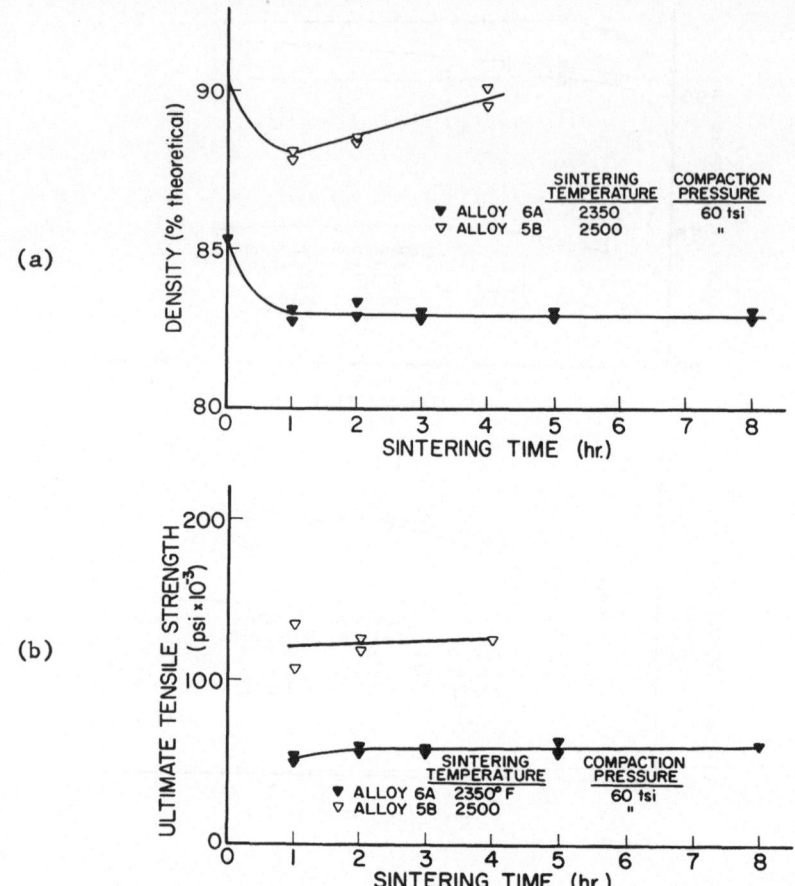

Figure 5. (a) Dependence of density on sintering time; (b) dependence of tensile strength on sintering time; both for Alloys 5 and 6.

In order to confirm the validity of the age hardening treatment
used, considered typical for wrought materials, an experiment was
conducted on Alloy 5 compacted at 60 tsi, sintered at 2350°F for 5
hours, and solution annealed at 1500°F for 1 hour. The hardness
versus time data given in Fig. 6 indicate that the standard 3 hour
aging time used is quite valid for the P/M system.

SUMMARY AND DISCUSSION

The present study has demonstrated the feasibility of making
18% Ni maraging steel P/M parts by conventional press and sinter
techniques. Using specific partially prealloyed powders and a fine
particle size base iron powder promotes compressibility and diffu-

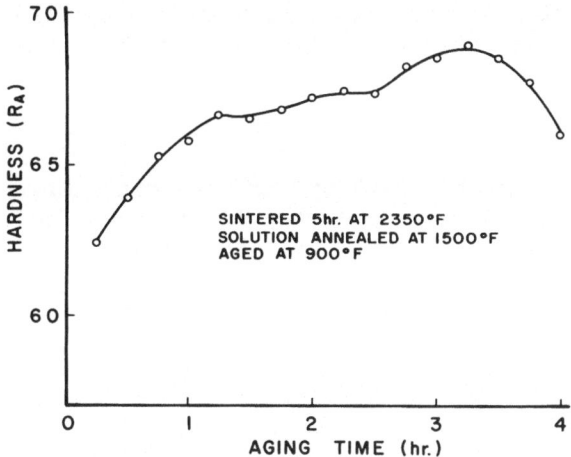

Figure 6.   Age hardening curve for Alloy 5 in annealed condition.

sional alloying.   Alloys 2 and 5 yielded the highest strengths.
Using both a carbonyl iron base and an iron-cobalt powder, instead
of pure cobalt, should produce even better results.   As in previous
studies, although high strengths can be obtained, very limited ductility was found.   Until this problem is solved the process and
property advantages of this P/M steel will not be realized.

    Improved results (higher strength and ductility) could also
be obtained by evaluating the following:

    (a) optimizing thermal treatments to avoid retained austenite;

    (b) increasing densities by developing optimum lubrication and
repressing procedures;

    (c) using a more reducing atmosphere to reduce impurities and
embrittlement;

    (d) using longer sintering times to increase alloying;

    (e) using a duplex aging treatment[6];

    (f) using higher purity powders; reducing the carbon content
to less than 0.005% can greatly improve fracture toughness[7];

    (g) adopting hot forming (pressing or forging) techniques to
increase density;

(h) adjusting composition to compensate for incomplete homogenization at reasonable sintering temperatures and times

ACKNOWLEDGEMENT

The support of this work by Burgess-Norton Mfg. Co., Amsted
Industries, Geneval, Ill. is gratefully acknowledged. Patent
application based on this work has been made.

REFERENCES

(1) J. J. Fisher, Int. J. Powder Met., vol. 2, no. 4, pp. 37-42,
     1966.

(2) S. W. McGee, U.S. Patent 3,303,066, February 7, 1967.

(3) T. Kinoshita, Y. Tokunaga, H. Kobayashi and I. Taniguchi,
     J. Japan Soc. for Powder Met., vol. 13, no. 5, pp. 228-235,
     pp. 236-242, 1966; vol. 14, no. 5, pp. 213-220, 1967.

(4) S. N. Kumar and J. S. Hirschhorn, University of Wisconsin,
     unpublished research.

(5) V. A. Tracey and R. S. K. Raman, Powder Met., vol. 12, no. 23,
     pp. 131-156, 1969.

(6) W. A. Spitzig, J. of Materials, vol. 5, no. 1, pp. 140-156,
     1970.

(7) R. P. M. Procter and H. W. Paxton, Spring Meeting TMS-AIME,
     Las Vegas, May, 1970.